Soil Carbon Sequestration
and The Greenhouse Effect

Soil Carbon Sequestration and the Greenhouse Effect

Proceedings of a symposium sponsored by Divisions S-3, S-5, and S-7 of the Soil Science Society of America at the 90th Annual Meeting in Baltimore, MD, 18–22 October 1998.

Editor
Rattan Lal

Organizing Committee
Rattan Lal and Kevin McSweeny

Editorial Committee
Rattan Lal and Kevin McSweeny

Editor-in-Chief SSSA
Warren A. Dick

Managing Editor
J.M. Bartels

SSSA Special Publication Number 57

Soil Science Society of America, Inc.
Madison, WI, USA

2001

Cover design: Patricia Scullion

Rows of no-till cotton were planted directly into this unplowed cornfield. Note the abundant residue between rows that helps prevent moisture loss and weed growth. Photo provided by ARS Photo Unit.

Soil Science Society of America, Inc.
677 S. Segoe Rd., Madison, WI 53711

Library of Congress Catalog Card Number: 00 136275

CONTENTS

FOREWORD

There is concern worldwide about increases in greenhouse gases (CO_2, CH_4, and N_2O) and their potential effects on global climate change. Carbon dioxide is increasing in the atmosphere at a rate of about 3.4 Pg yr^{-1} due to fossil fuel combustion, land use change, and tropical deforestation. The atmospheric concentration of CO_2 has increased by about 30% (280–370 ppmv) from the initiation of the Industrial Revolution (circa 1850) to the present. Soils represent a major component in the C cycle and afford a significant sink for C in the biosphere. It is estimated that about 1100 to 1600 Pg C is sequestered in soils worldwide (Izaurralde et al., 2000). This represents more than twice the amount of C in living vegetation (560 Pg) and in the atmosphere (750 Pg). The potential to sequester meaningful C via crop and soil management and reclamation of degraded land is pronounced. For example, it is estimated that 40 to 80 Pg of C can be sequestered in cropland soils over the next 50 to 100 years (Houghton et al., 1996).

However, further research is needed. Without question, one of the major frontiers in soil science in the next decade will be further elucidation of C dynamics in soils and ways to enhance C sequestration in terrestrial ecosystems. This timely publication reviews our current understanding of C sequestration and dynamics in soils and suggests many future research needs. I am confident that this will be a valuable resource for years to come to scientists, students, and policy-makers.

REFERENCES

Houghton, J.T., L.G. Meira Filho, B.A. Callander, N. Harris, A. Kattenberg, and K. Maskell. (ed.) 1996. The science of climate change. IPCC, Work. Group Rep. no. 1. Cambridge Univ. Press, New York.

Izaurralde, R.C., N.J. Rosenberg, and R. Lal. 2000. Mitigation of climate change by soil carbon sequestration: Issues of science, montoring, and degraded lands. Adv. Agron. 70:1–75.

Donald L. Sparks, *President*
Soil Science Society of America

PREFACE

The atmospheric concentration of several greenhouse gases has steadily increased since the onset of the industrial revolution around 1850. The present rate of enrichment of atmospheric CO_2 at 3.3 Pg C yr^{-1} is due to fossil fuel combustion, cement manufacture, deforestation and land use change, and agricultural activities. The magnitude of C emission due to deforestation and agricultural activities is estimated at about 1.6 ± 0.5 Pg C yr^{-1}. Several attempts at balancing the global C budget have pointed out the so-called "missing C" of the order of 1.8 to 2.0 Pg C yr^{-1}. Some have attributed this "missing C" or "fugitive C" to absorption by the terrestrial ecosystems, primarily in North America. It is apparent, therefore, that terrestrial ecosystems in general, but world soils in particular, play an important role in the global C cycle. Depending upon land use, farming or cropping system, tillage methods, and other soil management practices, soil can be a major source or sink for the atmospheric CO_2. There are several land uses, farming systems, and soil management practices that render soil as a net sink for the atmospheric CO_2. Such land uses and soil management practices are specific to soils and ecoregions, and need to be validated and adapted for site-specific situations. Further, the rates of soil C sequestration with recommended management practices also differ among soil types and climate conditions, and need to be determined. The importance of the strategy of soil C sequestration cannot be overemphasized. At present, it is the most cost-effective short-term option of reducing the emission of CO_2 into the atmosphere. Further, enhancement of soil organic C has numerous ancillary benefits, including improvement in soil structure, increase in soil buffering and water and nutrient retention capacities, decrease in risks of soil erosion, and increase in agronomic productivity. Soil C sequestration is a byproduct of adopting recommended agricultural practices for achieving food security. Hence, it is a truly win–win strategy.

It is with this background that a symposium was organized at the annual meetings of the Soil Science Society of America held in Baltimore, MD, on 20 October 1998. This one-day symposium was organized into two sessions. A total of 13 papers were presented, six in the morning session and seven in the afternoon session. The organization of the symposium was made possible because of the sponsorship by Dr. G. Petersen, SSSA President, and Dr. K. McSweeny, Chairman of the S-5 division of SSSA. The symposium was jointly sponsored by Divisions S-5, S-3, and S-7.

This Special Publication, comprising 16 chapters, includes research information on soil C sequestration from croplands, range lands, forest lands, and set-aside or CRP lands in the USA and Canada. This volume is a state-of-the-knowledge compendium. All authors made an outstanding effort to document and present their information on the current understanding of soil processes and the C cycle in a timely fashion. Their efforts have highlighted the importance of pedospheric processes in soil C sequestration. A unique feature of this volume is inclusion of up-to-date information on soil organic C and soil inorganic C, agricultural soils (croplands and grazing lands) and forest soils, and soil organic C dynamics in CRP land vis-a-vis the natural ecosystems.

Publication of this volume was made possible by the support from Ms. Brenda Swank of the Ohio State University who handled all the correspondence and organized the flow of manuscripts. Special thanks are due to Meg Pansegro, Jon Bartels, and Lisa Al-Amoodi of the SSSA Headquarters in Madison, WI, for their help and support. Their efforts in publishing this information on time to make it available to the scientific community are greatly appreciated.

Rattan Lal, *editor*
The Ohio State University
Columbus, Ohio

CONTRIBUTORS

Berc, J.L.	Natural Resource Manager, USDA-Natural Resources Conservation Service, South Ag. Bldg., Rm. 6250, P.O. Box 2890, Washington, DC 20013-2890
Birdsey, Richard A.	Program Manager, USDA Forest Service, Northeastern Research Station, 11 Campus Blvd., Suite 200, Newtown Square, PA 19073
Bluhm, George	USDA-Natural Resources Conservation Service, Dep. of Land, Air, and Water Resources, University of California, Davis, CA 95616
Campbell, C.A.	Agriculture and Agri-Food Canada, Eastern Cereal Oilseed Research Center, Central Experimental Farm, Ottawa, ON, Canada K1A-0C6
Cipra, Jan	Research Associate, Natural Resource Ecology Laboratory, Colorado State University, Fort Collins, CO 80523-1499
Desjardins, R.L.	Senior Research Scientist, Agric. & Agri-Food Canada, Central Experimental Farm, 960 Carling Ave., Ottawa, ON, Canada K1A 0C6
Drees, L.R.	Research Scientist, Dept. of Soil and Crop Sciences, Texas A&M University, College Station, TX 77843
Dumanski, Julian	Agric. & Agri-Food Canada, Central Experimental Farm, Ottawa, ON, Canada K1A 0C6
Elliott, E.T.	Director, School of Natural Resource Sciences, Biochemistry Hall, Room 302, University of Nebraska-Lincoln, Lincoln, NE 68583-0758
Eve, M.D.	Soil Scientist, USDA-Agriculture Research Service, c/o NREL, Colorado State University, Fort Collins, CO 80523-1499
Follett, R.F.	Research Leader, Soil Plant Nutrient Research, USDA-ARS, P.O. Box E, 301 S. Howes, Fort Collins, CO 80522-0470
Goss, D.W.	Research Scientist, Texas Agricultural Experiment Station, Blackland Research Center, 808 East Blackland Rd., Temple, TX 76502
Grant, Brian	Agricultural Technician, Agric. & Agri-Food Canada, Central Experimental Farm, 960 Carling Ave., Ottawa, ON, Canada K1A 0C6
Gregorich, E.G.	Agriculture and Agri-Food Canada, Eastern Cereal Oilseed Research Center, Central Experimental Farm, Ottawa, ON, Canada K1A-0C6
Heath, Linda S.	USDA Forest Service, Northeastern Research Station, 271 Mast Rd., P.O. Box 640, Durham, NH 03824-0640
Janzen, Henry H.	Agriculture and Agri-Food Canada, Box 3000, Lethbridge Research Center, Lethbridge, Alberta, Canada T1J 4B1
Kellogg, R.L.	Natural Resource Analyst, USDA-Natural Resources Conservation Service, South Ag. Bldg., Rm. 6250, P.O. Box 2890, Washington, DC 20013-2890
Killian, Kenrick	Natural Resource Ecology Laboratory, Colorado State University, Fort Collins, CO 80523-1499
Kimble, J.M.	USDA-NRCS, National Soil Survey Center, Fed. Bldg. Room 152 MS41, 100 Centennial Mall N., Lincoln, NE 68508-3866
Lal, Rattan	Professor, School of Natural Resources, The Ohio State University, 2021 Coffey Road, Columbus, OH 43210
LeCain, Daniel R.	Plant Physiologist, USDA-ARS Rangeland Resources Research, Crops Research Laboratory, 1701 Center Ave., Fort Collins, CO 80526

Liang, B.-C.   Agriculture and Agri-Food Canada, SPARC, P.O. Box 1030, Swift Current, SK S9H 3X2

McConkey, B.G.   Research Scientist, Agriculture and Agri-Food Canada, SPARC, P.O. Box 1030, Swift Current, SK S9H 3X2

McSweeny, Kevin   Dept. of Soil Science, Univ. of Wisconsin, 1525 Observatory Dr., Madison, WI 53717-1299

Morgan, Jack A.   Research Leader, Plant Physiologist, USDA-ARS Rangeland Resources Research, Crops Research Laboratory, 1701 Center Ave., Fort Collins, CO 80526

Nordt, L.C.   Assistant Professor, Dept. of Geology, Baylor University, P.O. Box 97354, Waco, TX 76798-7354

Paustian, K.   Senior Research Scientist, Natural Resource Ecology Laboratory, Colorado State University, Fort Collins, CO 80523-1499

Pruessner, E.G.   Research Technician, Soil Plant Nutrient Research, USDA-ARS, P.O. Box E, 301 S. Howes, Fort Collins, CO 80521-0470

Reeder, Jean D.   Soil Scientist, USDA-ARS Rangeland Resources Research, Crops Research Laboratory, 1701 Center Ave., Fort Collins, CO 80526

Roloff, Glaucio   Departamento de Solos, Universidade Federal do Paraná, Brazil

Samson-Liebig, Susan E.   Soil Scientist, USDA-NRCS, National Soil Survey Center, Fed. Bldg. Room 152 MS41, 100 Centennial Mall N., Lincoln, NE 68508

Sanabria, Joaquin   Assistant Research Scientist, Texas Agricultural Experiment Station, 808 East Blackland Rd., Temple, TX 76502

Schuman, Gerald E.   Soil Scientist, USDA-ARS, High Plains Grasslands Research Station, Rangeland Resources Research, 8408 Hildreth Road, Cheyenne, WY 82009

Smith, Jeffrey L.   Soil Biochemist, USDA-Agricultural Research Service, 215 Johnson Hall, Washington State University, Pullman, WA 99164-6421

Smith, W.N.   Agricultural Engineer, Agric. & Agri-Food Canada, Central Experimental Farm, 960 Carling Ave., Ottawa, ON, Canada K1A 0C6

Tarnocai, Charles   Soil Scientist, Agric. & Agri-Food Canada, Central Experimental Farm, 960 Carling Ave., Ottawa, ON, Canada K1A 0C6

Unger, Paul W.   Soil Scientist; now Soil Scientist (retired, collaborator), USDA-Agricultural Research Service, Conservation & Production Research Lab, P.O. Drawer 10, Bushland, TX 79012-0010

Van Pelt, R. Scott   Soil Scientist, USDA-ARS-CSRL, Wind Erosion and Water Conservation Research Unit, 302 W. I-20, Big Spring, TX 79720

Waltman, Sharon W.   USDA-NRCS, National Soil Survey Center, Fed. Bldg. Room 152 MS41, 100 Centennial Mall N., Lincoln, NE 68508-3866

Wilding, Larry P.   Professor of Pedology, Dept. of Soil and Crop Sciences, Texas A&M University, College Station, TX 77843-2474

Zentner, R.P.   Agriculture and Agri-Food Canada, SPARC, P.O. Box 1030, Swift Current, SK S9H 3X2

Zobeck, Ted M.   Soil Scientist, USDA-Agricultural Research Service, Wind Erosion & Water Cons. Res. Unit, 3810 4th St., Lubbock, TX 79415

Conversion Factors for SI and non-SI Units

Conversion Factors for SI and non-SI Units

To convert Column 1 into Column 2, multiply by	Column 1 SI Unit	Column 2 non-SI Units	To convert Column 2 into Column 1, multiply by
Length			
0.621	kilometer, km (10^3 m)	mile, mi	1.609
1.094	meter, m	yard, yd	0.914
3.28	meter, m	foot, ft	0.304
1.0	micrometer, μm (10^{-6} m)	micron, μ	1.0
3.94×10^{-2}	millimeter, mm (10^{-3} m)	inch, in	25.4
10	nanometer, nm (10^{-9} m)	Angstrom, Å	0.1
Area			
2.47	hectare, ha	acre	0.405
247	square kilometer, km² (10^3 m)²	acre	4.05×10^{-3}
0.386	square kilometer, km² (10^3 m)²	square mile, mi²	2.590
2.47×10^{-4}	square meter, m²	acre	4.05×10^3
10.76	square meter, m²	square foot, ft²	9.29×10^{-2}
1.55×10^{-3}	square millimeter, mm² (10^{-3} m)²	square inch, in²	645
Volume			
9.73×10^{-3}	cubic meter, m³	acre-inch	102.8
35.3	cubic meter, m³	cubic foot, ft³	2.83×10^{-2}
6.10×10^4	cubic meter, m³	cubic inch, in³	1.64×10^{-5}
2.84×10^{-2}	liter, L (10^{-3} m³)	bushel, bu	35.24
1.057	liter, L (10^{-3} m³)	quart (liquid), qt	0.946
3.53×10^{-2}	liter, L (10^{-3} m³)	cubic foot, ft³	28.3
0.265	liter, L (10^{-3} m³)	gallon	3.78
33.78	liter, L (10^{-3} m³)	ounce (fluid), oz	2.96×10^{-2}
2.11	liter, L (10^{-3} m³)	pint (fluid), pt	0.473

Mass

Column 1 → Column 2	Column 1 SI Unit	Column 2 non-SI Unit	Column 2 → Column 1
2.20×10^{-3}	gram, g (10^{-3} kg)	pound, lb	454
3.52×10^{-2}	gram, g (10^{-3} kg)	ounce (avdp), oz	28.4
2.205	kilogram, kg	pound, lb	0.454
0.01	kilogram, kg	quintal (metric), q	100
1.10×10^{-3}	kilogram, kg	ton (2000 lb), ton	907
1.102	megagram, Mg (tonne)	ton (U.S.), ton	0.907
1.102	tonne, t	ton (U.S.), ton	0.907

Yield and Rate

Column 1 → Column 2	Column 1 SI Unit	Column 2 non-SI Unit	Column 2 → Column 1
0.893	kilogram per hectare, kg ha⁻¹	pound per acre, lb acre⁻¹	1.12
7.77×10^{-2}	kilogram per cubic meter, kg m⁻³	pound per bushel, lb bu⁻¹	12.87
1.49×10^{-2}	kilogram per hectare, kg ha⁻¹	bushel per acre, 60 lb	67.19
1.59×10^{-2}	kilogram per hectare, kg ha⁻¹	bushel per acre, 56 lb	62.71
1.86×10^{-2}	kilogram per hectare, kg ha⁻¹	bushel per acre, 48 lb	53.75
0.107	liter per hectare, L ha⁻¹	gallon per acre	9.35
893	tonne per hectare, t ha⁻¹	pound per acre, lb acre⁻¹	1.12×10^{-3}
893	megagram per hectare, Mg ha⁻¹	pound per acre, lb acre⁻¹	1.12×10^{-3}
0.446	megagram per hectare, Mg ha⁻¹	ton (2000 lb) per acre, ton acre⁻¹	2.24
2.24	meter per second, m s⁻¹	mile per hour	0.447

Specific Surface

Column 1 → Column 2	Column 1 SI Unit	Column 2 non-SI Unit	Column 2 → Column 1
10	square meter per kilogram, m² kg⁻¹	square centimeter per gram, cm² g⁻¹	0.1
1000	square meter per kilogram, m² kg⁻¹	square millimeter per gram, mm² g⁻¹	0.001

Pressure

Column 1 → Column 2	Column 1 SI Unit	Column 2 non-SI Unit	Column 2 → Column 1
9.90	megapascal, MPa (10^6 Pa)	atmosphere	0.101
10	megapascal, MPa (10^6 Pa)	bar	0.1
1.00	megagram, per cubic meter, Mg m⁻³	gram per cubic centimeter, g cm⁻³	1.00
2.09×10^{-2}	pascal, Pa	pound per square foot, lb ft⁻²	47.9
1.45×10^{-4}	pascal, Pa	pound per square inch, lb in⁻²	6.90×10^3

(continued on next page)

Conversion Factors for SI and non-SI Units

To convert Column 1 into Column 2, multiply by	Column 1 SI Unit	Column 2 non-SI Units	To convert Column 2 into Column 1, multiply by
Temperature			
$1.00\ (K - 273)$	kelvin, K	Celsius, °C	$1.00\ (°C + 273)$
$(9/5\ °C) + 32$	Celsius, °C	Fahrenheit, °F	$5/9\ (°F - 32)$
Energy, Work, Quantity of Heat			
9.52×10^{-4}	joule, J	British thermal unit, Btu	1.05×10^3
0.239	joule, J	calorie, cal	4.19
10^7	joule, J	erg	10^{-7}
0.735	joule, J	foot-pound	1.36
2.387×10^{-5}	joule per square meter, J m^{-2}	calorie per square centimeter (langley)	4.19×10^4
10^5	newton, N	dyne	10^{-5}
1.43×10^{-3}	watt per square meter, W m^{-2}	calorie per square centimeter minute (irradiance), cal cm^{-2} min^{-1}	698
Transpiration and Photosynthesis			
3.60×10^{-2}	milligram per square meter second, mg m^{-2} s^{-1}	gram per square decimeter hour, g dm^{-2} h^{-1}	27.8
5.56×10^{-3}	milligram (H_2O) per square meter second, mg m^{-2} s^{-1}	micromole (H_2O) per square centimeter second, μmol cm^{-2} s^{-1}	180
10^{-4}	milligram per square meter second, mg m^{-2} s^{-1}	milligram per square centimeter second, mg cm^{-2} s^{-1}	10^4
35.97	milligram per square meter second, mg m^{-2} s^{-1}	milligram per square decimeter hour, mg dm^{-2} h^{-1}	2.78×10^{-2}
Plane Angle			
57.3	radian, rad	degrees (angle), °	1.75×10^{-2}

Electrical Conductivity, Electricity, and Magnetism

To convert Column 1 into Column 2, multiply by	Column 1 SI Unit	Column 2 non-SI Unit	To convert Column 2 into Column 1, multiply by
10	siemen per meter, S m⁻¹	millimho per centimeter, mmho cm⁻¹	0.1
10^4	tesla, T	gauss, G	10^{-4}

Water Measurement

To convert Column 1 into Column 2, multiply by	Column 1 SI Unit	Column 2 non-SI Unit	To convert Column 2 into Column 1, multiply by
9.73×10^{-3}	cubic meter, m³	acre-inch, acre-in	102.8
9.81×10^{-3}	cubic meter per hour, m³ h⁻¹	cubic foot per second, ft³ s⁻¹	101.9
4.40	cubic meter per hour, m³ h⁻¹	U.S. gallon per minute, gal min⁻¹	0.227
8.11	hectare meter, ha m	acre-foot, acre-ft	0.123
97.28	hectare meter, ha m	acre-inch, acre-in	1.03×10^{-2}
8.1×10^{-2}	hectare centimeter, ha cm	acre-foot, acre-ft	12.33

Concentrations

To convert Column 1 into Column 2, multiply by	Column 1 SI Unit	Column 2 non-SI Unit	To convert Column 2 into Column 1, multiply by
1	centimole per kilogram, cmol kg⁻¹	milliequivalent per 100 grams, meq 100 g⁻¹	1
0.1	gram per kilogram, g kg⁻¹	percent, %	10
1	milligram per kilogram, mg kg⁻¹	parts per million, ppm	1

Radioactivity

To convert Column 1 into Column 2, multiply by	Column 1 SI Unit	Column 2 non-SI Unit	To convert Column 2 into Column 1, multiply by
2.7×10^{-11}	becquerel, Bq	curie, Ci	3.7×10^{10}
2.7×10^{-2}	becquerel per kilogram, Bq kg⁻¹	picocurie per gram, pCi g⁻¹	37
100	gray, Gy (absorbed dose)	rad, rd	0.01
100	sievert, Sv (equivalent dose)	rem (roentgen equivalent man)	0.01

Plant Nutrient Conversion

To convert Column 1 into Column 2, multiply by	Elemental	Oxide	To convert Column 2 into Column 1, multiply by
2.29	P	P_2O_5	0.437
1.20	K	K_2O	0.830
1.39	Ca	CaO	0.715
1.66	Mg	MgO	0.602

1 Soils and the Greenhouse Effect

Rattan Lal

The Ohio State University
Columbus, Ohio

ABSTRACT

Atmospheric concentration of CO_2 has increased by about 32% [from 0.55 g m^{-3} (280 ppm) in 1700–0.727 g m^{-3} (370 ppm) in 2000]. Three principal sources of the atmospheric increase in CO_2 include fossil fuel combustion, deforestation and soil cultivation, and industrial manufacture of cement and fertilizers/lime materials. World soils comprise the third largest among active global C pools (1550 Pg of organic C and 750 Pg of inorganic C to 1-m depth), which is 3.2 times the atmospheric C pool (720 Pg) and 4.1 times the biotic pool (560 Pg). The soil organic carbon (SOC) pool has lost about 66 to 90 Pg C to the atmosphere due to conversion of natural to agricultural ecosystems and soil cultivation. Most agricultural soils have lost 25 to 75% of their original pool, and severely degraded soils have lost 70 to 80% of the antecedent pool. Restoration of eroded and degraded soils, and adoption of recommended agricultural practices can lead to sequestration of 60 to 80% of the historic soil C loss. There also is some potential for sequestration of soil inorganic carbon (SIC) by formation of secondary carbonates and leaching of carbonates in the groundwater. Realization of the soil C sequestration potential depends on development and adoption of appropriate soil management technology based on understanding of the soil processes and biogeochemical cycles of C, N, P and S.

Atmospheric concentration of carbon dioxide (CO_2) and other greenhouse gases (GHGs), (N_2O, CH_4, etc.) has increased steadily since the on-set of the Industrial Revolution around 1850 (Etheridge et al., 1996). The atmospheric concentration of CO_2 was about 0.51 g m^{-3} (260 ppmv) to 0.55 g m^{-3} (280 ppmv) from 4000 BC to 1700 AD and 0.55 g m^{-3} (280 ppmv) from 1700 to 1750 AD (Table 1–1). Since then, the concentration has steadily increased to about 0.727 g m^{-3} (370 ppmv) in 2000, and is currently increasing at the rate of about 0.5% yr^{-1}. If the current rate of increase in concentration of GHGs continues, significant climate changers are projected to occur during the 21st century. The business-as-usual (BAU) projection of CO_2 shows an increase in CO_2 concentration to 1.179 g m^{-3} (600 ppmv) in 2100, 2.553 g m^{-3} (1300 ppmv) in 2200 and leveling off at 3.536 g m^{-3} (1800 ppmv) in the year 2400 (Harvey, 2000). The projected increase, according to the plausible BAU scenario, may have drastic impact on the climate and natural and managed ecosystems (Rosenzweig & Hillel, 1998).

Table 1–1. Troporpheric concentration of CO_2 over 6000 yr (Edmonds, 1999).

Year	CO_2 concentration	
	ppmv	g m^{-3}†
4000 BC–1700 AD	260–280	0.51–0.55
1750	280	0.55
1800	285	0.56
1850	291	0.572
1900	300	0.589
1950	312	0.613
1959	316	0.621
1990	340	0.668
1994	358	0.703
1998	367	0.721
2000	370	0.727

† Concentration in SI units is computed on the basis of 44 g CO_2 combined in 22.4 L of CO_2 at standard temperature and pressure. The gaseous concentration in milligrams per liter is equal to grams per cubic meter

There are three principal anthropogenic activities that have disturbed the global C cycle, leading to atmospheric enrichment of CO_2 and CH_4. A major source of CO_2 is the fossil fuel (coal, oil and natural gas) combustion. Fossil fuel currently supplies about 85% of the world's total energy. The CO_2 emission factor is 22.2 to 26.6 kg C GJ^{-1} of energy for coal, 17 to 20 kg C GJ^{-1} of energy for oil and 13.5 to 14.0 kg C GJ^{-1} cf energy for natural gas (Harvey, 2000). Therefore decarbonizing the energy source would have an obvious impact on reducing the CO_2 emissions into the atmosphere.

The second industrial source of CO_2 emission is the production of cement, lime and ammonia. The emission factor in relation to these three industrial processes are 0.08 to 0.26 Mg C Mg^{-1} for cement, 0.215 Mg C Mg^{-1} for lime, and 0.431 Mg C Mg^{-1} for NH_3 (Harvey, 2000). Of these three industrial products, lime and ammonia are two important agricultural inputs used as plant nutrients and soil amendments.

The third source of CO_2 emission involves a wide range of agricultural activities. Important among these are deforestation and biomass burning leading to conversion of natural to agricultural ecosystems (Table 1–2). These activities affect both biotic and soil C pools leading to flux of GHGs (CO_2, CH_4 and N_2O) from soil to the atmosphere. The objective of this chapter is to discuss the importance of soil C pool in the global C cycle, to assess the impact of soil management practices on soil C pool and dynamics, and describe the potential of soil C sequestration to reduce the emission of CO_2 and other GHGs into the atmosphere.

SOIL CARBON POOL AND THE GLOBAL CARBON CYCLE

There are several active pools or temporary holding areas of C, and C cycles continuously between these pools. The C cycling process is complex, and the dynamics of some pools is not well understood. The complexity of the C cycling is attributed to numerous causes as follows: (i) C occurs in different chemical forms in different pools, (ii) the fluxes among pools are governed by numerous and com-

Table 1–2. The density of organic C in soil of different ecoregions (modified from Batjes, 1999).

Biome	SOC density	
	0.3-m depth	1-m depth
	— kg m^{-2} —	
A. Tropics		
Warm humid	5.2–5.4	10.0–10.4
Warm seasonally dry	3.6–3.8	7.0–7.3
Cool	4.4–4.7	8.4–8.9
Arid	2.0–2.2	3.7–4.1
B. Subtropics		
Summer rains	4.5–4.7	8.6–9.1
Winter rains	3.6–3.9	7.2–8.0
C. Temperate		
Oceanic	5.8–6.4	11.7–12.9
Continental	5.6–5.9	10.8–11.3
D. Boreal	9.8–10.2	23.1–24.0
E. Polar and Alpine	7.0–7.8	20.6–23.2

plex processes, and (iii) the atmospheric concentration of CO_2 responds to change in climate on a time scale ranging from day to millennia.

For example, C in the soil pool occurs in forms ranging from simple gas (CO_2, CH_4) to complex organomineral complexes and live biomass. The flux of C between the pedosphere (soil) and the atmosphere is mediated by numerous processes occurring in the pedosphere, atmosphere, biosphere, hydrosphere and the lithosphere (Fig.1–1). The complexity of C cycling among pools is enhanced by several positive and negative feedback mechanisms.

The analyses of the data in Fig. 1–1 show that the total amount of C in the upper 1-m depth of soil solum estimated at 2300 Pg (SOC = 1550 Pg and SIC = 750 Pg) is about: (i) 3.2 times the C in the atmospheric pool, (ii) 4.1 times the C in the biotic pool, and (iii) the soil C pool is directly linked with the atmospheric pool and change of 1 Pg of SOC pool is equal to 0.92×10^{-3} g m^{-3} (0.47 ppmv) of the atmospheric pool.

The C cycling between soil, atmosphere and the biotic pools is mediated by photosynthesis and respiration. The return of C from soil is mostly in the form of CO_2 but also as CH_4 from depressional sites, wetlands and termite mounds. Some of the soil C also is transferred to the groundwater as bicarbonates and others to the river and aquatic ecosystems along with eroded sediments (Lal, 1995). Net gain in the atmospheric pool is occurring at the rate of about 3.4 Pg yr^{-1}. This gain is primarily at the expense of the geologic pool (fossil fuel combustion), biotic, and the soil or the pedologic pool. Because land use and soil cultivation have contributed substantially to the atmospheric enrichment of CO_2, it is important to understand the processes governing the magnitude and properties of the soil C pool.

Soil Carbon Pools

Carbon in soil, biologically active upper layer of the earth crust to about 1-m depth, occurs in at least two forms, SOC and SIC. The SOC pool comprises highly

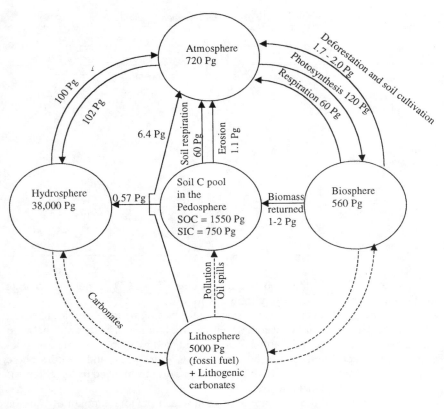

Fig. 1–1. Principal global C pools and fluxes between them. The number in the circle refers to the magnitude of the pool and arrows to the flux or the rate of C transfer between two pools. The direction of the arrow indicates the direction of the transfer of the C.

active humus and relatively inert charcoal C. In general terms, soil organic matter of which C constitutes about 58% of the total mass, includes a wide range of organic substances in soil. These substances comprise a mixture of plant and animal residues at various stages of decomposition and of the bodies of live microorganisms and small animals and their decomposing products, and of substances synthesized microbiologically or chemically from these products. These products may include: (i) soluble organic compounds such as sugars, proteins and other metabolic products, (ii) amorphous organic compounds such as humic acids, fats, waxes, oils, lignin and polyuronides, and (iii) organomineral complexes that involve hybrid compounds of organic molecules attached to clay particles (Schnitzer, 1991). The magnitude of SOC pool is estimated at 684 to 724 Pg in the 0- to 30-cm layer, 1468 to 1548 Pg in 0- to 100-cm layer and 2376 to 2456 Pg in 0- to 200-cm layer (Batjes, 1996).

The density of SOC varies widely among soils of different ecoregions (Table 1–2). Within a biome, the SOC density increases with increase in precipitation. Between biomes, the SOC density is higher in cool and moist than in warm and dry

Table 1–3. Anthropogenic activities leading to C emission from soil to the atmosphere.

SOC dynamics		SIC dynamics	
Activity	Flux/emissions	Activity	Flux/emissions
Deforestation	CO_2, N_2O	Irrigation	Bringing carbonates to the surface
Biomass burning	CO_2, CO, CH_4, N_2O	Fertilizer and liming	Acidification
Plowing	CO_2	Plowing & exposure of caliche	Acid rain
Fertilizer use	N_2O		
Manure application	CH_4, N_2O		
Drainage	CO_2		

biomes. The total amount of SOC pool to 1-m depth is 384 to 403 Pg in the tropics compared with 1078 to 1145 Pg in other biomes. The amount of SOC to 2-m depth is 616 to 640 Pg in the tropics vs. 1760 to 1816 Pg in other biomes (Batjes, 1999). Soils of the tropics, therefore, contribute 26% of the global SOC pool to both 1- and 2-m depths.

The SIC pool, mostly in the form of carbonates, can be significant in soils of the arid and semiarid regions (Lal & Kimble, 2000; Scharpenseel et al., 2000). The global estimate of SIC pool ranges from 222 to 245 Pg for 0- to 30-cm depth and 695 to 748 Pg to 1-m depth. The total amount of SIC pool to 1-m depth is 203 to 218 Pg in soils of the tropics compared with 492 to 530 Pg in soils of other regions (Batjes, 1999). Soils of the tropics, therefore, contribute 29% of the global SIC pool to 1-m depth. The SIC pool comprises two components: lithogenic inorganic carbon (LIC) and pedogenic inorganic carbon (PIC). The latter comprises secondary carbonates that are formed through the decomposition/dissolution of LIC or carbonate bearing minerals, and re-precipitation of weathering products, or through precipitation of atmospheric CO_2 with Ca^{2+} and Mg^{2+} or as other salts in soils. Formation of secondary carbonates and leaching of carbonates and bicarbonates are important processes in arid and semiarid regions leading to sequestration of inorganic C in soil and aquatic ecosystems (Nordt et al., 2000).

SOIL CARBON DYNAMICS AND THE ACCELERATED GREENHOUSE EFFECT

The SOC pool is very active, and plays an important role in the global C cycle. Principal anthropogenic activities that lead to depletion of SOC pool and emission of GHGs from soil to the atmosphere are listed in Table 1–3. Important activities that reduce SOC pool and accentuate emissions of GHGs include deforestation and biomass burning, plowing, drainage and indiscriminate use of fertilizers and lime. Conversion of natural to agricultural ecosystems have caused depletion of the global SOC pool by 55 Pg (IPCC, 1996) to 66 to 90 Pg (Lal, 1999). The annual losses from SOC pool are estimated in the range of 1 to 2 Pg C yr^{-1} (Bouwman & Sombroek, 1990). Principal processes that lead to depletion of SOC pool include mineralization, oxidation, methanogenesis and leaching. The rate and magnitude of SOC depletion is accentuated by soil degradative processes (e.g., accelerated erosion, nutrient depletion and salt imbalance in the root zone etc). Schematic of these processes

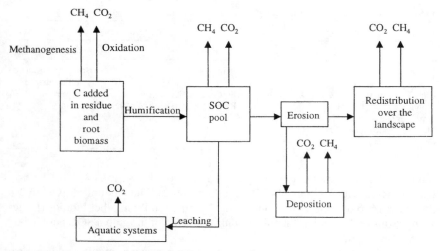

Fig. 1–2. Processes affecting soil C dynamics.

is shown in Fig. 1–2. Soil degradation accentuates emission of GHGs from soil to the atmosphere.

The dynamics of SIC is even less understood than that of SOC. Some anthropogenic activities may accentuate emission of CO_2 from the SIC pool (Table 1–3). These activities include irrigation with groundwater rich in carbonates, acidification caused by indiscriminate use of fertilizers and manures, and plowing or physical perturbation of the caliche or calciferous material exposing it to the climatic elements. The amount of SIC released into the atmosphere by anthropogenic activities is not known.

Soil Carbon Sink and Sequestration Potential

Carbon sequestration implies removal of carbon (as CO_2) from the atmosphere through photosynthesis and dissolution and the storage of C in soil as organic matter or secondary carbonates. Carbon sequestration as SOC is based on the proportion of NPP returned to the soil and conversion of biomass into humus. Processes leading to SOC sequestration include humification (conversion of biomass into humus), aggregation (formation of organomineral complexes as secondary particles), and translocation into subsoil (of biomass by deep roots and bioturbation, and of SIC by leaching into the groundwater as bicarbonates). These processes are set in motion by choice of an appropriate land use, restoration of degraded soils and ecosystems, and agricultural intensification through adoption of recommended agricultural practices (Table 1–4). Improved practices leading to agricultural intensification include no till or conservation tillage, soil-water conservation and water table management, judicious use of fertilizers in conjunction with the precision farming, use of improved varieties and cropping systems, growing cover crops and elimination of summer fallow (Lal, 1999). The potential of SOC sequestration through adoption of these practices is large, ranging from 75 to 208 Tg C yr^{-1} for the U.S.

Table 1–4. Anthropogenic activities leading to C sequestration in soil.

Activity	Process of soil C sequestration
Soil restoration	Humification, aggregation, calcification, translocation into subsoil
Conservation tillage	Aggregation, humification, translocation into subsoil
Residue/biomass mulching	Humification, aggregation
Integrated nutrient management	Humification, aggregation
Soil conversion	Aggregation, calcification, leaching of carbonates
Irrigation and water conservation	Calcification, humification, leaching of carbonates
Agricultural intensification	Humification, calcification

cropland (Lal et al., 1998) and 0.73 to 0.87 Pg C yr^{-1} for the world cropland (Lal & Bruce, 1999), 0.9 to 1.9 Pg C yr^{-1} through desertification control (Lal et al., 1999) and 1.2 to 2.6 Pg C yr^{-1} through management of world soils (Lal, 1999). Realization of this potential over the next 25 to 50 yr may have an important impact on the magnitude of emissions of CO_2 into the atmosphere.

Carbon sequestration as SIC is based on three processes: calcification, formation of secondary carbonates, and leaching of carbonates into the groundwater. The rate of C sequestration as SIC in the form of caliche and secondary carbonates is estimated at 10 to 11 Tg C yr^{-1} (Scharpenseel, et al., 1995; Schlesinger, 1997). A major mechanism of SIC sequestration, however, is through leaching of carbonates into the ground water (Nordt et al., 2000).

CONCLUSIONS

Soil C constitutes a large and an active pool. It is widely recognized that anthropogenic activities have caused depletion of the organic component of the soil C pool by 66 to 90 Pg C leading to emissions of GHGs into the atmosphere. Soil C emission is the second largest source of C emission (next to the fossil fuel) into the atmosphere. Most cultivated soils have lost 30 to 50% of their original SOC pool, whereas severely eroded soils may have lost 70 to 80% of the original pool. Soil C emissions are accentuated by deforestation and biomass burning undertaken for conversion of natural to agricultural or managed ecosystems, plowing, drainage of wetlands, and use of subsistence or soil fertility mining practices based on low external input.

Depending on land use and soil management, world soils can be a source or sink for atmospheric CO_2. The potential of world soils for C sequestration to mitigate the greenhouse effect is now being widely recognized. Restoration of degraded soils and ecosystems can lead to sequestration of 60 to 80% of the C lost due to historic mismanagement. Similarly, adoption of recommended agricultural practices can lead to increase in SOC pool in cropland and grazing lands. In such cases, the SOC pool stabilizes at a new equilibrium level because the increase in input of biomass exceeds the output. The SOC contents in agricultural soils, because of severe depletion, are now below their potential levels. Thus, there is a vast opportunity to increase soil C pool through adoption of recommended agricultural practices.

There is some scope, albeit lower than that of the SOC, for C sequestration as SIC. The potential may be realized through adopting improved systems of irrigation water management, and application of biosolids and other amendments that add Ca^{2+}, Mg^{2+} and other cations from outside the ecosystem.

Realization of the soil C sequestration potential requires a thorough understanding of the soil processes and properties that affect dynamics of SOC and SIC pools.

REFERENCES

Batjes, N.H. 1996. Total carbon and nitrogen in soils of the world. Eur. J. Soil Sci. 47:151–163.

Batjes, N.H. 1999. Management options for reducing CO_2 concentrations in the atmosphere by increasing Carbon sequestration in the soil. ISRIC Tech. Pap. 30. ISRIC, Wageningen, Holland.

Bouwman, A.F., and W.G. Sombroek. 1990. Inputs to climate change by soils and agriculture related activities: Present status and possible future trends. p. 15–30. In H.W. Scharpenseel et al. (ed.) Soils on a warmer earth. Develop. Soil Sci. Ser. Elsevier, Amsterdam.

Edmonds, J.A. 1999. Beyond Kyoto: Toward a technology greenhouse strategy. Consequences 5:17–28.

Etheridge, D.M., L.D. Steele, R.L. Langenfelds, R.J. Francey, I.M. Barnola, and V.I. Morgan. 1996. Natural and anthropogenic changes in atmospheric CO_2 over the last 10,000 years from air in Atlantic ice and fern. J. Geograph. Res. 101:4115–4128.

Halman, M.M., and M. Steinbeg. 1999. Greenhouse gas: Carbon dioxide mitigation. Science and technology. Lewis Publ., Boca Raton, FL.

Harvey, L., and D. Danney. 2000. Global warming: The hard science. Prentice Hall, Pearson Educ. Ltd., Harlow, UK.

Intergovernmental Panel on Climate Change. 1996. Climate change 1995. Work. Group 1. IPCC, Cambridge Univ. Press, Cambridge.

Lal, R. 1995. Global soil erosion by water and carbon dynamics. p. 131–142. In R. Lal et al. (ed.) Soils and global change. CRC/Lewis Publ., Boca Raton, FL.

Lal, R. 1999. Soil management and restoration for C sequestration to mitigate the accelerated greenhouse effect. Progr. Environ. Sci. 1:307–326.

Lal, R., and J.M. Kimble. 2000. Pedogenic carbonates and the global C cycle. p. 1–25. In R. Lal et al. (ed.) Global climate change and pedogenic carbonates. CRC/Lewis Publ., Boca Raton, FL.

Lal, R., and J.P. Bruce. 1999. The potential of world cropland soils to sequester carbon and mitigate the greenhouse effect. Environ. Sci. Policy 2:177–185.

Lal, R., H.M. Hassan, and J.W. Dumanski. 1999. Desertification control for C sequestration to mitigate the greenhouse effect. p. 83–107. In N.J. Rosenberg et al. (ed.) Carbon sequestration in soils: Science, monitoring and beyond. Battelle Press, Columbus, OH.

Lal, R., J.M. Kimble, R.F Follett, and C.V. Cole. 1998. The potential of U.S. cropland to sequester carbon and mitigate the greenhouse effect. Ann Arbor Press, Chelsea, MI.

Nordt, L.C., L.P. Wilding and L.R. Drees. 2000. Pedogenic carbonate transformations in leaching soil systems: Implications for global carbon cycle. p. 43–63. In R. Lal et al. (ed.) Global climate change and pedogenic carbonates. CRC/Lewis Publ., Boca Raton, FL.

Rosenzweig, C., and D. Hillel. 1998. Climate change and the global harvest: Potential impacts of the greenhouse effect on agriculture. Oxford Univ. Press, Oxford, UK.

Scharpenseel, H.W., J. Freytag, and E.M. Pfeiffer. 1995. The carbon budget in drylands: Assessment based on C residence time and stable isotope formation. p. 79–93. In V.R. Squires et al. (ed.) Combating global climate change by combating land degradation. UNEP, Nairobi, Kenya.

Scharpenseel, N.W., A. Mtimet, and J Freytag. 2000. Soil inorganic carbon and global change. p. 27–42. In R. Lal et al. (ed.) Global climate change and pedogenic carbonates. CRC/Lewis Publ., Boca Raton, FL.

Schlesinger, W.H. 1997. Biogeochemistry: An analysis of global change. Acad. Press, San Diego, CA.

Schnitzer, M. 1991. Soil organic matter: The next 75 years. Soil Sci. 151:41–58.

2   Myths and Facts About Soils and the Greenhouse Effect

Rattan Lal

The Ohio State University
Columbus, Ohio

ABSTRACT

It is not widely recognized that agricultural soils can be a sink for atmospheric C. Whereas land misuse and soil mismanagement can make soil a source of CO_2, CH_4 and N_2O, adoption of recommended agricultural practices can re-sequester C in soil making it a net sink. Agricultural practices that lead to emission of greenhouse gases (GHGs) from soil include conversion of natural to agricultural ecosystems, deforestation, biomass burning, indiscriminate use of fertilizers and manures, and excessive tillage. Emission of GHGs is exacerbated by soil degradation caused by accelerated erosion, acidification, salinization, and reduction in soil biodiversity. The historic loss of C from soil with attendant emission into the atmosphere is estimated at 66 to 90 Pg. In contrast, restoration of degraded soils and adoption of recommended agricultural practices can reverse the degradative trends and make world soils a net sink for atmospheric C. Soil C sequestration, removal of C from atmosphere via photosynthesis and its storage in soil as humus, is achieved through: (i) restoration of degraded soils, (ii) conversion of agriculturally marginal lands into restorative land uses, and (iii) management of prime soils with the recommended agricultural practices. The potential of soil C sequestration through such measures may be 50 to 70 Pg over a 25- to 50-yr period. However, the realization of this vast potential of agricultural soils for C sequestration depends on replacement of popular myths by facts through improvements in the knowledge base and strengthening of the data bank on soil properties at the pedon level.

Soil C pool is one of the five active global pools: (i) oceanic pool comprising carbonates estimated at 38 000 Pg (1 Pg = petagram = 10^{15} g = 1 billion t), (ii) geologic pool comprising fossil fuel at 5000 Pg and containing 4000 Pg of coal and 500 Pg of each of oil and gas, (iii) soil C pool at 2300 Pg to 1-m depth containing 1550 Pg of soil organic carbon (SOC) and 750 Pg of soil inorganic carbon (SIC), (iv) the atmospheric pool at 720 Pg, and (v) the biotic pool at 550 Pg. The atmospheric C pool has been steadily increasing since the industrial revolution with CO_2 concentration of 0.55 g m^{-3} (280 ppmv) around 1850 to about 0.727 g m^{-3} (370 ppmv) in 2000. The atmospheric C pool is increasing at the rate of 3.5×10^{-3} g m^{-3} (1.8 ppmv), 0.5% yr^{-1} or 3.3 Pg C yr^{-1} (Etheridge et al., 1996). An increase by about 30% in the atmospheric pool has occurred due to depletion of geologic pool (fossil fuel emissions currently at the rate of 6.4 Pg C yr^{-1}) and the soil/biotic pools (about 1.7–2 Pg C yr^{-1}) due to land use change and soil cultivation. Emissions from the soil and biotic pools comprised the principal source of the atmospheric increase in CO_2 until the mid-20th century. During the past 150 yr from 1850 to 2000, approximately 270 ± 30 Pg of C has been emitted into the atmosphere from fossil fuel burning and cement production, and about 136 ± 56 Pg C from land use change (IPCC, 2000). Both SOC and SIC constituents are among the active pools, influenced by anthropogenic activities, and are directly linked with the atmospheric pool. Change in 1 Pg of soil C pool is approximately equal to 0.47 ppmv of atmospheric CO_2.

Despite its significance in the global C cycle and the impact on atmospheric CO_2, the importance of soil C in the global C cycle is not widely recognized. Consequently, there are numerous myths about soil C pool in relation to the observed increase in atmospheric concentration of CO_2. Some common myths about soil C pool are discussed below in view of the current state-of-the-knowledge.

COMMON MYTHS AND SCIENTIFIC FACTS
ABOUT THE SOIL CARBON POOL

There are several common myths about soil C and its role in the accelerated greenhouse effect. These myths perpetuate misunderstanding and must be replaced by facts.

Myth 1: Emission From Soils and Biotic Pools Have Made Minor Contributions to Atmospheric CO_2 Enrichment.

A considerable amount of the 32% increase in atmospheric CO_2 concentration is attributable to depletion of the soil and biotic C pools. Until 1930s, as much as 90% of the increase in atmospheric CO_2 came from land use change and soil cultivation. During the 1990s, CO_2 emission from deforestation, land use change and soil cultivation has been estimated at 1 to 2 Pg C yr^{-1}. Since the dawn of settled agriculture, however, 136 Pg of C has been emitted from soil/biotic pools into the atmosphere that constitutes about 50% of the cumulative emission from fossil fuel combustion. The soil C emission is estimated at 66 to 90 Pg (Lal, 1999).

Myth 2: Emissions of Green House Gases from the Soil and Biotic Pools Have Been Significant Only Since the 1950s

The soil C emissions were probably the most significant during the period of rapid expansion of cropland area. In North America, the period of a rapid expansion of agricultural land was from about 1850 to 1950. Rozanov et al. (1990) estimated the emission from soil C pool at the rate of 15 Tg yr^{-1} (1 Tg = teragram $= 1 \times 10^{12}$ g = 1 million t) since agriculture began 10 000 yr ago, 200 Tg yr^{-1} for 300 yr from 1650 to 1950, and 440 Tg yr^{-1} from 1950 to 2000. Based on these estimates, historic loss of soil C may be as much as 230 Pg.

Myth 3: The Historic Loss of Soil Carbon Is Too Small to Warrant Strategic Planning for Carbon Re-Sequestration

Estimates of the historic loss of soil C to the atmosphere range widely. The historic loss has been estimated at 40 Pg (Batjes, 1996), 55 Pg (IPCC, 1996), 66 to

90 Pg (Lal, 1999), 150 Pg (Bohn, 1978), 500 Pg (Wallace, 1994), and 537 Pg (Buringh, 1984). These estimates vary by a factor of 13 to 14 and indicate the strong need for obtaining credible estimates of the historic loss of C due to change in land use and soil cultivation. The research information provides a reference point about the potential of soil C sequestration through adoption of an appropriate land use and recommended soil/crop management practice.

Myth 4: Soil Erosion Merely Leads to Carbon Redistribution Over the Landscape

Depletion of the soil C pool by agricultural practices is exacerbated by accelerated erosion due to wind and water. Water erosion is an important factor affecting soil C pool on sloping lands, whereas wind erosion is active on flat lands in dry regions. Accelerated erosion accentuates C depletion due to its adverse effects on both SOC and SIC pools, but especially on depletion of the SOC pool. The SOC is concentrated in the surface layer and, because of its low density (1.2–1.5 Mg m^{-3}), is preferentially removed. Consequently, the eroded sediments contain high concentration of SOC and clay contents. In fact, the enrichment ratio, the relative concentration of SOC in eroded sediments compared to that of the original soil, can be 1 to 30 for both water and wind erosion (Lal, 1976; Cihacek et al., 1993; Zobeck & Fryrear, 1986; Wan & El-Swaify, 1997; Palis et al., 1997).

The fate of soil C redistributed over the landscape by erosional processes depends on land use, properties of the depositional site, and hydrological characteristics of the landscape. Erosion leads to breakdown of aggregates, and SOC hitherto encapsulated is now exposed to microbial processes and climatic elements. Therefore, some of the SOC redistributed over the landscape may be readily mineralized. In contrast, the SOC brought in with the sediments also may enhance aggregation and reduce the rate of mineralization (Gregorich et al., 1998).

Myth 5: Sediment Deposition in Depressional Sites and Aquatic Ecosystems Lead to Carbon Sequestration

Some of the C buried in depressional sites and aquatic ecosystems is buried and taken out of circulation. However, the fate of SOC buried in depositional sites also depends on soil and hydrological characteristics. Some of the C carried into depressional sites may be emitted as CO_2 or CH_4 depending upon the degree of anoxia. Lal (1995) estimated that burial of C transported by soil erosion leads to sequestration of 0.57 Pg C yr^{-1}. In contrast, 1.14 Pg C yr^{-1} from that redistributed over the landscape is emitted into the atmosphere and 3.92 Pg remains within the landscape. In case of reducing conditions, methanogenesis also can set in leading to emission of CH_4. Stallard (1998) estimated that 1 Pg C yr^{-1} deposited in aquatic ecosystems and depositional sites may be sequestered and taken out of circulation. Similar to C, the fate of N translocated with the sediments also is not known. Some of the N may undergo denitrification leading to the emission of N_2O and NO_x.

Because of its importance to both biomass productivity and environment quality, conducting systematic study of the fate of C and N translocated by erosion is a high priority. Such a study needs to be done in conjunction with assessment of soil quality, soil moisture and temperature regimes, and soil/crop management.

Myth 6: The Principal Adverse Off-Site Effect of Soil Erosion and the Attendant Degradation Is that on Water Quality

Soil degradation, decline in soil quality due to human misuse, has severe on-site and off-site effects. On-site, soil degradation affects farm operations (e.g., frequency and intensity of plowing, time of seedbed preparation and planting, rate and formulation of fertilizer amendments) and biomass productivity. Depending on its severity, soil degradation can affect current and potential productivity. Decline in potential productivity is mainly due to reduction in the topsoil depth and the attendant loss in the available water capacity of the soil. Off-site, soil degradation adversely impacts water quality through transport of dissolved and suspended loads into natural waters. In addition, soil degradation is the source of GHGs second only to the fossil fuel combustion. Soil degradation exacerbates the emission of GHGs through the following:

1. Soil physical degradation leads to breakdown of aggregates and exposure of C to the microbial processes. In addition, soils with degraded structure in the surface horizon lose their capacity to oxidize CH_4 and may become a net source of CH_4. The rate of denitrification of inherent and applied N also is high on soils with poor structural attributes.
2. Soil chemical degradation leads to nutrient imbalance, acidification and salinization. Low biomass productivity and less biomass returned to the soil causes depletion of the SOC pool. Soil sodication causes formation of massive structure that accentuates methanogenesis and denitrification.
3. Soil biological degradation affects both the quality and quantity of SOC pool and can exacerbate emission of GHGs from soil to the atmosphere.

Myth 7: Subsistence Farming and Low-Input or Resource-Based Agriculture Are Environmentally Friendly

Agricultural practices that are based on mining soil fertility produce low returns and adversely impact the environment. Risks of soil erosion are exacerbated by soil management practices that produce less ground cover and return little if any biomass to the soil. Soil organic matter is depleted by low-input systems that accentuate the mineralization whereby CO_2 is emitted into the atmosphere while mineral elements (N, P, K, Ca, Mg, etc.) are absorbed by crop plants. The rate of nutrient (e.g., N, P, K) depletion in sub-Saharan Africa (SSA) is estimated at 30 kg ha^{-1} yr^{-1} (Smaling et al., 1993, 1996), thereby making soils of the region a major source of CO_2 and other GHGs. In contrast, conversion of biomass C into stable humus also requires additional input of N, P, S and other micronutrients. Himes (1998) calculated that to sequester 10 Mg of C from biomass in humus requires 833 kg of N, 200 kg of P, and 143 kg of S. The addition of this amount of humus to the soil will increase the SOC content of the plow layer by 0.7%. However, agricultural practices based on no external input of fertilizer accentuate the reverse process of mineralization leading to emission of 10 Mg ha^{-1} of C and release of 833 kg of N, 200 kg of P and 143 kg of S assuming C/N of 12:1, C/P of 50:1 and C/S of 70:1 (Himes, 1998).

With a continuous increase in population pressure and the attendant decrease in per capita land area, agricultural intensification is the environment-friendly choice. Agricultural intensification implies use of best soils with the state-of-the-art technology to produce optimum yield so that marginal soils can be converted to natural ecosystems for nature conservancy and environment enhancement.

Myth 8: Application of Nitrogenous Fertilizers Leads to Carbon Emission Due to Fossil Fuel Used in Their Manufacture, Transport and Application

Production and use of nitrogenous fertilizer requires use of energy. Over and above the energy needed for Haber-Bosch process for the production of NH_3, IPCC (1996) estimated that 1 kg of N used as fertilizer requires 0.86 kg of C in energy input involving manufacture, transport and application of fertilizer. Therefore, a careful analysis of input and output is needed to assess the energy balance. Application of N fertilizer is essential for enhancing agronomic production to achieve global food security, especially in developing regions of the world such as SSA where food shortage is associated with net depletion of soil fertility being caused by the lack of input of fertilizers at the recommended rates.

A judicious application of nitrogenous fertilizers can lead to a positive C balance even in commercial agriculture. For example, desirable corn (*Zea mays* L.) grain yield in midwestern USA is 11.3 Mg ha^{-1} (180 bu acre^{-1}). At 15.5% grain moisture content, grain/stover of 1:1 and stover/root biomass ratio of 1:0.5, the total dry matter produced by this yield is 23.9 Mg ha^{-1} containing 10.8 Mg C ha^{-1} (at C content of 45%). For an average corn grain yield of 2.5 Mg ha^{-1} during the 1950s, this production from 1 ha would have required 4.5 ha of land. Thus, improvement in agricultural technology has caused net saving of 3.5 ha of land for the same level of production [the land area saved by agricultural intensification is estimated at about 200 million ha (Mha) in USA and 50 Mha in India]. If all the residue (stover and roots) is returned to the soil, it would add 14.3 Mg ha^{-1} of biomass or 6.4 Mg C ha^{-1}. But, 244 kg N fertilizer is equivalent to 210 kg of C emission to the atmosphere. Therefore, for N fertilizer to cause a positive soil C balance, 6.4 Mg C ha^{-1} returned to the soil must sequester at least 210 kg ha^{-1} of C at humification efficiency of 3.3%. Many experiments from around the world have demonstrated the SOC increase efficiency of 5% or more, especially when fertilizer is used in conjunction with conservation tillage.

Judicious fertilizer use and saving land is a preferred option (over extensive agriculture) even if the efficiency of conversion of biomass returned into SOC were less than 5%. The N use efficiency may be further increased by precision farming and other agronomic practices (Matson et al., 1998). The strategy of integrated nutrient management, judicious use of chemical fertilizers with organic manure and biological N fixation, enhances nutrient use efficiency. While increasing productivity and enhancing soil quality, manuring also sequesters C in soil (Jenkinson et al., 1987), which would have otherwise been released into the atmosphere.

It is important to realize that application of N fertilizers to agricultural soils is made for enhancing food production. Soil C sequestration is an additional benefit of this practice essential to achieving food security.

Myth 9: Manuring Is Not a Viable Option for Soil Carbon Sequestration, Because It Takes Three to Four Hectares of Grazing Land to Collect Manure Enough to Apply on One Hectare of Cropland

The practice of adding animal and biomass manures to the soil to restore fertility dates back to several millennia. There is no question that SOC contents of the plots that receive large doses of manure regularly for a long time can exceed that of the natural ecosystem (Jenkinson et al., 1987). The question is regarding the net storage of C in the landscape from which manure is produced on grazing land and recycled again on cropland. The practice may lead to declining SOC content on large areas of off-site lands (Schlesinger, 1999). It is important to realize, however, that manure is an agricultural by-product and its judicious use, similar to that of the nitrogenous fertilizer, can enhance productivity while improving soil quality and increasing SOC pool. Smith and Powlson (2000) estimated that in Europe 820 million t of manure is produced every year, of which only 54% is applied to farmland. They estimated that if all manure produced is used on arable land in the European Union, there would be net sequestration equivalent to 0.8% of the 1990 CO_2-C emission from the same geographical area.

Myth 10: Net Effect of Irrigation on Soil Carbon Sequestration is Negative Because of the Power Use in Lifting Irrigation Water and Release of Carbon Dioxide from Carbonates Brought to the Surface with the Groundwater

Just as fertilizer, irrigation is another important productivity enhancing technology in semiarid and arid regions. Some 40% of world food is produced on 17% (255 Mha) of the cropland that is irrigated (Postel, 1999). Food security of some densely populated countries (India, China, Pakistan, Egypt) depends on irrigated agriculture. Similar to fertilizer use, irrigation is needed to meet the demands of food, feed and fiber production. It increases biomass by two to three times compared with the rainfed system and also leads to sequestration of SOC (Dormaar & Carefoot, 1998; Coneth et al., 1998). Despite its benefits to enhancing production, there is a strong need to improve water use efficiency of the traditional systems so that increase in productivity and enhancement in SOC pool can be achieved with minimal risks of salinization.

A judicious use of irrigation also can lead to sequestration of SIC. The role of SIC in C sequestration is less well understood than that of the SOC. In irrigated systems, a major mechanism of sequestration of SIC is via movement of HCO_3 into groundwater. This mechanism has implications when groundwaters unsaturated with $Ca(HCO_3)_2$ are used for irrigation (Wilding, 1999; Nordt et al., 2000). In addition, enhanced primary productivity of the vegetation and adoption of salinity control strategies (e.g., use of gypsum amendments and biosolids) can result in increased leaching of $Ca(HCO_3)_2$ under irrigation if the irrigation waters are not already saturated with respect to bicarbonates.

Myth 11: Incorporation of Crop Residue and Biomass in Soil by Plowing Under Is Essential to Enhancing the Soil Organic Carbon Content

Plowing and other disturbances cause a major perturbation in the soil ecosystem. Plowing increases soil temperature, decreases soil moisture, and disrupts aggregates. These alterations accentuate the rate of mineralization of humus and increase the release of CO_2 to the atmosphere (Reicosky, 1998). In contrast, leaving the crop residue and other biomass as mulch decreases the maximum soil temperature, increases soil moisture content, enhances soil biodiversity, increases elemental recycling, and increases SOC content (Duiker & Lal, 1999). Because of an intense biotic activity (or bioturbation) of earthworms and other soil fauna (Tomlan et al., 1995), crop residue left on the soil surface gets incorporated into the soil. Therefore, incorporation of crop residue by plowing can be counterproductive and leads to depletion of SOC content by enhanced rate of mineralization and high soil erosion risks.

Myth 12: Conversion from Plow Till to a Conservation Till or No Till System Always Leads to Soil Organic Carbon Enrichment

In general, use of conservation tillage enhances SOC sequestration. However, adoption of conservation tillage on a land that has been managed by a plow till system for a long time may not lead to SOC sequestration, especially during the first 3 to 5 yr. In some soils, conversion to conservation tillage may produce low yields due to soil compaction, poor drainage, heavy weed infestation, and high incidence of pests. The response to conservation tillage also may be negative in soils containing high clay or silt content. Any soil-related constraints that cause low production relative to plow till may lead to either no gains in SOC pool or to even depletion of the SOC pool. It is important, therefore, to develop a soil guide toward choice of appropriate tillage methods (Lal, 1985, 1989).

In suitable soils and agroecosystems where conservation tillage is applicable, however, it increases the SOC pool. Enhancement of SOC pool by conversion to conservation tillage have been reported for a wide range of soil and ecosystems including those in Australia (Dalal et al., 1991; So et al., 1999), Argentina (Costanza et al., 1996), Brazil (Resck et al., 1999), North America (Dick et al., 1998; Lal et al., 1994) and the tropics (Lal, 1989).

Myth 13: Rate of Soil Organic Carbon Sequestration of Recommended Agricultural Practice (e.g., No Till) Is Uniform Among Regions

The SOC pool and fluxes vary widely across space and time. The SOC content of the plow layer may vary widely even within a seemingly homogeneous soilscape. Within a landscape under a uniform management system, however, the SOC content depends on soil profile characteristics (e.g., topsoil depth, available nutrient and water capacities, internal drainage, horizonation), terrain attributes (e.g., slope gradient, length and aspect), landscape position (shoulder slope, side slope or foot slope) and drainage density. The SOC content is high in soils of depressional sites, with slow internal drainage and north-facing aspect. Jenny (1980) observed a strong relationship between SOC pool and annual precipitation.

Myth 14: Returning Crop Residue to Soils of Low Inherent Fertility Can Increase Soil Organic Carbon Pool

The humification efficiency of C returned to the soil is low and often less than 10%. The efficiency depends on climate (temperature and moisture regimes), composition of crop residue (C/N/P/S) including the lignin and suberin contents, and soil properties (clay content, moisture retention, drainage) including nutrient availability. Humification efficiency may be extremely low in soils deficient in essential nutrients. Over and above the nutrients (N, P, S, etc.) required for crop growth and yield, nutrients also are needed to convert C-rich residue into humus. Most residue of cereals have a ratio of 80 to 100 for C/N and much wider ratio for C/P and C/S. The available C supplied in crop residue cannot be converted into humus in soils with severe deficiency of N, P, S and other nutrient elements. Therefore, application of deficient nutrients is essential to realizing the SOC sequestration potential of soils of low inherent fertility. Realization of the vast potential of SOC sequestration in soils of SSA that are devoid of essential plant nutrients depends on making investment towards application of fertilizers and other amendments. In such cases, the realizable potential is the net storage of C in soils after the input has been balanced with the output.

Myth 15: Roots, Crop Residue and Other Identifiable Plant Material is Carbon Sequestration

The global annual rate of photosynthesis is 120 Pg C. However, 50% of the C photosynthesized is returned back to the atmosphere as plant respiration and the remaining 50% as soil respiration. The turnover time of the easily decomposable plant biomass is very short. Thus, the amount of biomass produced and the residue returned per se do not constitute C sequestration. The amount of C in the biomass/residue remaining at the end of 1 yr, however, may be considered sequestration. The amount of residue remaining in soil at any time is dependent on the rate of decomposition (Eq. [1]).

$$R_t = R_o e^{-kt} \qquad [1]$$

Where R_t is the amount of C remaining in the residue at any time t, R_o is the initial concentration and k is a constant with units of per unit time (e.g., ha^{-1}, d^{-1}, or yr^{-1}) (Paul & Clark, 1988, p. 117–132). Rearrangement of this equation shows that a plot of $\ln(R_t/R_o)$ vs. t is linear with a slope of $-k$ (plot of log 10 will give slope equals $-k/2.303$). Half of the initial C in residue is given by the $t_{1/2}$ as shown in Eq. [2].

$$t_{1/2} = 0.693/k \qquad [2]$$

The mean residence time (turnover time) is $1/k$ for the first-order reactions and R_o/k for the zero-order reactions. It is generally agreed that the amount of C remaining at the end of the 1st yr is considered as input into the SOC pool.

In view of the above analyses, soil sampling and sample preparation are important considerations. All identifiable plant material (IPM) should be carefully removed from the soil, weighed and documented separately.

Myth 16: Science of Soil Organic Carbon Sequestration in Agricultural Soils is a "Work in Progress" Because of Difficulties in Monitoring and Verification of Soil Organic Carbon Pool

Standard methods are available to quantitatively assess SOC pool and fluxes at scales ranging from pedon or soilscape to ecosystems, national and global scales. There is an inherent variability in SOC content at spatial and temporal scales, and this variability can be addressed by adopting appropriate sampling techniques and use of GIS and geostatistical analyses.

However, there are numerous challenges in obtaining reliable data on soil C pool on global scale. The challenges arise due to the following factors that need to be carefully considered.

1. There are different types of C pools. The SOC pool comprises three principal components: (i) labile pool (turnover time of $<10^0$ or 1 yr), (ii) intermediate pool (turnover time of 10^1–10^2 yr), and (iii) passive pool (turnover time of 10^2–10^4 yr). The charcoal C also can be a significant component, especially in fire-dependent ecosystems. The relative amount of charcoal C may range from 0 to 50%. Despite its importance, there may be no obvious relationship between charcoal C and total soil C.
2. The depth of assessment of soil C pool is to be carefully considered. While the SOC pool should be measured at least to 2-m depth, it is rarely measured below the plow layer for most agricultural ecosystems. The most recent global estimates of SOC pool are made to 1-m depth, yet the amount of C in soil below 1-m depth constitutes a significant part of the total C pool.
3. Laboratories around the world use different methods. There is a wide range of methods used for determination of the SOC content among laboratories around the world. There is a strong need to standardize methods so that results are easily comparable.
4. Supporting data are often unavailable. There is often lack of supporting data on soil bulk density and other soil properties (e.g., clay content, CEC).
5. Scaling techniques need to be developed. There are difficulties involved in scaling the data from a point source to ecosystem or regional scale.

These problems outlined above are not unique to the soil C pool and also are encountered in assessing the biotic C pool. The important consideration is to recognize and overcome these problems by standardizing procedures for physical sampling depth, sample preparation and handling for different types of C pools, measurement of soil bulk density, assessment of total system C and scaling from point source to large scales. Attempts have been made to standardize these procedures (Lal et al., 2000).

Soil scientists and agronomists have studied changes in soil organic matter in relation to soil fertility and plant nutrition since at least 1900. It is the right time to also assess soil C dynamics in relation to the environmental issues of water quality and the accelerated greenhouse effect.

Myth 17: Attributes of Carbon in Air, Live or Dead Biomass and Soil Are Identical

While the elemental C in different pools is the same, its attributes differ widely among pools because of its interaction with other constituents specific to the pool. For example, C in humus differs from that in biomass because of the differences in composition of the constituents. The SOC in humus has considerably narrow C/N/P/S than that in the biomass. Consequently, the residence time of C in soil humus is different than that in the biomass. Important among other attributes of soil C are its reactivity and affinity towards H_2O. Humic substances have high surface area, and high charge density (anion and cation exchange capacity). Thus, humic substances form organomineral complexes that stabilize soil structure (Tisdall, 1996; Kay, 1998). Because of high charge density, soil humus holds anions (NO_3^-, HPO_4^-) and cations (Ca^{2+}, Mg^{2+}, K^+) and make them available to plants. Soils with high SOC content have a relatively high available water-holding capacity, because organic matter holds water at high water potential (e.g., field capacity) and releases it at low water potential (in the vicinity of the wilting point).

Realization of these differences in C in the air, biomass and soil would be helpful in determination of the value of soil C. Soil C in the form of humus must have a different monetary value than that in the biomass because of differences in composition of its constituents and its functions in soil with regards to productivity and environment quality.

Myth 18: Soil's Capacity to Sequester Carbon Is Infinite

The steady-state level of SOC pool depends on climate and soil properties. Important climatic factors include rainfall and temperature. Within the same isotherm, the SOC content increases with increase in rainfall regime. For the same isohyet, the SOC content increases with the decrease in mean annual temperature (Jenny, 1980). Within the same landscape unit, the SOC pool increases with increase in clay content and available water-holding capacity of the root zone.

Conversion of natural to agricultural ecosystems depletes the SOC pool. The magnitude of depletion is accentuated by accelerated soil erosion and other degradative processes. Most agricultural soils have lost a large proportion of their antecedent SOC pool ranging from 20 to 60 Mg C ha^{-1} (25–75% of the antecedent pool). It is possible to resequester 75% of the SOC pool lost by adopting at least two strategies: (i) restoration of degraded soils by taking these lands out of agricultural production (e.g., conservation reserve program, afforestation), and (ii) agricultural intensification through adoption of recommended agricultural practices on primeland.

The magnitude of SOC sequestration in both options outlined above depends on the extent to which soil's storage capacity has been depleted. The response of SOC pool to change in land use and/or management is generally a linear increase with increase in C input through root biomass, litter and detritus material, crop residue and other biomass. Soils with high SOC content, close to its storage capacity when under the climax vegetation, often do not respond to increased C inputs. The

SOC storage capacity of the natural ecosystems can be increased under managed ecosystems only if there are specific soil-related constraints to biomass production. These constraints include Al toxicity and P deficiency in Oxisols of acid savannas in South America. Alleviation of these constraints through liming, P application and introduction of adapted species can enhance the SOC pool (Fisher et al., 1994, 1995). Similarly, introduction of irrigation in areas of dryland farming can lead to increase in SOC pool (Varvel, 1994).

Myth 19: Why Sequester Carbon in Soil When the Jury Is Still Out on the Global Warming Debate?

The SOC sequestration is an important issue for its own merits and needs to be pursued regardless of the real or perceived risks of global warming. The global warming is a "century scale" and a "global commons" problem. Depletion of SOC pool has exacerbated soil and environmental degradation leading to reduction in agronomic productivity, increase in inputs (of fertilizers, seed, tillage or irrigation) required to produce the same yield, increase in actual and potential risks of soil erosion, and increase in transport of sediments and sediment-borne pollutants to natural waters. Increase in SOC content of degraded soils can lead to increase in crop yield. For a low-fertility soil in Thailand, Petchawee and Chaitep (1995) observed that maize grain yield increased at the rate of 2.9 Mg ha^{-1} for each 1% increase in soil organic matter content.

Estimates of global productivity loss due to soil degradation are alarming. Lal (1998) estimated global yield reduction in 1995 due to soil erosion at 10% in cereals, 5% in soybean (*Glycine max* L.) and pulses, and 12% in roots and tubers. Total production losses on a global scale were estimated at 272 million Mg yr^{-1} comprising 190 million Mg of cereals, 6 million Mg of soybean, 3 million Mg of pulses and 73 million Mg of roots and tubers. Oldeman (1998) estimated productivity losses due to erosion from World War II to 1995 (45-yr period) at 25% for Africa, 13% for Asia, 37% for Central America, and 13% for the world.

The importance of SOC sequestration on productivity will be enhanced in the event that the projected global warming were to occur. There may be a tremendous shift in rainfall amount and distribution with adverse impact on agronomic productivity, acceleration of soil degradation, and further depletion of the SOC pool (Rosenzweig & Hillel, 1998). In that event, the need and urgency for soil C sequestration will be much greater.

Myth 20: The Soil Organic Carbon Sequestration to Mitigate the Accelerated Greenhouse Effect Is a Pie in the Sky

The potential of SOC sequestration is large: 75 to 208 Tg C yr^{-1} for U.S. cropland (Lal et al., 1998), 30 to 112 Tg C yr^{-1} for U.S. grazing land (Follett et al., 2000), 0.75 to 1.0 Pg C yr^{-1} for world cropland (Lal & Bruce, 1999), 0.9 to 1.9 Pg C yr^{-1} for desertification control (Lal et al., 1999), and 3.0 Pg C yr^{-1} for restoration of degraded soils in the world (Lal, 1997). This finite global potential, of the magnitude of 50 to 75 Pg C, can be realized over a 25- to 50-yr period. The maximum potential rate of SOC sequestration of 3 Pg C yr^{-1} is high enough to almost nullify the annual increase in atmospheric concentration of CO_2 at 3.4 Pg C yr^{-1}. Realization of this potential, however, requires a coordinated effort at local, regional, national and global scale to: (i) identify soils and ecoregions (ecological hot spots) with high and economically feasible potential of soil C sequestration, (ii) validate soil/crop management technologies and quantify the rate of SOC sequestration, (iii) identify policy issues that facilitate adoption of recommended practices, and (iv) assess value of soil C.

There are numerous options of C sequestration (Halman & Steinberg, 1999) of which SOC sequestration is a strategic option. It is a cost effective and an interim option. It buys us time of 25 to 50 yr during which other energy related options take effect.

Myth 21: For Every Ten Atoms of Carbon Fixed in Trees, Only One is Fixed in Soil

The partitioning of net primary productivity (NPP) into SOC and biomass C in a forest ecosystem depends on many site-specific factors. In a degraded soil converted from agricultural land use to forestry, 80 to 100% of the NPP may go into the above- and below-ground biomass for the initial 5 to 10 yr. Once the canopy cover is established and there is a sufficient amount of leaf litter and detritus returned to the soil, 25 to 50% of the net primary productivity (NPP) may be converted into the SOC pool. The growth rate and hence the C sequestration potential of forest diminishes as trees approach maturity. Forests accumulate C in young and middle age, but the rate declines to zero as forest matures (Batjes, 1999). The above-ground net primary productivity (ANPP) may even decline in some aging forest stands (Gower et al., 1996). In a mature forest, there may be no net gain in the biomass production and most of the litter fall and detritus material may go into the SOC pool. Once the SOC and the vegetation have attained a steady state, the net accumulation of C in the forest ecosystem may be negligible. At steady state, the magnitude of SOC pool may be 50 to 75% of that of the biomass pool. Changes in below-ground C and N cycles during stand development are very poorly understood (Gower et al., 1996). Some argue that if the flux of C from soil to vegetation increases because of the fertilization effect of increase in atmospheric concentration of CO_2, forest soils may become net sink of C for decadal time scales (Bird et al., 1996).

Myth 22: Enhancing the Biotic Carbon Pool Is a Solution to Mitigating the Accelerated Greenhouse Effect

The biotic C pool at 560 Pg is the smallest of the five active pools. Nonetheless, the biotic pool is very dynamic, with an annual flux of 120 Pg C, and mean residence time of less than 5 yr. The photosynthetic activity of the world's flora is the basis of terrestrial life. However, the biotic C pool is only 78% of the atmospheric pool and merely 24% of the pedologic pool. The residence time of C in the biotic pool is about 4.7 yr compared with that of 25.8 yr in the soil. Although enhancing the magnitude of both pools is important, the strategy of soil C sequestration has numerous ancillary benefits that must be carefully considered.

Myth 23: On-Site Benefits of Soil Organic Carbon Sequestration Primarily Determine the Market Value of Carbon for Trading Credits

The SOC sequestration has both on-site and off-site benefits. The on-site benefits are those related to improvements in soil quality that enhance current and future productivity. There are numerous ancillary or off-site benefits of SOC sequestration. These include improvement in water quality, improvements in soil quality with attendant decrease in soil erosion and sediment transport rates, low risk of damage to infrastructure and waterways, and decrease in the emissions of GHGs to the atmosphere.

The market value of C thus should reflect both the on-site and off-site benefits. The value is appropriately determined by an interdisciplinary team comprising biophysical and social scientists.

Myth 24: The Soil Inorganic Carbon Is Not an Important Factor in the Global Carbon Cycle and in Influencing Atmospheric Concentration of Carbon Dioxide

The SIC pool, estimated at 750 Pg to 1-m depth, is about 50% of the SOC pool (1550 Pg to 1-m depth). In some soils (e.g., Aridisols), the SIC pool is more significant than the SOC pool. The annual rate of SIC sequestration can be substantial (Nordt et al., 2000) and may range from 10 to 12 Tg C yr^{-1} (Scharpenseel, et al., 1995, 2000).

Despite its importance in the global C cycle, however, the dynamics of SIC in relation to land use, soil and crop management, and farming systems, is not known. The SIC pool is influenced by irrigation, drainage and leaching, plowing, use of acidifying material on Aridisols and of liming materials on Ultisols and Oxisols, and exposure of caliche to the surface by erosion and other anthropogenic and natural perturbations. The dynamics of SIC also is influenced by the partial pressure of CO_2 in the soil air as moderated by the activity of soil fauna and flora. Yet, the impact of these factors and processes on SIC pool and fluxes are not known. The lack of knowledge about the SIC pool should not be confused with the lack of its importance to the global C cycle.

Myth 25: It Is Difficult to Increase the Soil Organic Carbon Pool, Especially so in Soils of the Tropics Where the Rate of Increase is Barely Above the Natural Variability

Enhancing SOC pool requires a thorough understanding of the interacting processes, properties and practices that affect its dynamics. Merely altering a crop rotation, tillage method, or rate of fertilizer application may not be enough to significantly influence the SOC pool over a short period of 1 to 2 yr. It is important to identify soil-related constraints to SOC sequestration. Simply adding crop residue or biosolids without alleviating the specific constraints may not enhance the SOC pool. Predominant soil-related constraints in the tropics include: (i) lack of sufficient quantity of nutrients (e.g., N, P, S) for converting biomass C into humus, (ii) high soil temperature regime, and (iii) soil water imbalance. There is a lack of sci-

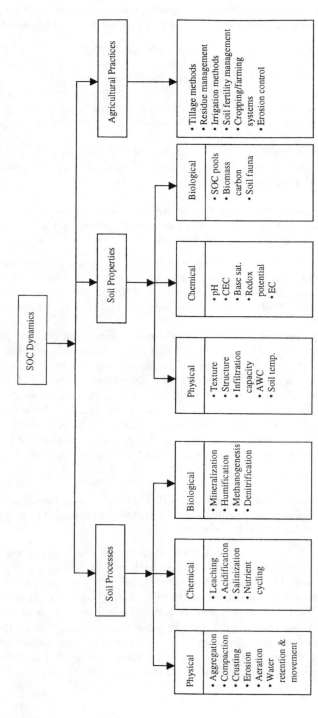

Fig. 2–1. Soil processes and properties, and agricultural management practices that influence the SOC dynamics.

entific data on the types of SOC pool, and their turnover time in relation to management practices.

With the knowledge of soil properties and processes, adoption of appropriate soil and crop management practices can lead to increase in SOC pool over a 5-yr period, even in the harsh environments of the tropics. The rate of SOC sequestration is high in cool and moist climates than in warm and dry ecoregions (Lal, 1999). The SOC sequestration potential is more in soils of the tropics and subtropics than in those of the temperate climates. The potential also is high in degraded soils and ecosystems. However, the realization of the potential is more challenging in the tropics and subtropics and in developing countries than in developed economies in the temperate climates.

CONCLUSIONS

Several misunderstandings about the role of soil C in the global C cycle persist because of the lack of knowledge and general awareness about the pool, its properties and the impact of anthropogenic and natural factors on its dynamics. Soil scientists and agronomists have studied the SOC for most of the 20th century, but only in relation to soil fertility and plant nutrition. It is important that SOC pool and its dynamics also be evaluated in relation to their roles in environmental issues e.g., water quality, and the accelerated greenhouse effect. There is a need for creating a new paradigm in soil science for understanding the importance of SOC and SIC pools in relation to the environmental impacts. The new paradigm would involve assessing the soil C pool (both SOC and SIC components): (i) up to at least 1-m depth, with standardized methods of soil sampling and analyses, and (ii) in relation to soil bulk density and other soil properties and processes. Important processes to which SOC pool is to be related include soil erosion, aggregation, leaching, humification of the biosolids returned to the soil, aeration and its effects on the redox potential etc. (Fig. 2–1).

There also is a lack of awareness about the importance of SOC and SIC pools in relation to the global C cycle. Soil scientists need to be pro-active in creating and strengthening the channels of communication with the general public, policy makers, planners and land managers.

Soil C sequestration is a strategic option toward mitigation of the accelerated greenhouse effect. This option is cost-effective and feasible. It has a finite role to play within the next 25 to 50 yr (up to Year 2050). This option buys us time during which energy-related alternatives to fossil fuel can be developed to decarbonize the world's energy system.

REFERENCES

Batjes, N.H. 1999. Management options for reducing CO_2 concentrations in the atmosphere by increasing carbon sequestration in the soil. ISRIC Tech. Pap. 30. Wageningen, Holland.

Bird, M.I., A.R. Chivas, and J. Head. 1996. A latitudinal gradient in carbon turnover in forest soils. Nature (London) 381:143–146.

Bohn, H. 1978. On organic soil carbon and CO_2. Tellus 30:473–475.

Buringh, P. 1984. Organic carbon in soils of the world. p. 91–109. *In* G.M. Woodwell (ed.) The role of terrestrial vegetation in the global carbon cycle: Measurement by remote sensing. SCOPE 23. John Wiley & Sons, Inc., Chichester, UK.

Cihacek, L.J., M.D. Sweeney, and E.J. Diebert. 1993. Characterization of wind erosion sediments in the red river valley of North Dakota. J. Environ. Qual. 22:305–310.

Coneth, A., G.J. Blair, and I.J. Rochester. 1998. Soil organic carbon fraction in a Vertisol under irrigated cotton production as affected by burning and incorporating cotton stubble. Aust. J. Soil Res. 36:655–667.

Costanza, A., D. Cosentino, and A. Segat. 1996. Influence of tillage systems on biological properties of a Typic Argiudoll soil under continuous maize in central Argentina. Soil Tillage Res. 38:265–271.

Dalal, R.C., P.A. Henderson, and J.M. Glasby. 1991. Organic matter and microbial biomass in a Vertisol after 20 years of zero tillage. Soil Biol. Biochem. 23:435–441.

Dick, W.A., R.L. Blevins, W.W. Frye, S.E. Peters, D.R. Christenson, F.J. Pierce, and M.L. Vitosh. 1998. Impacts of agricultural management practices on C sequestration in forest derived soils of the eastern Corn Belt. Soil Tillage Res. 47:235–244.

Dormaar, J.F., and J.M. Carefoot. 1998. Effect of straw management and nitrogen fertilizer on selected soil properties as potential soil quality indicators of an irrigated dark brown Chernozemic soil. Can. J. Soil Sci. 78:511–517.

Duiker, S.W., and R. Lal. 1999. Crop residue and tillage effects on C sequestration in a Luvisol in central Ohio. Soil Tillage Res. 52:73–81.

Etheridge, D.M., L.P. Steele, R.L. Lengenfelds, R.J. Francey, I.M. Barnola, and V.I. Moran. 1996. Natural and anthropogenic changes in atmospheric CO_2 over the last 1000 years from air in Antarctic ice and fern. J. Geophys. Res. 101:4115–4128.

Fisher, M.J., I.M. Rao, M.A. Ayarza, C.E. Lascano, J.I. Sanz, R.J. Thomas, and R.R. Vera. 1994. Carbon storage by introduced deep-rooted grasses in the South American savannas. Nature (London) 371:236–238.

Fisher, M.J., I.M. Rao, M.A. Ayarza, C.E. Lascano, J.I. Sanz, R.J. Thomas, and R.R. Vera. 1995. Pasture soils as carbon sink. Nature (London) 376:472–473.

Follett, R.F., J.M. Kimble, and R. Lal (ed.) 2000. The potential of U.S. grazing land to sequester carbon and mitigate the greenhouse effect. CRC/Lewis Publ., Boca Raton, FL.

Gower, S., R.E. McMurtrie, and D. Murty. 1996. Above-ground net primary production decline with stand age: Potential causes. TREE 11:378–383.

Gregorich, E.G., K.J. Greer, D.W. Anderson, and B.C. Liang. 1998. Carbon distribution and losses: erosion and deposition effect. Soil Tillage Res. 47:291–302.

Halman, M.M., and M. Steinberg. 1999. Greenhouse gas carbon dioxide mitigation: Science and technology. Lewis Publ., Boca Raton, FL.

Himes, F.L. 1998. Nitrogen, sulfur and phosphorus and the sequestering of carbon. p. 315–319. *In* R. Lal et al. (ed.) Soil processes and the carbon cycle. CRC Press, Boca Raton, FL.

Intergovernmental Panel on Climate Change. 1996. Climate change 1995: Impacts, adaptation and mitigation of climate change: Scientific and technical analysis. Work. Group 1. Cambridge Univ. Press, UK.

Intergovernmental Panel on Climate Change. 2000. Special report on land use, land-use change and forestry. Cambridge Univ. Press, Cambridge, UK.

Jenkinon, D.S., P.B.S. Hart, J.H. Rayner, and L.C. Parry. 1987. Modeling the turnover of organic matter in long-term experiments at Rothamsted, U.K. INTECOL Bull. 15:1–8.

Jenny, H. 1980. The soil resource: Origin and behaviour. Springer Verlog, New York.

Kay, B.D. 1998. Soil structure and organic matter: A review. p. 169–198. *In* R. Lal et al. (ed.) Soil processes and the carbon cycle. CRC Press, Boca Raton, FL.

Lal, R. 1976. Soil erosion on Alfisols in western Nigeria. IV. Nutrient losses in runoff and eroded sediments. Geoderma 16:403–417.

Lal, R. 1985. A suitability guide for different tillage systems in the tropics. Soil Tillage Res. 5: 179–196.

Lal, R. 1989. Conservation tillage for sustainable agriculture. Adv. Agron. 42:85–107.

Lal, R. 1995. Global soil erosion by water and carbon dynamics. p. 131–141. *In* R. Lal et al. (ed.) Soils and global change. CRC/Lewis Publ., Boca Raton, FL.

Lal, R. 1997. Residue management, conservation tillage and soil restoration for mitigating greenhouse effect by CO_2-enrichment. Soil Tillage Res. 43:81–107.

Lal, R. 1998. Soil erosion impact on agronomic productivity and environment quality. Crit. Rev. Plant Sci. 17:315–464.

Lal, R. 1999. Soil management and restoration for C sequestration to mitigate the accelerated greenhouse effect. Prog. Environ. Sci. 1:307–326.

Lal, R., T.J. Logan, M.J. Shipitalo, D.J. Eckert and W.A. Dick. 1994. Conservation tillage in the Corn Belt of the United States. p. 73–115. *In* M.R. Carter (ed.) Conservation tillage in temperate agroecosystems. Lewis/CRC Publ., Boca Raton, FL.

Lal, R., J.M. Kimble, R.F. Follett, and C.V. Cole. 1998. The potential of U.S. cropland to sequester carbon and mitigate the greenhouse effect. Ann Arbor Press, Chelsea, MI.

Lal, R., and J.P. Bruce. 1999. The potential of world cropland soils to sequester C and mitigate the greenhouse effect. Environ. Sci. Policy 2:177–185.

Lal, R., H.M. Hassand, and J. Dumanski. 1999. Desertification control to sequester carbon and mitigate the greenhouse effect. p. 83–151. *In* N.J. Rosenberg et al. (ed.) Carbon sequestration in soils: Science, monitoring and beyond. Battelle Press, Columbus, OH.

Lal, R., J.M. Kimble, R.F. Follett, and B.A. Stewart (ed.) 2000. Assessment methods for soil C pools. CRC/Lewis Publ., Boca Raton, FL (In press.)

Matson, P.A., R. Naylor, and I. Ortiz-Monasterío. 1998. Integration of environmental, agronomic and economic aspects of fertilizer management. Science (Washington, DC) 280: 112–115.

Nordt, L.C., L.P. Wilding, and L.R. Drees. 2000. Pedogenic carbonate transformations in leaching soil systems: implications for the global C cycle. p. 43–64. *In* R. Lal et. al. (ed.) Global climate change and pedogenic carbonates. CRC/Lewis Publ., Boca Raton, FL.

Oldeman, L.R. 1998. Soil degradation: A threat to food security. ISRIC, Wageningen, The Netherlands.

Palis, R.G., H. Ghadiri, C.W. Rose, and P.G. Saffigna. 1997. Soil erosion and nutrient loss. III. Changes in the enrichment ratio of total nitrogen and organic carbon under rainfall detachment and entrainment. Aust. J. Soil Res. 35:891–905.

Paul, E.A., and F.E. Clark. 1988. Soil microbiology and biochemistry. Acad. Press, London, UK.

Petchawee, S., and W. Chaitep. 1995. Organic matter management for sustainable agriculture. p. 21–26. *In* R.D.B. Lefroy et al. (ed.) Organic matter management in upland systems. ACIAR, Canberra, Australia.

Postel, S. 1999. Pillars of sand: Can the irrigation miracle last? W.W. Norton and Co., New York.

Reicosky, D.C. 1998. Tillage methods and CO_2 loss: Fall versus spring tillage. p. 99–111. *In* R. Lal et al. (ed.) Management of carbon sequestration in soil. CRC/Lewis Publ., Boca Raton, FL.

Resck, D.V.S. 1998. Agricultural intensification systems and their impact on soil and water quality in the Cerrado of Brazil. p. 228–300. *In* R. Lal (ed.) Soil quality and agricultural sustainability. Ann Arbor Press, Chelsea, MI.

Rosenzweig, C., and D. Hillel. 1998. Climate change and the global harvest: Potential impacts of the greenhouse effect on agriculture. Oxford Univ. Press, Oxford.

Rozanov, B.G., V. Targulian, and D.S. Orlov. 1990. Soils. p. 203–214. *In* B.L. Turner et al. (ed.) The Earth as transformed by human action: Global and regional changes in the biosphere over the past 300 years. Cambridge Univ. Press, UK.

Scharpenseel, H.W., J. Freytag, and E.M. Pfeifer. 1995. The carbon budgets in drylands: assessments based on carbon residence time and stable isotope formation. p. 79–93. *In* V.R. Squires et al. (ed.) Combating global climate change by combating land degradation. UNEP, Nairobi, Kenya.

Scharpenseel, H.W., A. Mtimet, and J. Freytag. 2000. Soil inorganic carbon and global change. p. 27–41. *In* R. Lal et al. (ed.) Global climate change and pedogenic carbonates. CRC/Lewis Publ., Boca Raton, FL.

Schlesinger, W.H. 1999. Carbon sequestration in soils. Science (Washington, DC) 286:2095.

Smaling, E.M.A., and L.O. Fresco. 1993. A decision-support model for monitoring nutrient balances under agricultural land use (NUTMON). Geoderma 60:235–256.

Smaling, E.M.A., L.O. Fresco, and A. de Jager. 1996. Classifying, monitoring and improving soil nutrient stocks and flows in African agriculture. Ambio 25:492–496.

Smith, P., and D.S. Powlson. 2000. Considering manure and carbon sequestration. Science (Washington, DC) 287:428–429.

Stallard, R.F. 1998. Terrestrial sedimentation and the carbon cycle: Coupling weathering and erosion to carbon burial. Glob. Biogeochem. Cycles 12:231–257.

Tisdall, J.M. 1996. Formation of soil aggregates and accumulation of soil organic matter. p. 57–96. *In* M.R. Carter and B.A. Stewart (ed.) Structure and organic matter storage in agricultural soils. CRC/Lewis Publ., Boca Raton, FL.

Tomlan, A.D., M.J. Shipitalo, W.M. Edwards, and R. Protz. 1995. Earthworms and their influence on soil structure and infiltration. p. 159–183. *In* P.F. Hendrix (ed.) Earthworm ecology and biogeography in North America. Lewis Publ., Boca Raton, FL.

Varvel, G.F. 1994. Rotation and nitrogen fertilization effects on changes in soil C and N. Agron. J. 86:319–325.

Wallace, A. 1994. Soil organic matter must be restored to near original level. Commun. Soil Sci. Plant Anal. 25:29–35.

Wan, Y., and S.A. El-Swaify. 1997. Flow-induced transport and enrichment of erosional sediment from a well-aggregated and uniformly textured Oxisol. Geoderma 75:251–265.

Wilding, L.P. 1999. Comments on paper by R. Lal, H.M. Hassan and J.M. Dumanski. p. 146–149. *In* N.J. Rosenberg et al. (ed.) Carbon sequestration in soils: Science, monitoring and beyond. Battelle Press, Columbus, OH.

Woodwell, G.M. (ed.) 1984. The role of terrestrial vegetation in the global carbon cycle. SCOPE and J. Wiley & Sons, Inc., Chichester, UK.

Zobeck, T.M., and D.W. Fryrear. 1986. Chemical and physical characteristics of wind-blown sediment. Trans. ASAE 29:1037–1041.

3

Carbon Sequestration Under the Conservation Reserve Program in the Historic Grassland Soils of the United States of America

R. F. Follett and E. G. Pruessner

USDA-ARS
Fort Collins, Colorado

S. E. Samson-Liebig, J. M. Kimble, and S. W. Waltman

USDA-NRCS, NSSC
Lincoln, Nebraska

ABSTRACT

Future emissions of CO_2 to the atmosphere are expected to continue to increase and, along with other "greenhouse gases", contribute to the potential for global climate warming. Capture of atmospheric CO_2-C by photosynthesis and its subsequent sequestration in soil is likely the best long-term option for C storage in terrestrial ecosystems. Adverse impacts of ongoing soil erosion in the USA resulted in legislative authority to implement the Conservation Reserve Program (CRP). The CRP returns cultivated land to permanent plant cover and that potentially increases the atmospheric CO_2-C captured through photosynthesis and its storage as soil organic C (SOC). This study evaluates that potential. Sampling sites were selected across three soil temperature and three soil moisture regimes found in the "historic grasslands" region of the USA. The sites had been in the CRP for a minimum of 5 yr and were paired with cropped sites that were as similar as possible. Weights of SOC and identifiable plant material (IPM) by soil layer to a depth of about 2 m were calculated using thicknesses, bulk densities, and C analyses data. Estimates of annual rates of SOC sequestration by the CRP and differences in total amounts of IPM were made by subtracting the amount measured in cropland sites from that measured under their paired CRP sites. Our estimates across the 13-state region in this study are that the CRP sequesters about 570, 740, and 910 kg SOC ha^{-1} yr^{-1} in the 0- to 5-, 0- to 10-, and 0- to 20-cm depth increments, respectively. A significant difference of the SOC and IPM under the CRP and cropped sites was observed at only these depth increments. The IPM likely precedes the introduction of and provides the C for SOC sequestration. Average amount of IPM-C under the CRP was 2990, 3470, and 3930 kg C ha^{-1} greater than under the cropped sites. Total amounts of SOC sequestered by the CRP for the entire 5.6 million ha of land under the CRP within the soil temperature (T) × soil moisture (M) regimes included in this 13-state region are estimated as 3.19, 4.15, and 5.14 million metric tons of C (MMTC) yr^{-1} within the 0- to 5-, 0- to 10-, and 0- to 20-cm depth increments, respectively.

A continuing trend of increasing emissions of carbon dioxide (CO_2) and its rising concentrations in the atmosphere are occurring (EPA, 1998). The Intergovernmental Panel on Climate Change (IPCC) predicts in its "UN Medium-High Case (IS92f)" scenario (IPCC, 1995) that future emissions of CO_2 to the atmosphere will increase from the current level of about 6.0 Pg yr^{-1} of C in 1990 to 26.4 Pg yr^{-1} by 2100. The IPCC also predicts a doubling of atmospheric CO_2 concentration by 2050. Although the effects of increased CO_2 levels on global climate are uncertain, many scientists agree that a doubling of atmospheric CO_2 concentrations could have a variety of serious environmental consequences.

By comparison to amounts of CO_2-C projected to be released to the atmosphere, the C stored in the world's soils is large, with 350 Pg C stored in North American soil profiles alone as reported in association with the work by Waltman et al. (1997). Carbon sequestration in terrestrial ecosystems is accomplished when C is stored in living plants by the photosynthetic process and then relocated to the soil and transformed into SOC. Carbon sequestration is defined as the capture and secure storage of C that would otherwise be emitted to or remain in the atmosphere as CO_2 or methane (CH_4). Land placed in the CRP has the potential to sequester significant amounts of SOC (Bruce et al., 1999; Lal et al., 1999). This manuscript discusses the potential of the CRP within the "historic grasslands" region (HGR) of the USA for sequestering SOC and for offsetting U.S. emissions of CO_2. Storage of C in soil is likely the best long-term option for C storage in terrestrial systems. Soil C dating studies for native grasslands indicate that once sequestered and not disturbed, large quantities of SOC remain in soil profiles for hundreds to thousands of years. The SOC at a depth of about 2 m was measured to have a mean residence time (MRT) of 9 000 to 13 000 yr (Follett et al., 1997). Paul et al. (1997) reported a MRT range of between 400 to 2100 yr for soils collected within the 0- to 15- or 15- to 30-cm depths under native and cultivated conditions for a number of sites within the U.S. Great Plains.

Increasing the storage of C in vegetation and soil potentially offers significant accompanying benefits including: improved soil quality, sustainable productivity, decreased pollution of surface and groundwaters by agricultural chemicals, reduced soil erosion, and less-overall off-site environmental degradation. Increases in SOC alone can provide significant benefits by allowing time for other possible C management options and strategies to be evaluated and implemented. The practical upper limit for these increases in SOC sequestration may be 50 yr (Lal et al., 1998), thus allowing time for adjustments in the world's energy production system to stabilize the amounts of CO_2 that are released to the atmosphere.

To increase the sequestration of C in terrestrial ecosystems, either increased photosynthetic C fixation and/or decreased losses (decomposition) of soil organic matter are necessary. Carbon that is stored below-ground is more permanent than plant biomass; however it too can be easily lost by the adoption of unsuitable soil management practices (Bowman et al., 1999; Karlen et al., 1998). Historically there has been little emphasis given to the development or implementation of strategies for C sequestration. Rather, C sequestration was not considered at all or had a low priority and losses of SOC occurred along with the release of large quantities of C to the atmosphere as CO_2.

As background to this study, numerous Federal programs and factors led to the creation of the Conservation Reserve Program (CRP) in the Food Security Act of 1985. As reported by Berg (1994), Federal legislation in 1956 authorized land to be set aside as part of a "Soil Bank" in what might be considered a previous version of the CRP. The primary purpose of the Soil Bank was to divert cropland from production to decrease agricultural inventories; a secondary benefit was the establishment of protective vegetative cover on land that would benefit from conservation practices. At its peak, the Soil Bank program had 11.6 million ha of cropland under contract. In the short-term, the Soil Bank resulted in significant reductions in soil erosion and substantial increases in wildlife habitat (Berg, 1994; Follett, 1998). However, because of unprecedented foreign grain sales, all USDA-subsidized cropland-retirement programs were suspended by 1973 and eventually few enduring resource conservation benefits remained (Berg, 1994). The conduct of national resources inventories by the USDA began in about 1972 and implementation of the Soil and Water Conservation Act (RCA) of 1977 (P.L. 95-192) resulted in the collection of data showing that soil erosion had increased to levels that were perhaps worse than those during the Dust Bowl years of the 1930s. Soils that were once adequately treated or retired from cultivation were plowed and planted to annual crops. Continued recognition of the adverse impacts of ongoing soil erosion finally resulted in legislative authority in P.L. 98-198 in 1985 to implement a nationwide conservation reserve. Thus, by 1992, there were just over 13.8 million ha of land under CRP contract and of those 7.2 million ha were identified by 1982 (Berg, 1994), before the CRP was implemented, as highly erodible cropland (HEL). Gomez (1995) estimates there are about 6 million ha of HEL in the conterminous USA. It is critical for the USA to continue to assess the role of the CRP and other conservation practices that may contribute to environmental quality issues including global climate change, soil quality and productivity, sediment-related nonpoint source pollution, groundwater quality, and others.

This chapter addresses the impacts of the CRP on C sequestration as it relates to global climate change. Although now greater, the amount of land in the CRP program in 1998 was 10.6 million ha (sign-ups 10–19, excluding 18), while about 66% (7.0 million ha) was within the 13-state region (Texas, New Mexico, Colorado, Wyoming, Montana, Oklahoma, Kansas, Nebraska, South Dakota, North Dakota, Missouri, Iowa, and Minnesota) that encompasses much of the HGR of the USA as generalized from Kuchler (1985). This study was conducted within the HGR (Fig. 3–1). Answers are needed about the beneficial effects that have accrued by placing (and/or keeping) land under permanent grass cover, such as with the CRP, and the benefits that may be lost if land currently in the CRP is returned to cultivation or unnecessarily lost by a change in management and cropping practice. Current literature documents rates of SOC sequestration under the CRP by the use of models (Paustian et al., 1995). Such estimates indicate the rates of C sequestration for the western and central USA are between <100 to 400 kg ha^{-1} yr^{-1} of soil organic matter and 250 to 1350 kg ha^{-1} yr^{-1} of total below-ground C, including roots (Bruce et al., 1999). Estimates by Lal et al. (1998, 1999) suggest that about 500 and 650 kg C ha^{-1} yr^{-1} are sequestered under the CRP as SOC in the 0- to 5- and 0- to 10-cm depths, respectively. Work by Gebhart et al. (1994) at five sites across

Texas, Kansas, and Nebraska indicated that about 800 and 1100 kg C ha^{-1} yr^{-1} were sequestered in the 0- to 40- and 0- to 300-cm depths under the CRP. Research by Robles and Burke (1998) indicated that returning cultivated fields in southeastern Wyoming to perennial grasses would restore SOC by increasing labile pools that might include coarse particulate C and mineralizable C, however they observed only a slight increase in SOC itself after 6 yr in the CRP. Thus, there is a considerable range reported in the literature for the amounts of SOC that can be sequestered under the CRP.

The rate at which SOC pools change under the CRP as compared to cropped soils is likely a function of climate, previous cropping history, type of plants seeded

Study Sites and Historic Natural Grasslands

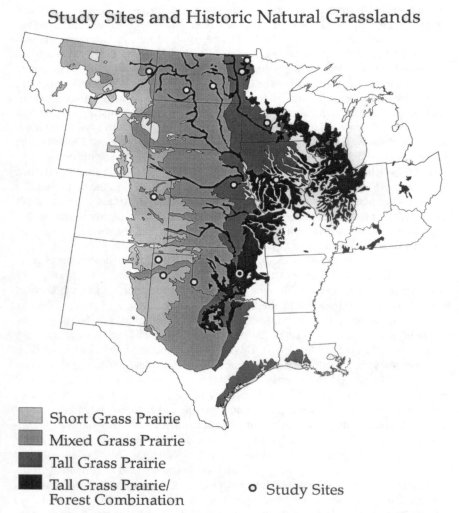

Short Grass Prairie
Mixed Grass Prairie
Tall Grass Prairie
Tall Grass Prairie/   o Study Sites
Forest Combination

Fig. 3–1. The historic grasslands region of the USA based on potential natural vegetation as identified by Kuchler (1985) and the locations of the sites sampled.

on the CRP, topographic location, soil texture and mineralogy, and time. Information obtained from studying SOC sequestration under the CRP can provide valuable insights concerning processes and mechanisms of C sequestration and soil quality. Benefits derived from placing or maintaining land in the CRP in relation to climate change and soil quality are among those effects needing better understanding. Such information is expected to increase our awareness of the relative vulnerability of various C pools and perhaps will allow insights about ways in which the rate of movement of C into long-term, recalcitrant SOC storage pools can be increased. This study was designed to provide broad regional information about the potential of using the CRP as a means to sequester atmospheric CO_2-C in soil as SOC and to provide some relative estimate of the importance of the use of the CRP within the USA as a management option to address the issue of climate change.

METHODS AND MATERIALS

Sampling sites were selected across a matrix of three soil temperature regimes and three soil moisture regimes (Soil Surv. Staff, 1998) that are found in the HGR of the USA, except within the area of the "tall grass prairie/forest combination" located in Wisconsin and Illinois (Kuchler, 1985). A paired-plot design was used where the sites chosen included fields that had been in the CRP program for a minimum of 5 yr. The CRP sites were then paired with cropped sites that were as similar as possible. Sites were selected on the same geomorphic unit, mapped series, slope, and aspect. However, soil erosion sometimes changed the series classification; for example, as with the loss of the mollic epipedon.

Soil samples were collected from the 0- to 5- and 5- to 10-cm depth and by genetic soil horizon thereafter to a depth of about 2 m in pits that were dug by backhoe or in a few cases by hand. The time that the sites were in the CRP, the temperature and moisture regimes, and soil profile properties (i.e., texture and taxonomic classification) are shown in Table 3–1. Soil bulk densities (33 kPa of moisture tension) were determined for each soil layer on clods collected from the face of the excavation and coated with Saran F-310[1] for transport (Soil Surv. Lab. Staff, 1992). To separate and measure identifiable plant material (IPM), samples from each soil layer were screened through a 2-mm sieve to separate the larger plant fragments and if necessary, gently washed to remove soil material. A subsample of the sieved soil was then picked free of recognizable plant and root fragments under 20× magnification. The IPM, thus collected, was oven dried (65 °C) weighed and analyzed for C using a Carlo Erba C/N analyzer (Haake Buchler Instruments, Saddle Brook, NJ) interfaced to a Tracer mass isotope-ratio mass spectrometer (Europa Scientific Ltd., Crewe, England). Soil samples were delimed as described by Follett et al. (1997) and Follett and Pruessner (2000), prior to SOC analyses by use of the C/N analyzer. The weights of SOC and IPM by soil layer were calculated using thicknesses, bulk densities, and C analyses data for IPM and SOC. Data could then be used directly for the 0- to 5- and 5- to 10-cm depth increments. Recalculation from the horizon data was necessary to statistically analyze and report data for discrete depth increments that were deeper than 10 cm.

---

[1] Trade and company names are included for the benefit of the reader and do not imply endorsement or preferential treatment of the product by the authors or the USDA.

Table 3–1. Years in the CRP, temperature and moisture regime numbers, and soil properties by location.

Location	Time in CRP (yr)	Temperature regime[†]	Moisture regime[‡]	Soil series	Surface texture Classification	Taxonomic classification
Bushland, Texas	8	T_1	M_1	Pullman	Clay loam	Fine, mixed, superactive, thermic Torrertic Paleustolls
Dalhart, Texas§	8	T_1	M_1	Dallam	Fine sandy loam	Fine-loamy, mixed, mesic Aridic Haplustalfs
Vinson, Oklahoma	10	T_1	M_2	Madge	Loam	Fine-loamy, mixed, superactive, thermic Typic Argiustolls
Boley, Oklahoma	10	T_1	M_3	Stephenville	Loamy fine sand	Fine-loamy, siliceous, active, thermic Ultic Haplustalfs
Akron, Colorado	6	T_2	M_1	Weld	Silt loam	Fine-loamy, smectitic, mesic Aridic Argiustolls
Lincoln, Nebraska	6	T_2	M_2	Crete	Silt loam	Fine, smectitic, mesic Pachic Argiustolls
Columbia, Missouri	7	T_2	M_3	Mexico	Silt loam	Fine, smectitic, mesic Aeric Vertic Epiaqualfs
Indianola, Iowa	8	T_2	M_3	Macksburg	Silty clay loam	Fine, smectitic, mesic Aquic Argiudolls
Glencoe, Minnesota	9	T_2	M_3	Nicollet	Clay loam	Fine-loamy, mixed, superactive, mesic Aquic Hapludolls
Sidney, Montana	5	T_3	M_1	Bryant	Loam	Fine-silty, mixed, superactive, frigid Typic Haplustolls
Mandan, North Dakota	10	T_3	M_2	Farnuf	Loam	Fine-loamy, mixed, superactive, frigid Typic Argiustolls
Medina, North Dakota	10	T_3	M_2	Barnes	Loam	Fine-loamy, mixed, superactive, frigid Calcic Hapludolls
Dorothy, Minnesota	7	T_3	M_3	Radium	Loamy sand	Sandy, mixed, frigid Aquic Hapludolls
Roseau, Minnesota	7	T_3	M_3	Percy	Loam	Coarse-loamy, mixed, superactive, frigid Typic Calciaquolls

† Temperature regimes T_1, T_2, and T_3 refer to thermic, mesic, and frigid soil temperature conditions, respectively.

‡ Moisture regime M_1 refers to aridic ustic soil conditions within each of the temperature regimes. Moisture regime M_2 refers to typic ustic and udic ustic conditions, but includes some aquic soil conditions in each temperature zone. Moisture regime M_3 refers to udic soil conditions, but includes the remaining aquic soil conditions that are not included with moisture regime M_2.

§ An earlier taxonomic classification identified the Dallam soil series as thermic instead of mesic, this study retains the thermic designation.

Analysis of variance (ANOVA) was performed using General Linear Models routine of SAS (1988). This investigation was designed to evaluate the differences among the effects of three soil temperature (T) and three soil moisture (M) regimes on the sequestration of SOC (Table 3–1). The eighth edition of *Keys to Soil Taxonomy* (Soil Surv. Staff, 1998) defines the various classes of T and M regimes, including those of this study. The area of each T and M regime combination for the 13-state region within our study area, and nationally, were summarized from county CRP field enrollment soil data to allow a regional estimate of the annual amount of SOC that is sequestered under the CRP. The area of each temperature moisture regime combination was determined using the following method. The national CRP field enrollment acreage for sign ups 10 to 19 (excluding 18) were provided by USDA-Farm Services Agency (Harte, 1999; Paul Harte, 1999, personal communication). These data included the first three county soil survey map-unit symbols for each field for all sign-ups, except sign-ups 10 to 12 when only the first map unit was provided and for sign-ups 13 and 14 where no soils information was available from USDA-Farm Services Agency (approximately 14% of the total enrollment acreage). The map unit symbols were used to link to soil survey attribute data that included the dominant soil series name for each symbol (Soil Surv. Staff, 1999a). The dominant soil series name was used to link to the Soil Classification File (Soil Surv. Staff, 1999b) to obtain current taxonomic classification including soil temperature and moisture regime. For those CRP acreage where taxonomic information could not be recovered from the Soil Classification File, a county estimate of dominant soil temperature and moisture regime was substituted and verified using the Official Series Description Database (Soil Surv. Qual. Assur. Staff, 1994; Soil Surv. Staff, 1999c). For each soil survey symbol provided by FSA, a CRP enrollment acreage value, dominant soil series or miscellaneous land type, and soil temperature regime and moisture regime was determined (approximately 150 000 data records). These acreage values were then summarized for each temperature moisture combination for the nation and the 13-state study area and then converted to hectares.

RESULTS

Soil Organic Carbon Stocks

The data in this manuscript are presented as the cumulative amounts of SOC and IPM with depth. Amounts of SOC under the CRP and cropped conditions are presented in Table 3–2a,b. The ANOVA indicated that temperature regime (T) consistently affected the amount of SOC at $P > 0.05$ for the 0 to 5, 0 to 10, 0 to 20 cm, and 0 to the bottom of the profile (0–bot) depth increments under both the CRP and cropped treatments. The effect of T at the 0- to 60- and 0- to 100-cm depth increments under the CRP was significant at $P > 0.10$ (Table 3–2a). The effect of T on the amount of SOC under cropping for the 0- to 60-cm depth increment also was significant at $P > 0.10$ (Table 3–2b). The lack of significance for the 0- to 100-cm depth increment under cropping occurred because the total soil profile depth for the Boley, Oklahoma, site was only 84 cm; thus resulting in the data being missing for

Table 3–2. Amount of soil organic carbon (SOC) measured under (*a*) CRP and (*b*) cropped soils by temperature regime for soil depth increments of 0 to 5, 0 to 10, 0 to 20, 0 to 60, 0 to 100 cm and 0 cm to the bottom of the soil pit (~200 cm).

Temperature regime†	df	Depth increment (cm)‡					
		0–5	0–10	0–20	0–60	0–100§	0–Bot
				kg C ha⁻¹			
(*a*) SOC within CRP soils							
T_1		6 910b	11 130b	17 400b	38 590b	51 460b	60 190c
T_2		11 330a	20 140a	34 000a	63 690a	74 260b	91 640b
T_3		16 040a	29 740a	53 050a	99 670a	117 930a	185 780a
				ANOVA			
Source of variation				Pr > F			
Temperature (T)	2	0.038 8	0.040 7	0.027 8	0.087 6	0.083 1	0.004 1
Moisture (M)	2	0.235 4	0.261 8	0.237 3	0.613 5	0.799 3	0.348 4
T × M	4	0.245 5	0.321 8	0.355 0	0.649 0	0.615 1	0.098 7
(*b*) SOC within cropped soils							
T_1		2 730b	5 310b	11 130b	30 280b	45 680	48 090c
T_2		9 290a	18 620a	33 790a	67 410a	79 370	101 040b
T_3		11 380a	23 280a	39 690a	86 850a	110 470	157 060a
				ANOVA			
Source of variation				Pr > F			
Temperature (T)	2	0.014 5	0.005 4	0.045 8	0.092 6	0.155 7	0.021 5
Moisture (M)	2	0.560 7	0.328 1	0.625 5	0.964 9	0.849 4	0.282 2
T × M	4	0.105 2	0.040 0	0.393 2	0.530 9	0.305 7	0.193 4

† See footnotes in Table 3–1 for the definitions of the temperature and the moisture regimes.
‡ Within columns, means followed by the same letter are not significantly different according to LSD (0.05) or LSD (0.10) based upon ANOVA test.
§ Because the total soil profile depth for the cropped site at the Boley, Oklahoma, location was only 84 cm, then the df were decreased to 1 for T, M and T × M at the 0- to 100-cm depth increment.

the 0- to 100-cm depth and a corresponding loss in the number of degrees of freedom available for testing the statistical significance. In terms of the overall data trends for all of the data shown in Table 3–2, moisture regime (M) was not significant nor was the T × M interaction. There was an apparent T × M interaction for the cropped treatment at the 0- to 10-cm depth increment; however compared to the rest of the data in Table 3–2, this interaction is likely due to an anomaly in the data. The lack of a significant effect of M may have resulted from our driest sites not being located far enough into the "aridic" moisture regime, and our wettest sites not being located far enough into the "udic" moisture regime. Irrespective, the sites sampled are on major soil types that have been placed into the CRP and are thus representative of the conditions required to meet the objectives of this study.

The strong effect of T is readily apparent for each depth increment shown in Table 3–2a,b with the total stock of SOC being lowest in the thermic temperature regime (T_1). Where T was in either the mesic (T_2) or frigid (T_3) regime, SOC levels were much higher. Mesic and frigid temperature regimes generally were not significantly different from each other, except for the "0-to bot" depth increment. These data also show the very large SOC stocks that are contained with depth in grassland soils, even when extensively cropped for long periods of time. Land area enrolled in the CRP within the T and M regimes of this study (Table 3–1) has an area

Table 3–3. Difference in amount of SOC and identifiable plant material carbon (IPM-C) measured for CRP minus that measured for cropped soils within temperature and moisture regimes for soil depth increments of 0 to 5, 0 to 10, and 0 to 20 cm.

Tempera-ture Regime[†]	Moisture Regime[†]	df	SOC[‡]			Identifiable plant material (IPM-C)		
			Depth increment (cm)					
			0–5	0–10	0–20	0–5	0–10	0–20
			kg C ha^{-1} yr^{-1}					
T_1	M_1		839	1 224	1 217	2 239	2 243	2 732.5
T_1	M_2		328	427	454	11 152	13 096	14 465
T_1	M_3		255	339	454	−193	−12	304
T_2	M_1		274	170	−460	2 325	2 636	2 769
T_2	M_2		518	924	942	1 111	1 689	2 142
T_2	M_3		160	−310	−267	3 965	4 172	4 183
T_3	M_1		−32	−281	206	1 946	3 038	3 517
T_3	M_2		170	−113.5	593	−280	358	1 574
T_3	M_3		1 781	3 132	4 731	2 251	2 530	2 758
			ANOVA					
Source of variation			P > F					
Temperature (T)		2	0.229 3	0.513 0	0.066 1	0.188 8	0.166 6	0.135 4
Moisture (M)		2	0.092 8	0.400 3	0.143 3	0.387 8	0.207 9	0.094 9
T × M		4	0.006 1	0.044 8	0.030 9	0.050 1	0.031 7	0.025 2

† See footnotes in Table 3–1 for the definitions of the temperature and the moisture regimes.
‡ The numbers shown for change in SOC have been calculated to an annual basis by dividing the difference between the CRP and the cropped locations by the number of years that the sites had been in the CRP.

of 5.6 million ha out of the 7 million ha of CRP within this 13-state region and have SOC stocks that average 105 Mg C ha^{-1}. Therefore, it can be estimated that total SOC stocks to a 2-m depth under the 7 million ha of the CRP land within this region is about 0.74 Pg that is about half of the total U.S. emissions of 1.44 Pg and many times greater than U.S. agricultural emissions of 0.04 Pg (Lal et al., 1998).

Sequestration of Soil Organic Carbon by the Conservation Reserve Program

Estimates of the annual rates of SOC sequestration by the CRP were made by subtracting the SOC measured in cropland sites from that measured under their paired CRP sites (Table 3–3). Analyses of variance were determined for the same depth increments that were reported in Table 3–2. However, a significant response (as an interaction of T × M) was observed only for the 0- to 5-, 0- to 10-, and 0- to 20-cm depth increments. The range of observed rates of SOC sequestration were fairly wide and included some negative values. The wide range observed is expected because of potential differences among the climate regimes studied, difficulties associated with paired-sampling designs, and that CRP grass stands and cropped fields often had different ownership and likely management quality even though the soil factors were well matched. However, because of the statistical significance achieved and number of locations sampled, the collective values that were obtained are considered highly representative of the region covered by this study.

Table 3–4. Amount of SOC sequestered in the CRP based upon the difference in amount of SOC measured for the CRP minus that measured for the cropped soil within temperature and moisture regimes, then multiplied by the corresponding land area in the CRP for depth increments of 0 to 5, 0 to 10, and 0 to 20 cm.

Temperature regime†	Moisture regime†	Area	SOC‡		
			Depth increment (cm)		
			0–5	0–10	0–20
		ha	Mg C yr^{-1}		
T_1	M_1	877 444	736 176	1 073 553	1 067 850
T_1	M_2	564 135	185 036	240 886	256 117
T_1	M_3	86 842	22 145	29 439	39 426
T_2	M_1	1 127 574	308 955	191 688	−518 684
T_2	M_2	679 477	351 969	627 837	640 067
T_2	M_3	924 174	148 176	−286 494	−246 446
T_3	M_1	19 694	−630	−5 534	4 057
T_3	M_2	596 568	101 417	−67 710	353 765
T_3	M_3	749 652	1 334 755	2 347 910	3 546 604
	Total =	5 625 560	3 188 000	4 151 570	5 142 760
	kg ha^{-1} yr^{-1} =		567	738	914

† See footnotes in Table 3–1 for definitions of the temperature and the moisture regimes.
‡ The numbers shown for change in SOC have been calculated to an annual basis by dividing the difference between the CRP and the cropped location by the number of years that the sites had been in the CRP.

In addition to annual rates of SOC sequestration, we also determined the difference in amounts of IPM-C by subtracting the IPM-C measured under the cropped fields from the IPM-C measured under the paired CRP fields. These data, categorized by T and M regime also are shown in Table 3–3. Again, a significant response (the interaction of T × M) was observed only for the top three depths with a moderately wide range. In general, these data show that there is much more IPM where permanent grass cover has been established than under cropped conditions; likely because of the absence of annual tillage effects, presence of actively growing plants, larger below-ground root systems, lower decomposition rates of residual root and plant material, and an overall higher density of individual plants. Because the IPM must be transformed into soil organic matter by microbial and other processes, we have considered the rate of SOC sequestration and the IPM separately, even though both are forms of C in the soil. Irrespective, IPM is the precursor to the introduction of and the amounts of C that are available to be sequestered as SOC.

Another objective of this study is to provide a regional estimate of the annual amount of SOC sequestered under the CRP. To do so, the area of the CRP within each of the nine T × M regimes (Table 3–4) were multiplied by the annual rates of SOC sequestration under the CRP minus the cropped sites for each T × M regime (Table 3–3). The sums, by depth increment, across all of the T × M regimes, then provide an estimate of the total annual amounts of SOC sequestered by the CRP for the entire 5.6 million ha of land under the CRP within the T × M regimes included in this 13-state region. As shown in Table 3–4, the annual amounts of SOC sequestered are 3.19, 4.15, and 5.14 MMTC within the 0- to 5-, 0- to 10-, and 0- to 20-cm depth increments, respectively. A significant difference of the SOC under

Table 3–5. Amount of increased identifiable plant mater carbon (IPM-C) for the CRP based upon the difference in amount of IPM-C measured for the CRP minus that measured for the cropped soils within temperature and moisture regimes, then multiplied by the corresponding land area in the CRP for depth increments of 0 to 5, 0 to 10, and 0 to 20 cm.

| | | | Identifiable plant material (IPM) | | |
| | | | Depth increment (cm) | | |
Temperature regime†	Moisture regime†	Area	0–5	0–10	0–20
		ha		Mg C	
T_1	M_1	877 444	1 964 159	1 967 668	2 397 616
T_1	M_2	564 135	6 291 235	7 387 914	8 160 215
T_1	M_3	86 842	−16 761	−1 042	26 400
T_2	M_1	1 127 574	2 621 611	2 972 286	3 122 254
T_2	M_2	679 477	754 899	1 147 636	1 455 439
T_2	M_3	924 174	3 664 042	3 855 346	3 865 820
T_3	M_1	19 694	38 325	59 832	69 265
T_3	M_2	596 568	−167 039	213 571	938 998
T_3	M_3	749 652	1 687 092	1 896 620	2 067 540
	Total =	5 625 560	16 837 563	19 499 831	22 103 547
	kg C ha^{-1} =		2 993	3 466	3 929

† See footnotes in Table 3–1 for definitions of the temperature and the moisture regimes.

the CRP and cropped sites was observed at only those depth increments (Table 3–3). Similar calculations for the IPM across this 13-state region show that there were 16.84, 19.50, and 22.10 MMT more IPM-C under the CRP than under the cropped soils in 0- to 5-, 0- to 10-, and 0- to 20-cm depth increments, respectively (Table 3–5).

Dividing the total annual amounts of SOC sequestrated by the total land area represented by the T × M regimes studied within the 13-state region (Table 3–4) provide regional average-annual SOC sequestration rates. The rates thus determined are 567, 738, and 914 kg C ha^{-1} yr^{-1} for the 0- to 5-, 0- to 10-, and 0- to 20-cm depth increments, respectively. Regional average amounts of IPM-C also can be calculated in the same manner as were those for SOC. Such a calculation indicates that the average amount of IPM-C under the CRP was about 2990, 3470, and 3930 kg C ha^{-1} greater than under the cropped sites (Table 3–5).

DISCUSSION

Agriculture in the USA has previously been identified as being able to provide an extremely important opportunity to help mitigate climate change by sequestering atmospheric CO_2 as C in soil (CAST, 1992; Follett, 1993). Management of cropland soils and their potential to sequester C is discussed by Lal et al. (1998) but they identify the need to strengthen the data base, including the impact of the CRP, as this study was designed to do. Considerable information is available about C sequestration. However, the literature contains a fairly wide range of values for the rates of SOC sequestration under the CRP. The purpose of this study was to provide experimentally measured data about SOC sequestration on a broad regional basis that also would represent a large percentage of the total area in the USA in the CRP. Data from this study can be used for comparisons with other studies conducted to determine SOC sequestration rates.

Our estimates across three soil temperature (T) and three moisture (M) regimes within the historic grasslands region of the USA are that the CRP sequesters about 570, 740, and 910 kg SOC ha^{-1} yr^{-1} in the 0- to 5-, 0- to 10-, and 0- to 20-cm depth increments, respectively. Our estimates (from a total of 14 sites) compare very well with the estimates made by Gebhart et al. (1994) at five sites across Texas, Kansas, and Nebraska that indicated that about 800 and 1100 kg C ha^{-1} were sequestered in the 0- to 40- and 0- to 300-cm depths under the CRP. Of particular interest is that these two independent studies are based upon actual physical measurements from several sites and both result in similar rates of SOC sequestration. Other results that can be compared are those by Bruce et al. (1999), who indicate that typical rates of C gain under the CRP in the western and central USA are 100 to 400 kg C ha^{-1} yr^{-1} as soil organic matter and 250 to 1350 kg C ha^{-1} yr^{-1} as total below-ground C, including roots (Paustian et al., 1995). These results appear conservative based upon results from the current study. In addition to values for SOC sequestration, the present study included measurements of the IPM-C (including roots) content that amounted to about 2990, 3470, and 3930 kg C ha^{-1} more in the CRP than in cropped sites in the 0- to 5-, 0- to 10-, and 0- to 20-cm depth increments, respectively. However, we have no measurements of rates of IPM accumulation or breakdown on an annual basis, either under the CRP or cropped sites and are not able to make direct comparisons with those by Paustian et al. (1995).

The area represented by the nine T $\times$ M regimes in our study is 5.62 million ha or about 53% of the total 10.61 million ha of land that was enrolled in the CRP across the USA at the time of this study. Based upon both the study by Gebhart et al. (1994) and this study, actual average rates of SOC sequestration likely approach an annual rate of 1000 kg SOC ha^{-1} yr^{-1}. If we assume a sequestration rate near that which we measured in the 0- to 20-cm depth increment, or about 900 kg SOC ha^{-1} yr^{-1}, then 9.5 MMTC would be sequestered annually in 10.61 Mha of the CRP. All U.S. cropland agriculture is reported to emit 42.9 MMTC (Lal et al., 1998); thus, the CRP could offset about 20% of agriculture's CO_2 emissions. Lal et al. (1998) had assumed that C sequestration in the CRP averaged 500 kg SOC ha^{-1} yr^{-1}. If so, then at 100% enrollment of 14.73 million ha of land under the CRP, 17% of the CO_2 emissions for U.S. agriculture could be offset. However, if the rate of SOC sequestration is assumed to be 900 kg ha^{-1} yr^{-1} (instead of 500 kg ha^{-1} yr^{-1}) under 100% enrollment, then the CRP could offset over 30% (13.3 MMTC) of the CO_2 emissions from all U.S. cropland agriculture.

The CRP program provides many apparent benefits such as soil erosion control and wildlife habitat. Also, a benefit that was not recognized during the original implementation of the program was its potential to offset U.S. CO_2 emissions and to contribute to the overall role of soil as a C sink for the USA. Lal et al. (1998) estimated that SOC sequestration by U.S. cropland agriculture (including the CRP) could offset 75 to 208 MMTC (>7%+ of total U.S. emissions) of all U.S. greenhouse gases (GHG) and thus contribute greatly towards the USA meeting its Kyoto protocol requirements. Extrapolation of the data in this study to all the CRP land in the USA results in an amount of SOC sequestration of perhaps as much 13.3

MMTC annually. This amount represents a significant sink in comparison with total U.S. emissions of GHG at either 1485 MMTCE (0.9%) or 1709 MMTCE (0.8%). In summary, an added benefit of the CRP that was not anticipated when it was originally implemented is that of SOC sequestration and the offset to the emissions of CO_2 that occur from other sources within the USA. This benefit contributes directly and significantly to agricultural's role in helping to mitigate the greenhouse effect.

ACKNOWLEDGMENTS

The authors gratefully acknowledge and appreciate the assistance that was provided by local and/or state personnel of the Natural Resources Conservation Service (NRCS), often in collaboration with Agricultural Research Service (ARS) scientists within the states where sampling occurred, for their assistance in the selection of the sites samples and often as active participants in sampling and describing the soil profiles selected. The authors also express special appreciation to Dr. Gary Richardson, ARS Statistician, for statistical assistance, to Dr. Paul Harte, FSA statistician, and to Jule Roth for technical support.

REFERENCES

Berg, N.A. 1994. The genesis of the CRP. p. 7–12. *In* When conservation reserve program contracts expire: The policy options. Proc. Conf., Arlington, VA. 10–11 February. Soil Water Conserv. Soc. Ankeny, IA.

Bruce, J.P., M. Frome, E. Haites, H. Janzen, R. Lal, and K. Paustian. 1999. Carbon sequestration in soils. J. Soil Water Conserv. 54:382–389.

Bowman, R.A., M.F. Vigil, D.C. Neilsen, and R.L. Anderson. 1999. Soil organic matter changes in intensively cropped dryland systems. Soil Sci. Soc. Am. J. 63:186–191.

Council for Agricultural Science and Technology. 1992. Preparing U.S. agriculture for global climate change. Task Force rep. no. 119. CAST, Ames, IA.

Follett, R.F. 1993. Global climate change, U.S. agriculture, and carbon dioxide. J. Prod. Agric. 6:181–190.

Follett, R.F. 1998. CRP and microbial biomass dynamics in temperate climates. p. 305–322. *In* R. Lal et al. (ed.) Management of carbon sequestration in soil. CRC Press, Boca Raton, FL.

Follett, R.F., E.A. Paul, S.W. Leavitt, A.D. Halvorson, D. Lyon, and G.A. Peterson. 1997. Carbon isotope ratios of Great Plains soils and in wheat-fallow systems. Soil Sci. Soc. Am. J. 61:1068–1077.

Follett, R.F., and E.G. Pruessner. 2000. Interlaboratory carbon isotope measurements on five soils. *In* R. Lal et al. (ed.) Methods of carbon analyses. (In press.)

Follett, R.F., E.G. Pruessner, S.E. Samson-Liebig, J.M. Kimble, and S.W. Leavitt. 1997. Organic carbon storage in historic US-grassland soils. Agron. Abstr. 89:208.

Food Security Act. 1985. Public Law 98-198.

Gebhart, D. L., H.B. Johnson, H.S. Mayeux, and H.W. Polley. 1994. The CRP increases soil organic carbon. J. Soil Water Conserv. 49:488–492.

Gomez, B. 1995. Assessing the impact of the 1985 farm bill on sediment-related nonpoint source pollution. J. Soil Water Conserv. 50:374–377.

Intergovernmental Panel on Climate Change. 1995. WGII: Climate Change 1995. p. 21–53. *In* Impacts, adaptations and mitigations of climate change: Scientific-technical analyses. *In* R.T. Watson et al. (ed.) Contributions of working group II to the second assessment report of the Intergovernmental Panel on Climate Change. Cambridge Univ. Press, New York.

Karlen, D.L., J.C. Gardner, and M.J. Rosek. 1998. A soil quality framework for evaluating the impact of CRP. J. Prod. Agric. 11:56–60.

Kuchler, A.W. 1985. Potential natural vegetation. 1:7,500,000 scale map. Digital product. Univ. Kansas, 1966, Natl. atlas of the United States of America, Dep. Interior, U.S. Geol. Serv., Reston, VA.

Lal, R., R.F. Follett, J.M. Kimble, and C.V. Cole. 1999. Carbon sequestration in soils. J. Soil Water Conserv. 54:374–381.

Lal, R., J.M. Kimble, R.F. Follett, and C.V. Cole. 1998. The potential for U.S. cropland to sequester carbon and mitigate the greenhouse effect. Ann Arbor Press, Chelsea, MI.

Paul, E.A., R.F Follett, S.W. Leavitt, A. Halvorson, G. Peterson, and D. Lyon. 1997. Determination of the pool sizes and dynamics of soil organic matter: Use of carbon dating for Great Plains soils. Soil Sci. Soc. Am. J. 61:1058–1067.

Paustian, K., C.V. Cole, E.T. Elliott, E.F. Kelly, C.M. Yonker, J. Cipra, and K. Killian. 1995. Assessment of the contribution of CRP lands to C sequestration. Agron. Abstr. 87:136.

Robles, M.D., and I.C. Burke. 1998. Soil organic matter recovery on conservation reserve program fields in southeastern Wyoming. Soil Sci. Soc. Am. J. 62:725–730.

SAS Institute. 1988. SAS user's guide: Statistics. Version 6.03 ed. SAS Inst., Cary, NC.

Soil Survey Laboratory Staff. 1992. Soil survey laboratory methods manual. Soil Surv. Invest. Rep. 42. Version 2.0. USDA-SCS, Natl. Soil Surv. Center, Lincoln, NE.

Soil Survey Staff. 1998. Keys to soil taxonomy. 8th ed. USDA NRCS.

Soil Survey Quality Assurance Staff. 1994. Soil climate regimes of the United States. USDA–SCS, Natl. Soil Surv. Cent., Lincoln, NE.

Soil Survey Staff. 1999a. National map unit record (MUIR) database dated March 1999. USDA-NRCS Publ. Stat. Lab., Iowa State Univ., Ames, IA.

Soil Survey Staff. 1999b. Soil classification file database dated March 1999. USDA-NRCS Publ. Stat. Lab., Iowa State Univ., Ames, IA.

Soil Survey Staff. 1999c. Official soil series database dated April 1999. USDA-NRCS Publ. Stat. Lab., Iowa State Univ., Ames, IA. Available at http://www.statlab.iastate.edu/soils/osd/.

Soil and Water Resources Conservation Act. 1977. P.L. 95-192.

U.S. Environmental Protection Agency. 1998. Executive summary. p. 1–19. In Inventory of U.S. greenhouse gas emissions and sinks: 1990–1996. EPA, Washington, DC.

Waltman, S.W., B. Lacelle, C. Tarnocai, N.B. Bliss, and F. Orozco-Chavez. 1997. Soil organic carbon map and database for North America. Agron. Abstr. 89:259.

4

A National Assessment of Soil Carbon Sequestration on Cropland: Description of an Analytical Approach

D. W. Goss and Joaquin Sanabria

Texas Agricultural Experiment Station
Temple, Texas

R. L. Kellogg and J. L. Berc

NRCS
Washington, District of Columbia

ABSTRACT

Carbon sequestration in soil can contribute to greenhouse gas mitigation and potentially help the USA meet its target under the Kyoto Protocol of 1997. Reliable national estimates will be needed for international acceptance. This chapter presents a simulation modeling approach developed for this purpose. Field-level process models are combined with a national simulation model consisting of about 177 000 representative fields located throughout the USA. Each representative field is a sample point in the National Resources Inventory (NRI). The NRI statistical design determines the acres represented by each sample point. Soil organic matter models can be run for each representative field using data on soils and crops from the NRI and data on climate and management practices from other sources. The national simulation model can estimate a baseline as well as potential C accumulation assuming adoption of appropriate field management strategies designed to conserve C. Estimates can be updated with future NRIs to track progress.

Alarmed about the potential for global climate change, most of the countries of the world agreed in 1992 to reduce atmospheric concentrations of greenhouse gases originating from anthropogenic activities. The Kyoto Protocol was adopted in 1997, which included specific commitments by the industrialized nations to reduce atmospheric concentrations of carbon dioxide, methane, and nitrous oxide. This could be accomplished by either reducing emissions or sequestering and storing C in sinks. One possibility is to increase the C stored in agricultural soils, which can be accomplished through the adoption of reduced or no-till, use of cover crops, improved nutrition and yield enhancement, elimination of bare fallow, use of forages in crop rotations, use of improved varieties, and use of organic amendments (Bruce et al., 1998).

Reliable national estimates of the amount of C stored in agricultural soils, and how the levels change over time, will be needed for international acceptance. The purpose of this chapter is to present a conceptual approach for estimating a national C sequestration indicator for cropland that can be used to estimate baseline levels of C in agricultural soils for 1992, and to track changes from 1992 to the present and in future years. Work is presently underway to estimate the indicator for cropland in 1992 and 1997.

OVERVIEW OF THE ANALYTICAL FRAMEWORK

The analytical framework will be based on the NRI sample points for culti-vated cropland. The NRI is a national survey of private land use that is based on about 800 000 sample points throughout the 48 states, including cropland, pas-tureland, rangeland, forest land, urban land, and other uses of private land (Nusser & Goebel, 1997; Kellogg et al., 1994). The Inventory is conducted every 5 yr. At each NRI sample point, information is collected on nearly 200 attributes, includ-ing land use and cover, cropping history, conservation practices, potential cropland, highly eroding land, water and wind erosion estimates, wetlands, wildlife habitat, vegetative cover conditions, and irrigation. The NRI is linked to a national soil data-base (SOILS5), which provides data on soil characteristics needed to estimate C levels and potential changes in C levels. For the simulation, each NRI sample point is treated as a "representative field". The number of acres each "representative field" represents was determined from the survey design. Land use data for 1992 will be used to establish a baseline for C sequestration potential, and inventories for future years can be used to monitor progress.

The NRI has about 177 000 sample points with crops that can be used in the simulation. These include the following crops: corn (*Zea mays* L.), wheat (*Triticum aestivum* L.), soybean (*Glycine max* L.), cotton (*Gossypium hirsutum* L.), barley (*Hordeum vulgare* L.), sorghum (*Sorghum bicolor* L.), rice (*Oryza sativa* L.), oat (*Avena sativa* L.), hayland, peanut (*Arachis hypogaea* L.), sunflower (*Helianthus annuus* L.), potato (*Solanum tuberosum* L.), tobacco (*Nicotiana tabacum* L.), and sugar beet (*Beta vulgaris* L.). In 1992, these crops accounted for 110.6 million ha (273 million acres) of cultivated cropland. Additional cultivated cropland was at-tributed to summer fallow [9.7 million ha (24 million acres)], cropland not planted in 1992 including USDA set-aside acres [5.3 million ha (13 million acres)], other close grown crops [2.0 million ha (5 million acres)], other vegetables [1.2 million ha (3 million acres)], and other row crops [1.2 million ha (3 million acres)].

Ideally, potential C accumulation would be estimated for each of these points by running a plant growth and soil process model such as EPIC for characteristics unique to each sample point. At the present time, however, the information re-quirements to run EPIC at each sample point are too large. The approach taken here is to create a smaller number of soil groups and climate groups that will reasonably represent the 177 000 sample points, and produce estimates of C accumulation for each soil-climate combination. These estimates will then be linked back to the appropriate NRI sample points and aggregated to national and regional totals. A similar approach has been used by Goss et al. (1998) to derive national estimates of the potential for pesticide loss from farm fields.

ESTIMATING CARBON ACCUMULATION IN CROPPED SOILS

The importance of calcium carbonate in the total soil C accumulation is recognized. Currently the EPIC (Erosion/Productivity Impact Calculator) model does not account for this accumulation. There will not be an attempt to estimate results from this process.

EPIC (Erosion/Productivity Impact Calculator) will be used to estimate C accumulation. EPIC is a daily time-step model that simulates plant growth, erosion, runoff and leaching of water and nutrients, and includes a component for soil C (Sharpley & Williams, 1990). Simulations will be conducted for 50 yr to capture the potential for C accumulation over time. A national database will be established for combinations of soil and climate groups. For each soil-climate group, model results will be obtained for a variety of tillage practices, conservation practices (contour farming and terracing), and nutrient management practices.

These results will include a sufficient number of conditions to allow construction of a 1992 baseline as well as simulation of alternative practices that can help define the potential for C accumulation in cropland soils.

The initial experimental design for construction of the national database includes the following:

1. Irrigation in areas where irrigation is used.
2. Three tillage practices—no-till, minimum till, and conventional till.
3. Three conservation practices—straight row cultivation, contour farming, and terraces.
4. Three N application rates (low, medium, high) representing common practices for commercial N application in 1990 to 1995 (based on farmer survey results). Estimates are specific for irrigated and non-irrigated conditions, and separate for situations where manure is applied.
5. One P rate of application coinciding with each of the three N rates. The P rate is dependent on the N rate and based on farmer survey results.
6. Dominant times of application (single applications in the fall, spring, at plant, and after plant and double applications for combinations of the four application times).

Nutrient management practices can have a significant impact on the amount of C stored in soils. The 1990–1995 Cropping Practices Survey database assembled by the National Agriculture Statistics Service is being used to derive estimates of commercial N and P application rates and times that represent common practices needed to construct a baseline for early 1990s. Nutrient management practices vary by crop and region of the country. The Cropping Practices Survey database shows that nutrients may be applied multiple times per year. Sometimes there are N applications but no P applications. Phosphorus, however, is rarely applied more than once per year. Because it was found that commercial fertilizer rates tended to be lower when manure also was applied to the field, separate fertilizer rate estimates will be made from samples with manure application and samples without manure application. The surveys had information as to whether or not manure was applied, but did not include estimates of nutrients from the manure. Separate fertilizer rate estimates also will be made for irrigated and non-irrigated conditions. The percentage of acres treated with each nutrient management practice will be calculated, and a sufficient number of dominant application times will be selected so as to represent about 70% of the acres planted to each crop in each state.

MODELING WITH EROSION/PRODUCTIVITY IMPACT CALCULATOR

The goal for operating EPIC is to develop a database with sufficient breadth that many farming systems may be examined at each NRI point without running a 50-yr EPIC simulation for each situation (177 000 times).

The region selected was east of the Rocky Mountains (Fig. 4–1) and contained 90% of the cropland in the USA. This area was selected to reduce climatic and soil phase variation as compared to the entire USA. This area contains over 90% of the cropped land in the USA. Clustering techniques were used to place climates and soils into clusters with similar characteristics. The clustering of soils and climate was designed to reduce the number of EPIC simulations to produce a comprehensive set of data to satisfy the goal of sufficient breadth.

Fig. 4–1. Outline of modeling area with Thiessen weights of weather stations.

SOIL CLUSTERING

The soils selected for clustering are those soil phases from the NRI data that grow crops. This selection reduced the number of unique soil phases to about 30 000 from the 177 000 NRI points. The selected soil phases were clustered using a procedure after Sanabria and Goss (1997). Soil grouping was based in linear combinations of soil properties. The coefficients used in the linear combinations resulted from a multivariate factor analysis performed on the standardized matrix of soil characteristics. Soil properties associated with the highest coefficients in the linear combinations were: surface horizon sand content, silt content, dry bulk density and soil erosion factor; horizon with the minimum saturated conductivity sand content, silt content, dry bulk density and saturated conductivity; total soil water volume for field capacity and wilting point; total soil weight of organic C and calcium carbonate; total soil average pH and cation exchange capacity; and the average slope of the map-unit. Soils with carbonate were clustered separate from those without carbonate. The number of clusters selected for noncarbonate soils (110) reduced the variability within the clusters to a level the entire suite of soils in the cluster would expect to behave similarly when used in the EPIC model. One soil near the centroid of a soil cluster will be chosen to represent all the soils of the cluster when running EPIC. The carbonate soils have not been clustered at this time. The dominant noncarbonate soil cluster per eight-digit hydrologic unit is mapped in Fig. 4–2.

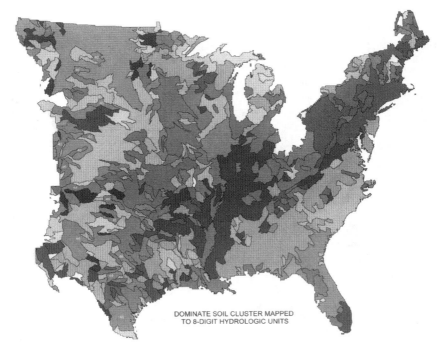

DOMINATE SOIL CLUSTER MAPPED
TO 8-DIGIT HYDROLOGIC UNITS

Fig. 4–2. Dominate noncarbonate soil clusters.

Table 4–1. Cluster statistics for two broadly spaced clusters.

	Cluster 19					Cluster 22			
Variable	N	Mean	SD†	Maxi-mum	Mini-mum	N	Mean	SD†	Maxi-mum	Mini-mum
Annual means										
Precipitation	26	87.272	8.311	102.983	69.017	22	86.856	15.109	105.375	57.958
% wet/dry	26	0.207	0.026	0.243	0.158	22	0.188	0.035	0.241	0.123
% wet/dry	26	0.417	0.037	0.484	0.347	22	0.419	0.042	0.487	0.327
MNRDAL	26	7.981	1.050	9.376	6.116	22	7.447	1.420	9.662	4.858
MI_5AL	26	23.046	1.740	26.633	19.892	22	26.227	1.321	28.625	23.008
Max. temp	26	20.728	0.900	22.233	19.231	22	22.939	1.033	25.408	20.909
Min. temp	26	7.711	0.951	9.568	6.469	22	9.920	0.898	11.706	8.433
Solar rad.	26	388.526	14.389	429.583	367.583	22	409.992	19.370	445.167	385.583
Dew point	26	7.945	1.077	9.768	5.817	22	9.485	1.354	11.802	7.391
SD†										
Precipitation	26	13.609	0.821	15.317	11.892	22	14.683	1.035	16.783	13.183
Mean monthly maximum temp										
Winter	26	9.279	1.693	12.080	6.033	22	12.333	2.017	16.220	7.997
Spring	26	20.700	1.044	22.520	18.670	22	22.992	1.245	25.680	20.710
Summer	26	31.059	1.049	33.230	29.120	22	32.650	0.939	34.483	30.877
Fall	26	21.873	0.806	23.593	20.683	22	23.781	0.960	26.147	22.013
Mean monthly minimum temp										
Winter	26	−2.950	1.724	−0.067	−5.997	22	−0.200	1.959	2.867	−3.847
Spring	26	7.131	0.902	8.880	5.667	22	9.462	0.883	11.277	7.853
Summer	26	18.326	0.888	19.893	16.370	22	19.987	0.628	20.883	18.497
Fall	26	8.337	1.031	10.367	6.937	22	10.430	0.822	11.863	9.113
Mean monthly precipitation										
Winter	26	65.394	24.561	99.733	25.133	22	74.983	35.768	118.367	21.000
Spring	26	94.174	8.962	115.733	77.867	22	100.100	11.958	116.133	72.567
Summer	26	106.51	6.360	116.767	94.600	22	95.780	13.181	114.500	66.567
Fall	26	83.009	6.914	94.100	68.667	22	76.561	11.407	102.333	58.733
Mean SD monthly precipitation										
Winter	26	9.942	1.538	13.233	6.933	22	11.753	2.068	14.300	7.700
Spring	26	12.382	1.359	15.333	9.833	22	14.635	1.426	17.133	12.033
Summer	26	16.004	1.817	18.733	12.367	22	15.398	1.503	17.633	12.267
Fall	26	16.108	0.967	18.200	14.367	22	16.945	2.235	22.167	13.467

(continued on next page)

CLIMATE CLUSTERING

Climatic station records with sufficient length and completeness have been developed from other modeling efforts. These approximately 680 stations east of the Rocky Mountains were clustered using a modified Sanabria and Goss (1997) method. The thiessen-weighted areas of the weather stations are shown in Fig. 4–1. Variables to cluster were selected using the following procedure. The year was divided into four seasons: December to February, March to May, June to August, and September to November. The variables available for evaluation are shown in the left column of Table 4–1. These values were calculated from the EPIC climatic data set. This set of variables was processed with a multivariate factor analyses and one or more strongly weighted variables were chosen from each factor from the clus-

Table 4–1. Continued.

		Cluster 19					Cluster 22			
Variable	N	Mean	SD†	Maxi-mum	Mini-mum	N	Mean	SD†	Maxi-mum	Mini-mum
Mean % wet/dry d										
Winter	26	0.196	0.048	0.267	0.123	22	0.188	0.063	0.277	0.087
Spring	26	0.237	0.018	0.263	0.200	22	0.210	0.021	0.253	0.163
Summer	26	0.225	0.030	0.277	0.177	22	0.208	0.043	0.270	0.130
Fall	26	0.169	0.019	0.207	0.133	22	0.148	0.020	0.173	0.110
Mean % wet/wet d										
Winter	26	0.393	0.052	0.473	0.293	22	0.414	0.058	0.507	0.293
Spring	26	0.428	0.031	0.480	0.360	22	0.424	0.036	0.500	0.357
Summer	26	0.431	0.045	0.523	0.343	22	0.425	0.050	0.490	0.320
Fall	26	0.418	0.034	0.497	0.353	22	0.415	0.038	0.510	0.337
Mean number of monthly raindays										
Winter	26	7.420	1.801	9.663	4.507	22	7.339	2.392	10.907	3.433
Spring	26	8.984	0.768	10.067	7.637	22	8.205	0.921	9.870	6.330
Summer	26	8.716	1.239	10.647	6.733	22	8.143	1.737	10.603	5.023
Fall	26	6.821	0.741	7.847	5.510	22	6.101	0.909	7.483	4.483
Mean monthly maximum 0.5-h precipitation										
Winter	26	11.674	2.703	17.800	7.200	22	12.782	3.234	19.700	8.567
Spring	26	21.594	3.926	28.467	16.500	22	26.973	3.064	31.967	17.533
Summer	26	34.278	2.114	37.400	29.633	22	37.023	3.129	42.900	31.900
Fall	26	24.640	4.916	33.200	15.567	22	28.129	2.912	34.100	22.667
Mean monthly solar radiation										
Winter	26	229.141	12.502	254.333	202.333	22	246.303	16.189	277.000	219.333
Spring	26	449.423	11.423	481.333	435.333	22	472.318	15.249	497.333	448.333
Summer	26	548.244	25.307	608.000	512.333	22	564.136	32.658	617.000	528.333
Fall	26	327.295	20.179	374.667	301.333	22	357.212	21.724	394.667	323.000
Mean monthly dew point										
Winter	26	−2.149	1.734	0.560	−5.560	22	0.039	1.936	3.393	−3.110
Spring	26	6.633	0.971	7.863	4.740	22	8.512	1.231	10.787	6.607
Summer	26	18.539	0.598	19.333	17.020	22	19.287	0.831	20.540	17.480
Fall	26	8.755	1.280	11.600	6.547	22	10.102	1.545	12.487	7.743

† SD = standard deviation.

ter analyses. These variables were: for each season the mean monthly dew point, mean monthly maximum and minimum temperature, and average standard deviation of the monthly precipitation and mean monthly precipitation. Also selected were mean monthly solar radiation for the spring and winter and mean and standard deviation of the annual precipitation.

The number of climate clusters selected (35) reduced variability within the clusters to a level the entire suite of climate stations within a cluster produce similar results when used in the EPIC model. One climatic station near the centroid of a climate cluster will be chosen to represent all climatic stations in the cluster when generating a 50-yr climate for EPIC. The climate clusters are mapped in Fig. 4–3. Note that the clusters are not contiguous across the mapped area. The statistics of two of these clusters (Clusters 19 and 22) that are widely separated east and west are presented in Table 4–1.

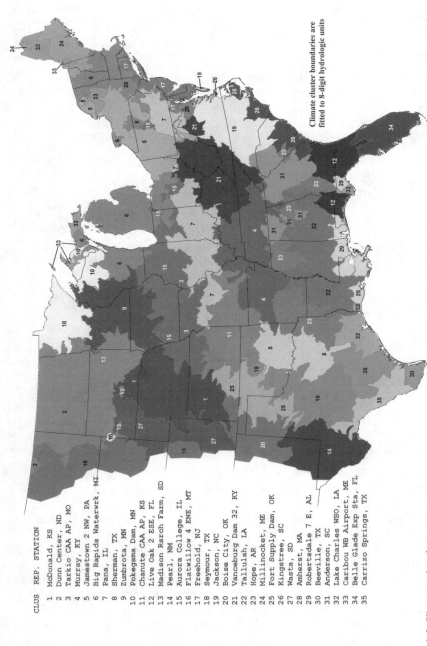

CLUS REP. STATION

1   McDonald, KS
2   Dunn Center, ND
3   Tarkio CAA AP, MO
4   Murray, KY
5   Jamestown 2 NW, PA
6   Big Rapids Waterwrk, MI
7   Pana, IL
8   Sherman, TX
9   Zumbrota, MN
10 Pokegama Dam, MN
11 Chanute CAA AP, KS
12 Live Oak 2 ESE, FL
13 Madison Rsrch Farm, SD
14 Pearl, NM
15 Aurora College, IL
16 Flatwillow 4 ENE, MT
17 Freehold, NJ
18 Seymour, TX
19 Jackson, NC
20 Boise City, OK
21 Vanceburg Dam 32, KY
22 Tallulah, LA
23 Hope, AR
24 Millinocket, ME
25 Fort Supply Dam, OK
26 Kingstree, SC
27 Wasta, SD
28 Amherst, MA
29 Robertsdale 7 E, AL
30 Beeville, TX
31 Anderson, SC
32 Lake Charles WSO, LA
33 Caribou WB Airport, ME
34 Belle Glade Exp Sta, FL
35 Carrizo Springs, TX

Fig. 4-3. Climate station clusters.

The soil and climate clusters provide a base to select climate/soil combinations for running EPIC. The soil clusters selected to use with each climate cluster were accomplished using Geographic Information Systems (GIS). One soil cluster occurred in two to nine-climate clusters.

DATABASE DEVELOPMENT

The soil-climate combinations are the bases to develop the extensive EPIC database. Each of these combinations will be input into 50-yr EPIC simulations with crop, fertilizer application irrigation, conservation and tillage varied by state. The results would be a large database that can be entered with NRI point data, nutrient and tillage data to determine long-term soil organic C and root C changes over time.

ESTIMATING A BASELINE FOR 1992

In order to estimate any increase in C sequestration for cropland soils, it is first necessary to establish a baseline for C stocks. The NRI provides 1982, 1987, and 1992 data needed for soil and crop EPIC input variables at each point, as well as information on the presence of conservation practices (contour farming and terraces). The EPIC model will be run for 50 yr using 1982 NRI data and conventional tillage, followed with 1982, 1987 and 1992 data for 5 yr each using tillage systems derived from other sources such as the Conservation Technology Information Center and National Agricultural Statistical Service. The tillage system will be distributed according to percentage of region covered. The resulting C from organic matter and roots will be the base line soil C from which future changes will be estimated. The resulting national and regional estimates will not be perfect, but they will provide the best possible estimate of a 1992 baseline, given existing data, and will represent much of the diversity in soils and land management that was actually present during that time.

ESTIMATING POTENTIAL SOIL ORGANIC CARBON CHANGES

The national database generated using EPIC contains the information to estimate the potential C changes that will be associated with each of the NRI cropland point for the next 50 yr. The national database will be entered using soil, climate, crop, conservation practice, nutrient management, and tillage practice. The NRI point data provides the soil, climate, crop and conservation practice. The Cropping Practices Survey database provides estimates of the nutrient management. Tillage data for each NRI cropland point is not available; however, NRCS conducted a pilot study in 1992 to determine the feasibility of collecting tillage data at NRI sample points, and sampled about 10% of the cropland sample points. A prediction equation will be developed from these data that will predict the probability associated with each of three tillage conditions—no-till, minimum till, and conventional till. Using this prediction equation, tillage types will be apportioned to each

NRI cropland point. Estimates of potential soil C changes over 50 yr will assume no changes in land use or land management.

TRACKING THE INDICATOR OVER TIME

The NRI is routinely conducted every 5 yr, providing an automatic opportunity to update the C sequestration indicator with each new NRI. The release of the 1997 NRI was in November 2000. Because of the increasing concern about nutrification of water bodies, it is expected that nutrient use databases in the future will be maintained, if not improved. A source for tillage data remains to be found, however. At present, there are no plans to include tillage as an NRI attribute in future inventories. The C database generated using EPIC and 1992 NRI data, according to the methodology described above, will be useful in analyzing future C and conservation policy impacts and scenarios. The most accurate nutrient and tillage data will be used.

It is also expected that refinements will be made in the C estimation procedures in EPIC that will improve its validity for this use. When significant changes in the model occur, the baseline for 1992 can be re-estimated, as well as estimates for subsequent years.

REFERENCES

Bruce, J.P., M. Frome, E. Haites, H. Janzen, R. Lal, and K. Paustian. 1998. Carbon sequestration in soils. J. Soil Water Conserv. 54:382–389.

Goss, D.W., R.L. Kellogg, J. Sanabria, and S. Wallace. 1998. The National Pesticide Loss Database: A tool for management of large watersheds. Available at http://www.nhq.nrcs.usda.gov/land/pubs/gosstext.html. (Verified 7 July 2000.)

Kellogg, R.L., G.W. TeSelle, and J.J. Goebel. 1994. Highlights from the 1992 National Resources Inventory. J. Soil Water Conserv. 49:521–527.

Nusser, S. M., and J.J. Goebel. 1997. The National Resources Inventory: A long-term Multi-resource monitoring programme. Environ. Ecol. Stat. 4:181–204.

Sanabria, J., and D.W. Goss. 1997. Construction of input for environmental simulation models using multivariate analysis. p. 217–234. In Proc. South Central SAS Users Meet., Houston, TX. November 1997.

Sharpley, A.N., and J.R. Williams. 1990. EPIC—erosion/productivity impact calculator: 1. Model documentation. USDA Tech. Bull. 1768.

U.S. Department of Agriculture, National Agricultural Statistics Service. 1995. Cropping practices survey database. Available on-line at http://usda.mannlib.cornell.edu. (Verified 7 July 2000.)

5 An Inventory of Carbon Emissions and Sequestration in United States Cropland Soils

M. D. Eve, K. Paustian, and R. F. Follett

USDA-ARS and Colorado State University
Fort Collins, Colorado

E. T. Elliott

University of Nebraska
Lincoln, Nebraska

ABSTRACT

Means of estimating, monitoring and modeling changes in terrestrial C storage at a national level are just beginning to be developed, although much has been accomplished through localized field studies, state and regional analyses. In the USA, an accurate and defensible estimate of C storage in cropland soils is critical for the development of effective agricultural and environmental policies and strategies. Furthermore, such estimates are needed to fulfill U.S. obligations under the Framework Convention on Climate Change (FCCC). Our research focuses on using a balance sheet approach to estimate net changes in soil C storage in U.S. croplands for the FCCC baseline year (1990). Using input data from a number of sources to obtain information on land use, soils, crops, tillage practices, and climate, we developed a series of spreadsheet models that estimate changes in C storage in cropland soils. Preliminary estimates indicate that changes in land use and management have resulted in mineral soils being a net sink of 11.31 million metric tons (MMT) of C yr^{-1}. Organic soils remain a source of atmospheric C, with emissions estimated at 6.03 MMT of C yr^{-1}. Mineral soils in the Warm Temperate, Moist climatic region of the eastern USA are sequestering the most C, while organic soils in the Sub-Tropical, Moist region have the largest total emissions. Additional modeling is being conducted to obtain estimates of the year-to-year changes in C storage, as well as the potential for increased C sequestration. Strengths and limitations of the approach are discussed.

Means of estimating, monitoring and modeling changes in terrestrial C storage at a national level are just beginning to be developed, although much has been accomplished through localized field studies and state and regional analyses. In the USA, an accurate and defensible estimate of C storage in agricultural cropland soils is critical for the development of effective agricultural and environmental policies and strategies. Furthermore, such estimates are needed to fulfill U.S. obligations under the FCCC. Utilizing analysis of long-term, field-scale soil C data and synthesis of literature on soil C, Lal et al. (1998, 1999) estimate that improving management on U.S. cropland has the potential to sequester between 75 and 208 MMT of C yr^{-1} for the next several decades. Bruce et al. (1999) estimate U.S. potential at 75 MMT yr^{-1} over the next 20 yr. Much of the estimated potential is the result of changes in land use and agricultural land management (Bruce et al., 1999; Lal et al., 1998, 1999; Paustian et al., 1997a,b). These types of changes are currently occurring (although slowly) in the USA (Kellogg et al., 1994; CTIC, 1998; Paustian et al., 1998). The development of an efficient and reliable method for verifying and tracking these changes in C stocks must be a high priority if we are to fully understand the potential to increase soil C stocks through further land use and management change in the future.

The objective of our research was to estimate net changes in soil C storage in U.S. croplands for the FCCC baseline year (1990). The estimate was derived utilizing application of a balance sheet approach to national greenhouse gas inventories that was developed by the Intergovernmental Panel on Climate Change (Houghton et al., 1997a,b,c). We used input data available for the USA from sources including USDA-NRCS (Natural Resources Conservation Service) National Resources Inventory (NRI), Conservation Technology Information Center (CTIC) Crop Residue Management Survey, and the PRISM (Parameter-elevation Regressions on Independent Slopes Model) climate mapping program. Using these data sources to obtain information on land use, soils, crops, tillage practices, and climate, we applied a series of spreadsheet models that estimate changes in C storage in cropland soils.

MATERIALS AND METHODS

Intergovernmental Panel on Climate Change Inventory Approach

The Intergovernmental Panel on Climate Change (IPCC) was established in 1988 by the World Meteorological Organization (WMO) and the United Nations Environment Programme (UNEP). Its role is to assess the scientific, technical and socioeconomic information relevant to the understanding of human-induced climate change risk, making this information available to WMO and UNEP members and the world scientific community. The IPCC played a key role in the development of the UN Framework Convention on Climate Change (FCCC). The IPCC provides assessments, reports, guidelines, and methodologies in support of the 1990 FCCC. More specifically, the IPCC has been responsible for developing and refining an internationally agreed upon technique that IPCC member countries and Parties to the FCCC can use in conducting and reporting greenhouse gas inventories (Houghton et al., 1997a).

The methods developed by IPCC for inventorying greenhouse gas emissions are comprehensive, covering the full range of anthropogenic influences on sources and sinks of greenhouse gasses, such as agricultural, industrial, energy, waste, and land use change related greenhouse gas fluxes. The section dealing with land use change accounts for changes in terrestrial C storage in plant biomass as well as soils.

Since CO_2 is exchanged between plant/soil systems and the atmosphere through the processes of photosynthesis and respiration, net changes in plant/soil C stocks can be equated to net changes in CO_2 emissions. The research reported here focuses on changes in soil C stocks primarily related to changes in land use and/or agricultural management practices. Land use change includes such activities as rangeland being converted into cropland, cropland being enrolled in the Conservation Reserve Program (CRP) and planted back to grass or trees, agricultural land being converted to urban development, or agricultural land being planted back into forest. Changes in agricultural management include changing cropping systems or tillage management practices. Changes in cropping systems would include activities such as shifting from a corn (*Zea mays* L.) and soybean (*Glycine max* L. Merrill) rotation to a corn/hay rotation or shifting from a wheat (*Triticum* spp.) fallow rotation to continuous wheat. Changes in tillage management would include changes in tillage practices such as shifting from conventional tillage to a no-till strategy where the soil surface is left undisturbed from harvest until planting the next crop. Documentation related to the inventory methods for land use and management change can be found in IPCC Workbook Module 5, "Land-Use Change and Forestry" (Houghton et al., 1997b) and Reference Manual Chapter on "Land-Use Change and Forestry" (Houghton et al., 1997c).

The method was developed as a practical first-order approach using simple assumptions about the effects of land use change on C stocks, and then applying those assumptions in order to estimate changes in C stocks due to land use change in the past 20 yr. Changes in soil C stocks as a function of land use and land management practices are estimated using a series of coefficients based on climate, soil type, disturbance history, tillage intensity, and productivity (C input rate) (Houghton et al., 1997b). The default method applies to the upper 30 cm of the soil profile only.

The IPCC method is very general in order to facilitate broad application by parties to the FCCC, but flexible enough that more detailed information and computation can be included if available. Information is laid out in a series of worksheets, each related to a different source of C flux. The worksheets contain the formulas necessary to compute soil C storage (Houghton et al., 1997b). The authors of the approach searched the literature to establish default values for native C levels and changes in C stocks under different land use change scenarios. For the most basic application of the inventory, the investigator needs only the estimated area under each land use/management system at the beginning and end of the inventory period. If more detailed information is available for the country being inventoried, C values and change factors can be adjusted to make the inventory as accurate as possible.

Input Data Sources

The primary data requirements for the IPCC method deal with the land use and land management changes over time. However, information also is needed to stratify these changes according to climate and soil type. Under the IPCC approach, climate is divided into eight distinct categories based upon average annual temperature, average annual precipitation, and the length of the dry season (Houghton et al., 1997c). Six climatic regions occur in the conterminous USA (Table

Table 5–1. Description of the IPCC climate categories that occur in the conterminous USA.

Climate zone	Annual average temperature (°C)	Annual average precipitation (mm)	Length of dry season (months)
Cold temperate, dry	<10	<PET†	NA
Cold temperate, moist	<10	≥PET†	NA
Warm temperate, dry	10–20	<600	NA
Warm temperate, moist	10–20	≥PET†	NA
Sub-Tropical, dry	>20	<1000	Usually long
Sub-Tropical, moist (w/short dry season)	>20	1000–2000	<5

† PET = potential evapotranspiration.

5–1). Soils are grouped by merging soil orders into one of six classes based upon texture, morphology, and ability to store organic matter (Table 5–2; Houghton et al., 1997c). High clay activity mineral soils are those having a large proportion of high activity clays (i.e., expandable clays such as montmorillonite) that are effective in long-term stabilization of soil organic matter (Houghton et al., 1997a). Low clay activity mineral soils are those dominated by low activity, relatively nonexpandable clays (such as kaolinite or gibbsite) that have less ability to stabilize soil organic matter (Houghton et al., 1997a). Sandy soils have less than 8% clay and greater than 70% sand, have poor structural stability, and poor C stabilization ability (Houghton et al., 1997a). Volcanic soils generally have allophane as the primary colloidal mineral and are rich in C and highly fertile (Houghton et al., 1997a). Aquic soils are mineral soils that have developed in wet sites with poor drainage. Under native conditions, these soils typically have high organic matter content and a reduced rate of decomposition (Houghton et al., 1997a). Organic soils form under water-saturated conditions with greatly reduced decomposition (Houghton et al., 1997a) and have very high organic matter content. The IPCC method provides default estimates of C contents for each of the five mineral soil classes under native (i.e., pre-agricultural) conditions.

The types of land use and/or land management change are defined specifically for the country being inventoried. It is important only that the systems identified capture the changes over the previous 20 yr (Houghton et al., 1997c).

For this type of inventory, the USA has better data available than many other parts of the world. We have utilized much data and conducted a detailed inventory. We started with the Major Land Resource Areas (MLRA; Anonymous, 1981) as our basic spatial unit. Major Land Resource Areas were originally delineated in the 1960s as a tool to assist land managers and land use planners, and have undergone occasional revisions since then. Each MLRA represents a geographic unit with relatively similar soils, climate, water resources, and land uses (Anonymous, 1981).

Table 5–2. Taxonomic soil orders grouped into the IPCC inventory categories.

IPCC inventory soil categories	USDA taxonomic soil orders
High clay activity mineral soils	Vertisols, Mollisols, Inceptisols, Aridisols, and high base status Alfisols
Low clay activity mineral soils	Ultisols, Oxisols, acidic Alfisols, and many Entisols
Sandy soils	Sand and loamy sand-textured Entisols
Volcanic soils	Andisols and Spodosols
Aquic soils	Soils denoted as Hydric (excluding Histosols)
Organic soils	Histosols

Climate in the USA is monitored through an extensive network of National Weather Service (NWS) cooperative weather stations. Other national agencies also maintain specific climate data bases such as the USDA-NRCS Snotel network and the Global Gridded Upper Air Statistics data base maintained by the National Climatic Data Center (NCDC). The PRISM (Parameter-Elevation Regressions on Independent Slopes Model) Climate Mapping Program has combined the 1961 to 1990 averages from each of these stations (point data) and sources with topographic information derived from digital elevation models (DEM, grid data) to generate gridded (4-km grid cells) estimates of temperature and precipitation for the USA (Daly et al., 1994, 1998). Average annual precipitation and average annual temperature were derived for each MLRA from PRISM model outputs. These averages were used to aggregate the nearly 180 MLRAs that make up the conterminous USA into the six prescribed IPCC climatic zones represented within the USA (Fig. 5–1).

Land use and land use change information was derived from the National Resources Inventory (NRI) database developed by the USDA-Natural Resources Conservation Service (NRCS). The NRI is a stratified two-stage area sample of over 800 000 points across the USA (Nusser & Goebel, 1997). Each point in the survey is assigned an area weight (i.e., expansion factor) based on other known areas and land use information so that each point has a statistically assigned area that it represents (Nusser & Goebel, 1997). An extensive amount of soils, land use, and land management data are collected each time all, or nearly all, sites are visited every 5 yr (Nusser et al., 1998). The National Resource Inventory was designed as a tool

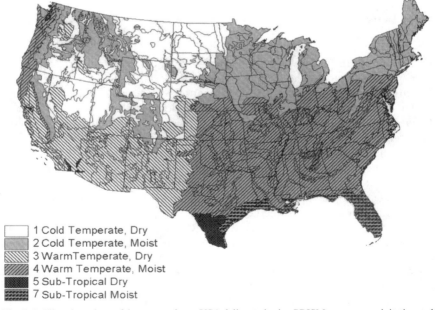

1 Cold Temperate, Dry
2 Cold Temperate, Moist
3 Warm Temperate, Dry
4 Warm Temperate, Moist
5 Sub-Tropical Dry
7 Sub-Tropical Moist

Fig. 5–1. Climatic regions of the conterminous USA delineated using PRISM average precipitation and temperature summarized by Major Land Resource Area (MLRA, USDA-NRCS) according to the IPCC climatic region definitions. The lines overlaying the climate zones are the state boundaries, the thin lines in the background are the MLRA boundaries.

Table 5–3. Land use and management categories grouped from the National Resources Inventory data.

General land use categories	Specific management related subcategories
Agricultural land	Irrigated crops
	Continuous row crops
	Continuous small grains
	Continuous row crops and small grains
	Row crop/fallow rotation
	Small grains/fallow rotation
	Small grains, small Grains then fallow rotation
	Row crops and small grains with fallow rotations
	Row crops in rotation with hay and/or pasture
	Small grains in rotation with hay and/or pasture
	Row crops and small grains in rotation with hay and/or pasture
	Vegetable crops (continuous or in a rotation with other crops)
	Rice (*Oryza sativa* L., continuous or in a rotation with other crops)
	Low residue annual crops [such as cotton (*Gossypium hirsutum* L.) or tobacco (*Nicotiana tabacum* L.)]
	Continuous hay
	Continuous pasture
	Perennial and/or horticultural crops
	Conservation reserve program (CRP)
Range land	
Forest	
Urban land	
Water	
Federal land†	
Miscellaneous non-cropland‡	

† NRI sample points falling on U.S. federally owned land are noted as such, and no further data is collected at those sites.
‡ Miscellaneous non-cropland includes such things as roads and highways and barren areas such as mines or beaches.

to assess conditions and trends for soil, water, and related natural resources primarily on non-federal lands of the USA (Kellogg et al., 1994; Nusser & Goebel, 1997). Specifically, NRI is intended to provide monitoring and status information in support of natural resource conservation policy development and program implementation (Nusser et al., 1998). Because the data points and the information collected have remained fairly constant since 1982, NRI is a useful source for much of the data required for the U.S. inventory. Land use and land management information for 1982 and changes from 1982 to 1992, enrollment in the Conservation Reserve Program (CRP), dominant soil order and other soil characteristics were obtained from the data set (Anonymous, 1994). For the purposes of our research, the land use information in NRI was merged into a combination of land use and management systems as listed in Table 5–3. Each NRI point was assigned to one of these systems based upon the land use data collected in the 1982 and 1992 surveys as well as the cropping history data recorded for the 3 yr prior to each survey (Anonymous, 1994). Each of the over 800 000 NRI points was assigned an aerial extent based upon the weighted expansion factors discussed earlier.

The dominant taxonomic soil order was obtained for each point by querying the soils database that accompanies the NRI data that is linked through a common pointer (Anonymous, 1994). A relatively small percentage of the points did not have an associated soil order record in the soils database. These points were mostly in water or urban areas and did not create a major problem. For NRI sites where soil information was needed and not available, the site was assigned to the low activity mineral soils IPCC category because that category would minimize potential over- or underestimation of C changes at that site.

Reliable estimates of the adoption of improved tillage practices such as conservation tillage or no-till are critical for our inventory because much of the gain in C storage in U.S. cropland is projected to come through the adoption of these practices (Bruce et al., 1999; Lal et al., 1999; Paustian et al., 1997b). The CTIC, a nonprofit information and data transfer facility that promotes environmentally and economically beneficial natural resource systems, conducts annual surveys of the adoption of conservation tillage management (CTIC, 1998). Each year they conduct a Crop Residue Management survey to estimate the portion of cropland managed under various tillage systems. These are annual surveys, and indicate acres of a specific crop planted utilizing that system for each survey year. This does not account for rotations where a crop may be planted no-till 1 yr and using tillage the following year. Nor does it account for producers who may try no-till for a few years and then revert back to some other tillage system. Carbon sequestration is a slow, multiyear process so estimation of the adoption of long-term changes in tillage systems is needed to accurately assess changes in soil C. Modification of the information collected by CTIC is required to estimate the adoption of no-till and reduced tillage systems by region for the USA. For our research, use of a specific system for 5 yr was considered adoption. The Conservation Technology Information Center provided these estimates for both 1982 and 1992 to coincide with the years of NRI data utilization (Table 5–4; D. Towery, 1999, personal communication).

Integration of Existing Data into the Intergovernmental Panel on Climate Change Spreadsheets

Once the appropriate data had been extracted from each of the sources through a series of database manipulations, a merging procedure was implemented to combine areas that were in the same climatic region and where the soil and 1992 land use/crop rotation were the same. For each possible climate-soil-land use category, the area was summed giving a total area by region for each category. The extent of each cultivated cropland category was further separated into relative proportions of conventional tillage, reduced tillage, and no-till hectares by utilizing the information provided by CTIC. Likewise, the same evaluation and merging was conducted utilizing the 1982 land use/tillage management information. This analysis resulted in the number of hectares of each soil-land use-management category within each climatic region for the period 1979 to 1982 and the period 1989 to 1992. For example, we knew the extent of continuous row crops grown under reduced tillage on high activity mineral soils within the warm temperate moist climatic region in 1982, and how much that area had changed by 1992. These estimates were the number of hectares for the beginning and end of the inventory period that were subsequently used in the IPCC spreadsheets.

Because our inventory period was 10 yr, and not the 20 yr outlined in the IPCC documentation (IPCC, 1997b), we computed change using the IPCC 20-yr defaults and then divided by two. We also modified the factor to convert from the total change in C to an annual rate of change to reflect the 10-yr inventory period. For each climate-soil-land use/management category, change in C stocks was computed as:

$$\Delta C = [(ha1992 \times SCUN \times BF \times TF \times IF)$$

$$- (ha1982 \times SCUN \times BF \times TF \times IF)] / 2 \qquad [1]$$

where: ΔC is the change in C stocks for that land use scenario over the 10-yr inventory period; $ha1992$ is the number of hectares in that land use in 1992; $ha1982$ is the number of hectares in that land use in 1982; $SCUN$ is the IPCC default estimate of soil C under native vegetation, a different default is provided for each climatic zone and soil type; BF is the IPCC base factor, or the relative percentage of soil C that has been lost historically by a site being used a particular way; TF is the IPCC tillage factor, which adjusts soil C levels based on the tillage system in place; IF is the IPCC input factor, which adjusts soil C levels based upon the level of inputs such as irrigation, or utilizing hay in the rotation. The total change in C stocks for a climatic region is the sum of the changes in C stocks for each land use category within that region. The total change was then converted to MMT of C yr^{-1} average for the 10-yr inventory period.

Organic soils are handled differently in the IPCC approach. Organic soils that are under native vegetation are excluded from the inventory under the assumption that they are not significantly affected by human activity. Organic soils that are intensively managed are assigned a default rate of C loss based on land use system and the climatic region where they are located (IPCC, 1997c). Only two types of managed systems are considered: cropland and introduced pasture/forest (IPCC, 1997b). Estimated C loss from croplands grown on organic soils is four times greater than loss from managed pasture and forests in the same climatic region (IPCC, 1997b).

Extrapolation and Scaling

All of the input data were aggregated and extrapolated to the climatic regions shown in Fig. 5–1. The NRI point data were extrapolated based upon the statistically derived expansion factors included in the data set (Anonymous, 1994). The climate data was computed as average values for each MLRA. Based upon these averages, each MLRA was assigned to a specific climate region. The tillage information was estimated as an average for each climate region. The IPCC inventory approach is designed as a national tool, and works well for analysis of the six climatic regions in the conterminous USA. Ongoing research is investigating the effects of applying the technique at different spatial scales and levels of aggregation.

RESULTS AND DISCUSSION

Changes in Land Use and Management

The NRI data show that changes in land use are occurring in the USA (Fig. 5–2; Kellogg et al., 1994). One notable change is the enrollment of cropland in the

Change from 1982 to 1992

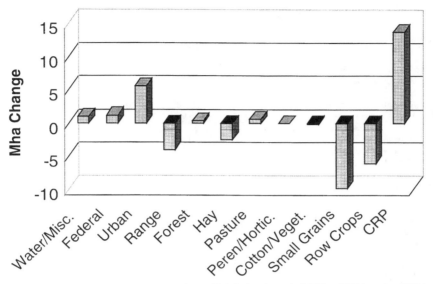

Fig. 5–2. Changes in land use in the conterminous USA during the period 1982 to 1992 based on USDA-NRCS National Resources Inventory data.

Conservation Reserve Program (CRP). The Conservation Reserve Program is a federal program of the 1985 U.S. Food Securities Act intended to take highly erodible cropped land out of agricultural production by planting it back to grass or trees for a 10-yr period. As of the 1992 NRI survey, nearly 14 Mha had been enrolled in CRP (Kellogg et al., 1994). Another notable land use shift between 1982 and 1992 is the loss of native range land (Fig. 5–2). Approximately 4 Mha was broken out of range and put into other uses. The most striking land use change is the 5.5 Mha increase in urban area in the USA, representing a 17% increase in the area of urban lands from 1982 to 1992. Urban development often occurs on high quality farmland, and often involves removal and sale of the topsoil. As neighborhoods are established, high-input lawns and trees are established. The fate of soil C during the process of urban establishment and the decades subsequent to development is not well researched. However, results of Groffman et al. (1995) suggest that urban forests (i.e., residential lawns, parks and golf courses) can accumulate soil C stocks that exceed those of rural forests.

Management of agricultural lands changed somewhat during the period 1982 to 1992 as well. With implementation of the 1985 farm bill, more producers shifted to reduced tillage and no-till systems (Table 5–4). No-till was rare in 1982. By 1992, in the warmer areas of the Corn Belt (Warm Temperate, Moist climatic region), about 10% of the area in continuous row crops was being managed under continuous no-till systems. In other crops and other areas, adoption has been lower (Table 5–4). Systems that reduce the intensity of tillage, however, have received more widespread adoption (Table 5–4).

Table 5–4. Tillage system adoption by percentage of cropland type for the beginning and end of the inventory period as estimated by CTIC (Dan Towery, personal communication).

| | 1982 | | | 1992 | | |
System	No till	Reduced tillage	Conventional tillage	No till	Reduced tillage	Conventional tillage
			%			
Sub-Tropical dry region						
Row crop rotations	0	2	98	0	4	96
Small grain rotations	0	0	100	0	2	98
Vegetables in rotation†	0	2	98	0	4	96
Low residue crops	0	2	98	0	4	96
Sub-Tropical moist region						
Row crop rotations	0	0	100	0	20	80
Small grain rotations	0	0	100	0	10	90
Vegetables in rotation†	0	2	98	0	3	97
Low residue crops	0	2	98	0	3	97
Warm temperate dry region						
Row crop rotations	0	0	100	0	10	90
Small grain rotations	0	2	98	0	15	85
Vegetables in rotation†	0	2	98	0	1	99
Low residue crops	0	2	98	0	1	99
Warm temperate moist region						
Row crop rotations	0	5	95	10	30	60
Small grain rotations	0	5	95	5	30	65
Vegetables in rotation†	0	8	92	1	10	89
Low residue crops	0	8	92	1	10	89
Cool temperate dry region						
Row crop rotations	0	2	98	2	25	73
Small grain rotations	0	5	95	4	25	71
Vegetables in rotation†	0	0	100	1	2	97
Low residue crops‡	0	0	100	1	2	97
Cool temperate moist region						
Row crop rotations	0	10	90	5	30	65
Small grain rotations	0	10	90	5	30	65
Vegetables in rotation†	0	0	100	1	2	97
Low residue crops‡	0	0	100	1	2	97

† Vegetables in rotation was assumed to follow an adoption pattern similar to the "Low residue crops" category.

‡ "Low residue crops" (primarily cotton) in the CTM and CTD region make up very few acres.

Changes in Carbon Stocks in the United States of America

Our preliminary estimates indicate that changes in land use and management during the 10-yr period in the USA have resulted in a net increase in soil C in mineral soils for every climatic region except the Sub-Tropical Moist region in the extreme southeastern USA (Table 5–5). The region contributing the most to the C storage is the Warm Temperate, Moist region. This region is the largest of the regions, with soils and climate that produce high-yielding crops. It also is the area where (as noted earlier) the rate of adoption of no-till is the highest in the country. Overall, changes in land use and agricultural management during the period of 1982 to 1992 have produced an annual increase in mineral soil C of just over 11.3 MMT yr^{-1}.

Table 5–5. IPCC inventory estimate of annual C emissions due to changes in land use and agricultural management practices during the period 1982 to 1992. (Negative values denote decreased emissions, and increases in soil C stocks.)

Climatic region	C from mineral soils	C from organic soils	Total C loss
		MMT yr^{-1}	
Subtropical moist	1.015	4.095	5.11
Subtropical dry	−0.046	0.000	−0.05
Warm temperate moist	−7.569	1.058	−6.51
Warm temperate dry	−0.728	0.471	−0.26
Cool temperate moist	−3.159	0.402	−2.76
Cool temperate dry	−0.828	0.000	−0.83
Totals	−11.31	6.03	−5.29

Organic soils, especially the large quantity of intensively managed organic soils in the southeastern USA (Sub-Tropical, Moist region), continue to lose soil C at an annual rate of about 6.03 MMT yr^{-1}. When the changes in C stocks are combined for the mineral and organic soils, the net effect is still an increase in soil C

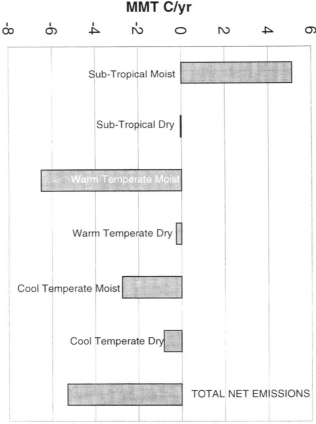

Fig. 5–3. Estimated annual carbon emissions by climatic region resulting from land use and management change between 1982 and 1992. Estimated using the IPCC inventory approach.

(Fig. 5–3, Table 5–5). In every climatic region except the Sub-Tropical Moist region, the mineral soils are storing enough C to compensate for the loss of C from the organic soils. The net annual change in soil C stocks in the conterminous USA related to changes in land use and agricultural management is estimated at 5.29 MMT (Table 5–5).

Limitations of the Intergovernmental Panel on Climate Change Approach

The IPCC inventory approach accounts for changes in C stocks resulting from changes in land use. It also takes into account agricultural land management such as cropping practices and tillage systems, and is sensitive to changes in these management practices. The approach was designed to address only "anthropogenic" sinks and sources—those occurring as a result of human management or impact on the landscape. The approach does not consider any changes in C stocks that result from natural processes or from nonagricultural change that occurred more than 20 yr prior (10 yr prior in our application). For example, some research indicates that native organic soils may continue to sequester C almost indefinitely (Armentano, 1980; Armentano & Menges, 1986). The IPCC inventory methods do not account for this type of increase in soil C. Another example, the approach assumes that native portions of the landscape such as forests have arrived at a long-term equilibrium of soil C and are in a steady-state, neither gaining or losing C stocks. Detailed land use change data tend to be available for the relatively recent past, making it difficult to incorporate more long-term trends in land use and land management. Finally, the IPCC inventory does not account for changes in soil C due to erosion and sedimentation. There is no estimate made in the method for C lost to the atmosphere during an erosion event or for C buried in sediments.

As with any analysis, there is a level of uncertainty inherent in the estimates. Each of the input datasets has an associated level of uncertainty that gets passed through the analysis and results in uncertainty in the final estimates. To minimize this uncertainty, we utilized data from tightly controlled data sets with rigid data collection procedures in place, and from highly respected sources. Ongoing research is aimed at identifying the sources and magnitude of uncertainty, and refining the approach to minimize that uncertainty. Some sources of uncertainty currently being addressed will be discussed below.

For our estimate of C stocks under native vegetation, as well as the base, tillage, and input factors, we utilized the default values contained in the IPCC inventory documentation (Houghton et al., 1997b). The IPCC authors determined these defaults after searching the relevant literature and experimental data. The defaults are set according to soil type and climatic region, and were intended for application anywhere in the world where better information is not readily available. The default values may carry with them a high degree of uncertainty. Many of the important default values are not well established or may be highly variable from region to region (Houghton et al., 1997a). This uncertainty can be reduced substantially for the USA by using field experiment data to adjust the default values used in the analysis. For this preliminary work, the default values have been evaluated and determined to be consistent with the available field data. More comparison with available field data and further refinements to the values may result in decreased uncertainty. Also, more field research and new experimental data will result in better inventory inputs and reduced uncertainty.

For this research, we aggregated the data by soil type and land use across each climatic region. Aggregating data over such a large area could result in some land use changes canceling each other, incorrectly showing no net loss or gain on those hectares. For example, within one climatic region, a conversion from forest to continuous row crop agriculture at one site could be offset by a conversion from continuous row crop agriculture back into forest at another site. This type of area offset, while expected to be small, would introduce error into the overall estimates. We are currently working on a modified approach where land use change and the subsequent changes in C stocks are evaluated independently at each NRI site for the inventory period. After C stock change has been estimated for each site, then the estimate will be summed over the climatic region, thus eliminating the uncertainty resulting from offsetting changes in land use.

Another possible source of uncertainty is in the spatial scaling of the data. The NRI data that we used was designed for regional and national analysis of natural resources, and implemented and tested primarily at the county level (Nusser & Goeble, 1997). The sampling framework will not be intensive enough to allow reliable estimation of parameters at more localized scales. The effects of analyzing the NRI and other data at various scales, from county level to the larger climatic regions, remains uncertain, and is part of our ongoing investigation. This will be as much a test of the sensitivity of the approach as it is a test of uncertainty.

CONCLUSIONS

Our initial application of the IPCC inventory approach in estimating changes in soil C stocks related to land use and agricultural land management changes in the USA indicates that mineral soils are sequestering atmospheric C in most regions of the country (although in some regions the rate of sequestration is very slight). Managed organic soils continue to release C. Even with the emissions from organic soils, all agricultural soils and land use change are resulting in a net sink of atmospheric C into the soil in every area of the conterminous USA except the Sub-Tropical Moist region of the Southeast. There are several explanations for this sink including: (i) most upland agricultural soils have been in production long enough that they are no longer a net source of atmospheric C (Cole et al., 1993), (ii) crop yields (and crop biomass) have increased substantially since the 1940s (Allmaras et al., 1998), (iii) the adoption of conservation tillage practices has increased (Kern & Johnson, 1993; Lal & Kimble, 1997), and (iv) cropland has been converted to grass and trees through the Conservation Reserve Program (CRP) (Gebhart et al., 1994; Lal et al., 1999).

The IPCC inventory approach for estimating changes in soil C stocks is efficient and can be accomplished with reasonable land use, soils and climate data inputs. The approach has limitations, and the resulting estimates will include a measure of uncertainty. However, many of the limitations can be overcome with minor modifications to the approach, or addressed with other modeling techniques. Further testing and refinement can minimize the level of uncertainty, and new techniques can be developed to more accurately measure the level of uncertainty. There is a continuing need for experimental research and information to refine and improve the values used as input factors to compute changes in C stocks. Also, additional factors may be needed that can accommodate new or improved agricultural technologies for enhancing soil C stocks. Research also is ongoing to assess the capability of the CENTURY ecosystem model in deriving national estimates of changes in C stocks. Use of a simulation model approach will require more detailed input data, but may facilitate more rigorous quantification of uncertainty.

These results indicate that with proper input data the IPCC approach can provide an estimate of changes in soil C stocks. The IPCC is a defensible technique, that can provide reliable base-line estimates for compliance with FCCC obligations. Furthermore, with additional validation and possible refinement, estimated changes in soil C stocks will be a useful input to planning U.S. agricultural and environmental policy. The results presented here confirm the findings of Lal et al. (1998, 1999) and Bruce et al. (1999) that there is considerable potential for further soil C sequestration through strategic changes in land use and agricultural management.

ACKNOWLEDGMENT

This work was funded by the USDA-ARS, with additional funding being provided by EPA. Appreciation is expressed to Dr. John Kimble and Sharon Waltman (USDA-NRCS Natl. Soil Surv. Lab.), and Dan Towery (Conserv. Technol. Inform. Center), for their input. Gratitude also is extended to Drs. Dennis Ojima and Vern Cole for their review of this manuscript. We also are grateful to the chairs, experts and authors who developed the IPCC inventory guidelines. K. Paustian was Co-Chair for the Land Use Change and Forestry section.

REFERENCES

Allmaras, R.R., D.E. Wilkins, O.C. Burnside, and D.J. Mulla. 1998. Agricultural technology and adoption of conservation practices. p. 99–158. *In* F.J. Pierce and W.W. Frye (ed.) Advances in soil and water conservation. Sleeping Bear Press, Inc., Chelsea, MI.

Anonymous. 1981. Land resource regions and major land resource areas of the United States. USDA Agric. Handb. 296. USDA, NRCS, NSSC, Lincoln, NE.

Anonymous. 1994. 1992 national resources inventory digital data. USDA, NRCS, Washington DC.

Armentano, T.V. 1980. Drainage of organic soils as a factor in the world carbon cycle. BioScience 30:825–830.

Armentano, T.V., and E.S. Menges. 1986. Patterns of change in the carbon balance of organic soil—wetlands of the temperate zone. J. Ecol. 74:755–774.

Bruce, J.P., M. Frome, E. Haites, H. Janzen, R. Lal, and K. Paustian. 1999. Carbon sequestration in soils. J. Soil Water Conserv. 54:382–389.

Cole, C.V., K. Flach, J. Lee, D. Sauerbeck, and B. Stewart. 1993. Agricultural sources and sinks of carbon. Water Air Soil Pollut. 70:111–122.

Conservation Technology Information Center. 1998. 1998 Crop residue management executive summary. CTIC, West Lafayette, IN.

Daly, C., G.H. Taylor, W.P. Gibson, T. Parzybok, G.L. Johnson, and P.A. Pasteris. 1998. Development of high-quality spatial datasets for the United States. p. I-512–I-519. *In* Proc. 1st Int. Conf. Geospatial Informat. Agric. Forest., Lake Buena Vista, FL. 1–3 June. ERIM Int., Inc., Ann Arbor, MI.

Daly, C., R.P. Neilson, and D.L. Phillips. 1994. A statistical-topographic model for mapping climatological precipitation over mountainous terrain. J. Appl. Meteorol. 33:140–158.

Food Security Act of 1985. 1985. P.L. 99-198.

Gebhart, D.L., H.B. Johnson, H.S. Mayeux, and W.W. Polley. 1994. The CRP increases soil organic carbon. J. Soil Water Conserv. 49:488–492.

Groffman, P.M., R.V. Pouyat, M.J. McDonnell, S.T.A. Pickett, and W.C. Zipperer. 1995. Carbon pools and trace gas fluxes in urban forest soils. p. 147–158. *In* R. Lal et al. (ed.) Advances in soil science: Soil management and the Greenhouse Effect. CRC Press, Inc., Boca Raton, FL.

Houghton, J.T., L.G. Meira Filho, B. Lim, K. Treanton, I. Mamaty, Y. Bonduki, D.J. Griggs and B.A. Callender (ed.). 1997a. IPCC guidelines for national greenhouse gas inventories reporting instructions. Vol.1. Rev. 1996. IPCC, OECD, IEA, U.K. Meteorol. Office, Bracknell, United Kingdom.

Houghton, J.T., L.G. Meira Filho, B. Lim, K. Treanton, I. Mamaty, Y. Bonduki, D.J. Griggs and B.A. Callender (ed.). 1997b. IPCC guidelines for national greenhouse gas inventories workbook Vol. 2. Rev. 1996. IPCC, OECD, IEA, U.K. Meteorol. Office, Bracknell, United Kingdom.

Houghton, J.T., L.G. Meira Filho, B. Lim, K. Treanton, I. Mamaty, Y. Bonduki, D.J. Griggs and B.A. Callender (ed.). 1997c. p. 5.1–5.74. IPCC guidelines for national greenhouse gas inventories reference manual. Rev. 1996. Vol. 3. IPCC, OECD, IEA, U.K. Meteorol. Office, Bracknell, United Kingdom.

Kellogg, R.L., G.W. TeSelle, and J.J. Goebel. 1994. Highlights from the 1992 National Resources Inventory. J. Soil Water Conserv. 49:521–527.

Kern, J.S., and M.G. Johnson. 1993. Conservation tillage impacts on national soils and atmospheric carbon levels. Soil Sci. Soc. Am. J. 57:200–210.

Lal, R., J. Kimble, R.F. Follett, and C.V. Cole. 1998. The potential for U.S. cropland to sequester carbon and mitigate the Greenhouse Effect. Sleeping Bear Press, Ann Arbor, MI.

Lal, R., R.F. Follett, J. Kimble, and C.V. Cole. 1999. Managing U.S. cropland to sequester carbon in soil. J. Soil Water Conserv. 54:374–381.

Lal, R., and J.M. Kimble. 1997. Conservation tillage for carbon sequestration. Nutr. Cycl. Agroecosyst. 49:243–253.

Nusser, S.M., and J.J. Goebel. 1997. The National Resources Inventory: A long-term multi-resource monitoring programme. Environ. Ecol. Stat. 4:181–204.

Nusser, S.M., F.J. Breidt, and W.A. Fuller. 1998. Design and estimation for investigating the dynamics of natural resources. Ecol. Appl. 8:234–245.

Paustian, K., O. Andren, H.H. Janzen, R. Lal, P. Smith, G. Tian, H. Tiessen, M. Van Noordwijk, and P.L. Woomer. 1997a. Agricultural soils as a sink to mitigate CO_2 emissions. Soil Use Manage. 13:230–244.

Paustian, K., H.P. Collins, and E.A. Paul. 1997b. Management controls on soil carbon. p. 15–49. *In* E.A. Paul et al. (ed.) Soil organic matter in temperate agroecosystems: Long-term experiments in North America. CRC Press, Boca Raton, FL.

Paustian, K., C.V. Cole, D. Sauerbeck, and N. Sampson. 1998. CO_2 mitigation by agriculture: An overview. Climate change. 40:135–162.

6

Carbon Dynamics and Sequestration of a Mixed-Grass Prairie as Influenced by Grazing

G. E. Schuman

High Plains Grasslands Research Station
Cheyenne, Wyoming

D. R. LeCain, J. D. Reeder, and J. A. Morgan

Crops Research Laboratory
Fort Collins, Colorado

ABSTRACT

Grazing rangelands can influence plant community structure, soil chemical and physical properties, and the distribution and cycling of nutrients within the plant-soil system. Studies at the High Plains Grasslands Research Station near Cheyenne, Wyoming, have shown that after 12 yr of season-long grazing, the total C mass of the plant-soil (0–60 cm) system was not affected when compared to a nongrazed treatment. However, significant increases in the mass of C in the primary root zone (0–30 cm) of the soil were evident in the grazed treatments. A gas-exchange chamber used to assess CO_2 exchange rates (CER) of these grazing treatments exhibited as much as two times greater CER from mid-April through the end of June for the grazed treatments compared to the nongrazed treatment. This increase in CER was best related to a green vegetation index, hence the grazed pastures exhibited a more vigorous system. We hypothesize that grazing removes and/or prevents the accumulation of dead plant material, and reduces litter, thereby resulting in a warmer soil and enhancing the illumination of green shoots in the spring. This results in earlier spring green-up and greater CER compared to the nongrazed treatment. These data indicate that grazing does not have any detrimental affects on the C balance of the ecosystem and may enhance soil C and the potential for soil C-sequestration on rangelands if managed appropriately.

The effects of grazing on soil organic C dynamics in rangeland have been variable and inconsistent among the ecosystems studied (Milchunas & Lauenroth, 1993; Smoliak et al., 1972; Bauer et al., 1987; Frank et al., 1995). Studies have shown that grazing affects the soil organic C levels, either through changes in plant community (Coupland et al., 1960; Coupland, 1992; Dormaar & Willms 1990), or by other more subtle and less discernable responses such as changes in the chemical composition of organic matter, amount and quality of root exudates, increased root biomass, and more rapid cycling of C (Smoliak et al., 1972; Christie, 1979; Naeth et al., 1991; Shariff et al., 1994; Detling et al., 1979; Schuster, 1964; Davidson, 1978; Holland & Detling, 1990; Dyer & Bokhari, 1976).

Research conducted at the High Plains Grasslands Research Station near Cheyenne, Wyoming, on a northern mixed-grass prairie evaluated the effects of three grazing intensities on the soil organic C, plant-soil system C balance (Schuman et al., 1999), soil respiration and photosynthesis (LeCain et al., 1999). The general objective of these studies was to better understand the role of grazing on C cycling and its impact on sustainability of the grassland ecosystem. This chapter summarizes the overall findings of this research and relates them to the effects of grazing management on the potential for C sequestration by a mixed-grass ecosystem.

SITE DESCRIPTION AND METHODS

The research was conducted on a native northern mixed-grass rangeland, near Cheyenne, Wyoming (41°11′ N lat, 104°54′ W long), with rolling topography, elevations ranging from 1910 to 1950 m, a 127-d growing season, and an average annual precipitation of 384 mm, of which 70% occurs from April 1 through September 30 (NOAA, 1994). Soils are Ascalon sandy loams (mixed, mesic, Aridic Argiustolls) (Stevenson et al., 1984). Vegetation is predominately grasses [55% cool-season (C_3) species and 23% warm-season (C_4) species], forbs, sedges and half-shrubs. Dominant grasses are: C_3, western wheatgrass [*Pascopyrum smithii* (Rydb.) A. Love] and needleandthread (*Stipa comata* Trin & Rupr.); and C_4, blue grama [*Bouteloua gracilis* (H.B.K.) Lag. Ex Steud.]. This study area had not been grazed by domestic livestock for over 40 yr prior to establishment of the grazing treatments in 1982.

Three treatments were imposed: (i) nongrazed 0.2-ha exclosures, EX; (ii) continuous season-long light grazing (0.16–0.23 steers ha^{-1}), CL; and (iii) continuous, season-long heavy grazing (0.56 steers ha^{-1}), CH. The CL and CH pastures were 41 and 9 ha, respectively. The experiment was organized in a randomized block design with two replicate blocks (pastures). The heavy stocking rate utilized about 45% of the annual production. Details of the grazing treatments and pasture design are given in Hart et al. (1988).

Vegetation field transects (50 m) were established in each replicate pasture on near-level sites of the Ascalon soil map unit. Depth of the A horizon and solum of the Ascalon soil is 15 cm ±2 and 100 cm ±7, respectively. In July 1993, soil and plant samples were collected at 10-m intervals along the transects to assess the C content of the various components of the plant-soil system. Soil samples were collected to a depth of 60 cm and segregated into the following depth increments: 0 to 3.8, 3.8 to 7.6, 7.6 to 15, 15 to 30, 30 to 45, and 45 to 60 cm. Separate soil cores were obtained for root biomass and C assessment. Bulk density data also were obtained and used to convert the soil C concentrations determined on a mass concentration basis (mg kg^{-1}) to mass of C on a land area basis (kg ha^{-1}). Bulk density was obtained using a 3.2-cm diam. core 3.8 cm long for the first two depth increments and 7.6 cm long for the lower four depths (Blake & Hartge 1986). Root biomass was obtained using a washing and screening technique described by

Lauenroth and Whittman (1971). Surface litter and standing dead plant biomass were estimated at 10-m increments along the transects within a 0.18-m^2 quadrat. Estimates of annual production were obtained from duplicate 0.18-m^2 quadrats clipped from three 1.5- by 1.5-m temporary exclosures located throughout each of the grazed pastures; in the exclosures five 0.18-m^2 quadrats were sampled along the 50-m transects.

Soil organic C was determined using the Walkley-Black dichromate oxidation procedure (Nelson & Sommers 1982) and plant component C was determined with an automated combustion analyzer.

In 1995, five 1-m^2 angle iron metal frames were driven into the soil about 3 m from and parallel to each 50-m transect for assessing canopy CER, a measure of photosynthetic activity and soil respiration. The CER measurements were only made on one replicate pasture. The carbon dioxide exchange rate was measured with a 40 cm high by 100 by 100 cm Lexan (Regal Plastics, Littleton, CO[1]) chamber. The air in the chamber was circulated by small fans and a sample of the air pumped to a portable infrared analyzer that assessed the CO_2 depletion within the chamber during a 2- to 3-min period. The CER measurements were taken from 1000 to 1300 h (Mountain Standard Time) approximately every 3 wk during the growing seasons in 1995 to 1997. Measurements made on this site showed that maximum daily photosynthesis, as measured by CER, occurred during this period of the day which agrees with Detling et al. (1979). Each time CER measurements were made, a 10-pin point frame was used to determine the relative photosynthetic surface area (green vegetation index, GVI) within each of the CER frames (Warren-Wilson 1963). Green leaves, stems, and sheaths were reccorded by the point frame since all three are significant contributors to photosynthesis (Caldwell et al., 1981). Standard methods of assessing leaf area index would not work on a plant canopy of this short stature, and destructive sampling was unacceptable to achieve repeated sampling over time on the same areas. Soil respiration was measured on small plots (82 cm^2) adjacent to each CER frame in which all vegetation had been removed; all CER data were adjusted for soil respiration by subtracting the CO_2 contributed by soil respiration from the CO_2 depletion measured (PP Systems, SRC-1, Hertfordshire, UK[1]). The CER and soil respiration data were calculated on a ground surface area basis.

Greater detail of these studies are reported by Schuman et al. (1999) and LeCain et al. (1999).

RESULTS AND DISCUSSION

Grazing of these pastures at the heavy stocking rate decreased the peak standing crop (PSC), Table 6–1, and shifted plant composition. Western wheatgrass represented 45% of the PSC (weight basis) in the CL and only 21% in the CH grazing treatments, while the percentage of blue grama increased from 17% under CL to 27% under the CH treatment (Schuman et al., 1999). Even though the PSC was similar for the EX and CL treatments, plant composition was quite different. Forbs were the dominant species group in the EX (33%) but only accounted for 16 and

[1] Brand names and company identification are made for the benefit of the reader only and in no way imply endorsement by USDA or the Agricultural Research Service.

Table 6–1. Total above and below ground vegetation biomass as affected by grazing, 1993 (modified from Schuman et al., 1999).

Component	Exclosure (EX)	Continuous light grazing (CL)	Continuous heavy grazing (CH)	LSD$_{(0.10)}$
		kg ha^{-1}		
Above-ground				
Live biomass	1 330	1 224	816	270
Dead biomass	3 344	2 139	1 271	1 054
Total biomass	4 673	3 363	2 087	1 176
Root biomass				
0–15 cm	31 474	21 695	27 319	6 500
15–30 cm	5 516	6 971	5 289	n.s.
30–60 cm	1 618	1 779	1 162	n.s.
Total roots	38 609	30 446	33 770	n.s.
Total plant biomass	43 281	33 809	35 857	n.s.

22% in the CL and CH pastures, respectively (Schuman et al., 1999). Litter and standing dead biomass together accounted for 72, 63, and 61% of the above-ground plant biomass in EX, CL, and CH pastures, respectively. Root biomass in the 0- to 15-cm soil depth was significantly lower in the CL than in the EX and CH treatments but no difference was found between the EX and CH grazing treatments. Coupland et al. (1960) and Dormaar and Willms (1990) found that grazing significantly altered plant community composition and they found that with a change from a cool-season dominated community to warm-season species like blue grama, root biomass in the surface 15 to 30 cm increased. However, our root biomass did not respond like that reported by Dormaar and Willms (1990). The lack of a similar response may be because of less intense grazing or greater variability in the root biomass and/or the grass species involved in our study.

Carbon distribution in the above-ground vegetation components (Table 6–2), as would be expected, responded like the vegetation component biomass (Table 6–1). Heavy grazing significantly reduced live biomass C, while both the CL and CH treatments reduced dead biomass C. Mass of root C in the 0- to 15-cm depth was greater in the EX than for either of the grazing treatments. The greater root C

Table 6–2. C mass of vegetation components as affected by grazing, 1993 (modified from Schuman et al., 1999).

Component	Exclosure (EX)	Continuous light grazing (CL)	Continuous heavy grazing (CH)	LSD$_{(0.10)}$
		kg ha^{-1}		
Above-ground				
Live biomass	587	535	355	119
Dead biomass	1 015	742	394	255
Total above-ground C	1 602	1 277	749	252
Roots				
0–15 cm	7 166	6 011	5 763	1 073
15–30 cm	1 244	1 646	1 312	n.s.
30–60 cm	379	504	346	n.s.
Total root C	8 790	8 160	7 421	n.s.
Total plant C	10 392	9 437	8 170	1 259

Table 6–3. Soil profile C mass and total system C mass of a grazed mixed-grass prairie, 1993 (modified from Schuman et al., 1999).

Soil profile	Exclosure (EX)	Continuous light grazing (CL)	Continuous heavy grazing (CH)	LSD$_{(0.10)}$
		kg ha^{-1}		
Depth increments				
0–3.8 cm	9 595	12 675	12 000	1 309
3.8–7.6 cm	5 906	7 457	8 478	660
7.6–15 cm	12 662	15 009	15 472	1 573
0–15 cm	28 162	35 141	35 950	2 188
15–30 cm	19 761	22 847	22 348	2 485
0–30 cm	47 923	57 998	58 298	2 463
30–45 cm	22 932	20 353	25 281	n.s.
45–60 cm	17 291	13 595	17 689	n.s.
0–60 cm	88 147	91 937	101 267	11 853
Total ecosystem C (includes plant components)				
0–30 cm	58 315	67 425	66 468	4 334
0–60 cm	98 539	101 374	109 437	n.s.

mass and the greater root biomass at this depth may be due in part to the greater proportion of forbs found in the EX. The forbs have much larger roots and represent greater root biomass than the grass roots. The organic soil C mass in the 0- to 30-cm soil depth was significantly lower in the EX than either of the grazed treatments (Table 6–3). Blue grama has been shown to partition more C below-ground in blue grama-dominated systems (Coupland & Van Dyne 1979) and in mixed-grass prairie (Frank et al., 1995). Even though we could not detect a significant increase in root biomass in the grazed plots, we did observe a slight increase in C/N of the root material from the grazed treatments (CL 27:1, CH 28:1) compared to the 24:1 for the EX. This supports the point that forb roots present in the EX represented less C than the more fibrous root systems present in the CL and CH pastures. Frank et al. (1995) showed a similar, but not significant, increase in the C/N of the soil-root composite under grazing compared to the nongrazed exclosure. Their data also showed that grazing resulted in an increase in blue grama and an increase in the δ^{13}C that further substantiates an increase in C$_4$ species contribution to the organic C pool. Dodd and Hopkins (1985) and Mutz and Drawe (1983) showed that simulated grazing stimulated greater above-ground production that resulted in greater root growth. Dyer and Bokhari (1976) also suggested that simulated grazing may stimulate root respiration and root exudation. Increased below-ground C in the grazed treatments was limited to the surface 30 cm where >90% of the root biomass exists in these ecosystems. About 89 to 93% of the ecosystem C was stored in the soil organic matter of the 0- to 60-cm soil depth, with less than 10% found in the vegetation components. The roots accounted for 85 to 91% of the vegetation component C. Therefore, the effect of grazing on the above-ground plant components has limited potential impact on the total C mass of the system. However, any effect it has on plant growth and resource transfer to below-ground components can significantly impact system C distribution. Above-ground vegetation components have a greater potential to photochemically oxidize (Coupland & Van Dyne 1979) and be lost from the system unless grazing breaks down the plant material and encourages soil contact and/or incorporation and decomposition.

Canopy CER, soil respiration rate, and GVI were measured for three growing seasons (1995–1997). The results were similar in all 3 yr (LeCain et al., 1999);
therefore for brevity, only the 1997 data are shown here. Early season CER of the
plant community in the grazed treatments was as much as two times that exhibited
by the plant community of the EX treatment (Fig. 6–1). This grazing induced increased CER was generally observed mid-April through the end of June, and was
closely related to the green vegetation index (Fig. 6–1) on the sites. The CER and
GVI data illustrate that grazing caused earlier and more vigorous growth that resulted in greater GVI, and hence greater early season photosynthesis. Even though
LeCain et al. (1999) showed that the growing season average CER was not signif-

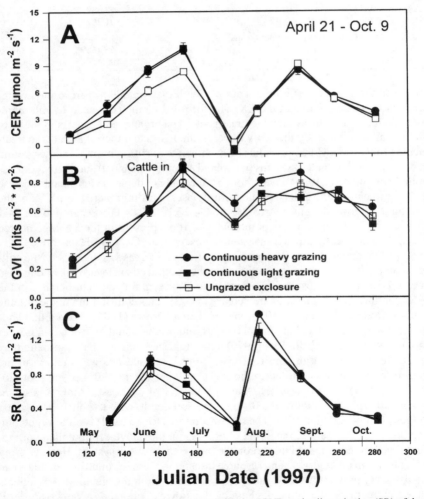

Fig. 6–1. CO_2 exchange rate (CER), green vegetation index (GVI), and soil respiration (SR) of three
mixed-grass prairie pastures subjected to different grazing intensities, 1997. Data are means of five
replications ± standard error (from LeCain et al., 2000).

icantly different between the grazing treatments, the potential for greater above-ground production is almost always greater during the early part of the growing season, when soil moisture and cool temperatures are more favorable for cool-season, C_3 dominated grasslands than during the latter warmer and typically drier portion of the growing season. Eighty to one-hundred percent of the PSC on the northern mixed prairie in southeastern Wyoming is typically achieved by 30 June (W.S. Johnson, 1984, unpublished data; D. Gasseling, 1998, unpublished data). Assuming that the accrual of new below-ground plant C corresponds to above-ground growth, then one also would assume that production of new below-ground tissues in the spring also dominates seasonal growth rates, and that below-ground production is limited in the latter half of the growing season. However, LeCain et al. (1999) found that substantial photosynthesis can occur in these grasslands in the second half of the growing season, assuming stored soil water does not become depleted. If that is true, then where does the C go when conditions are favorable and mid- to late-season photosynthesis rates are high? LeCain et al. (1999) speculated that much of the C may go to below-ground organs. As perennial plants prepare for winter dormancy, they tend to partition more photosynthate to below-ground storage compounds (Smith, 1973). The same mechanism apparently exists in cool-season grasses like western wheatgrass as they enter the dry, hot days of summer (Morgan et al., 1998). However, rates often are limited by low available soil water so the total amount of photosynthate available for translocation is usually limited. Further, since much of the C that goes below-ground at this time is converted to storage carbohydrates, most of it may be used in the regeneration of above-ground tissues, and so may not reside very long in the soil. For these reasons, we speculate that spring is generally the most important time for the production of plant C that has a significant residence time in the soil. The fact that photosynthesis rates were high in the grazed pastures in the spring could account for a portion of the increased C mass observed in the surface 30 cm of the soil profile.

There also was a trend for greater soil respiration in the grazed compared to the nongrazed treatment early in the season, particularly in the CH pasture, Fig. 6–1. Since the CER data are corrected for soil respiration, this is not a factor in the CER treatment differences. This indeed is another indication of greater biological activity during the early part of the season in the grazed pastures. The reduction of litter and standing dead plant material (Table 6–1) most likely allows better light penetration to the soil, creating warmer soil temperatures and a warmer micro-environment for plant growth and soil microbial activity.

Our research has shown that grazing does not adversely affect the C resources of this system, and in fact shows an increase in the soil C mass associated with the surface 30 cm of soil, which is the most active portion of the soil profile as it relates to root growth and microbial activity. We believe this increase in soil organic C is the result of several things: enhanced incorporation and decomposition of the litter and standing dead through hoof action of the grazing animals, transfer of net primary production to below-ground plant parts, increased plant vigor, improved plant community quality (fewer forbs more perennial grass), and increased plant vigor during early portion of the growing season (exhibited by increased CER and GVI). Increased soil C in the 0- to 30-cm zone has important implications to management and the potential C sequestration by these grasslands. These ecosystems developed under grazing by large herbivores and removing livestock from this system could over the long-term reduce soil C and the productivity and sustainability of the system (Milchunas et al., 1988).

Several questions require further study. We need to quantify the effects of grazing on root respiration, elongation, and exudation and plant C allocation. These factors have a major impact on the potential sequestration of C and its cycling, since these processes produce soil C that is readily available to the microbial cycling of nutrients.

ACKNOWLEDGMENTS

Contribution from the Rangeland Resources Research Unit, USDA, ARS. Authors would like to thank Ernie Taylor, Matt Mortenson, Jeremy Manley, and Pam Freeman for their able field and laboratory assistance on this research. We would also like to thank Drs. Rudy Bowman and Gary Peterson for their timely and thorough review of this manuscript.

REFERENCES

Blake, G.R., and K.H. Hartge. 1986. Bulk density-core method. p. 363–375. In A. Klute (ed.) Methods of soil analysis. Part 1. 2nd ed. Agron. Mongr. 9. ASA and SSSA, Madison, WI.

Bauer, A., C.V. Cole, and A.L. Black. 1987. Soil property comparisons in virgin grasslands between grazed and nongrazed management systems. Soil Sci. Soc. Am. J. 51:176–182.

Caldwell, M.M., J.H. Richards, D.A. Johnson, R.S. Nowak, and R.S. Dzurec. 1981. Coping with herbivory: Photosynthetic capacity and resource allocation in two semiarid *Agropyron* bunchgrasses. Oecologia 50:14–24.

Christie, E.K. 1979. Ecosystem processes in semiarid grasslands. II. Litter production, decomposition and nutrient dynamics. Aust. J. Agric. Res. 30:29–42.

Coupland, R.T. 1992. Mixed prairie. p. 151–182. In R.T. Coupland (ed.) Ecosystems of the world 8A. Natural grasslands: Introduction and Western Hemisphere. Elsevier, New York.

Coupland, R.T., N.A. Skoglund, and A.J. Heard. 1960. Effects of grazing in the Canadian mixed prairie. p. 212–215. In Proc. 8th Int. Grassland Congr., Univ. Reading, Reading, UK. 11–21 July. Alden Press, Oxford, UK.

Coupland, R.T., and G.M. VanDyne. 1979. Systems synthesis. p. 97–106. In R.T. Coupland (ed.) Grassland ecosystems of the world: Analysis of grasslands and their uses. Int. Biol. Programme 18. Cambridge Univ. Press, Cambridge, UK.

Davidson, R.L. 1978. Root systems—the forgotten component of pastures. p. 86–94. In J.R. Wiklson (ed.) Plant relations in pastures. CSIRO, East Melbourne, Australia.

Detling, J.K., M.I. Dyer, and D.T. Winn. 1979. Net photosynthesis, root respiration and regrowth of *Bouteloua gracilis* following simulated grazing. Oecologia 41:127–134.

Dodd, J.D., and H.H. Hopkins. 1985. Yield and carbohydrate content of blue grama as affected by clipping. Trans. Kansas Acad. Sci. 61:282–287.

Dormaar, J.F., and W.D. Willms. 1990. Effect of grazing and cultivation on some chemical properties of soils in the mixed prairie. J. Range Manage. 43:456–460.

Dyer, M.I., and U.G. Bokhari. 1976. Plant-animal interactions: Studies of the effects of grasshopper grazing on blue grama grass. Ecology 57:762–772

Frank, A.B., D.L. Tanaka, L. Hofmann, and R.F. Follett. 1995. Soil carbon and nitrogen of Northern Great Plains grasslands as influenced by long-term grazing. J. Range Manage. 48:470–474.

Hart, R.H., M.J. Samuel, P.S. Test, and M.A. Smith. 1988. Cattle, vegetation, and economic responses to grazing systems and grazing pressure. J. Range Manage. 41:282–286.

Holland, E.A., and J.K. Detling. 1990. Plant response to herbivory and belowground nitrogen cycling. Ecology 71:1040–1049.

Lauenroth, W.K., and W.C. Whittman. 1971. A rapid method for washing roots. J. Range Manage. 24:308–309.

LeCain, D.R., J.A. Morgan, G.E. Schuman, J.D. Reeder, and R.H. Hart. 2000. Carbon exchange of grazed and ungrazed pastures of a mixed grass prairie. J. Range Manage. 53:199–206.

Milchunas, D.G., and W.K. Lauenroth. 1993. Quantitative effects of grazing on vegetation and soils over a global range of environments. Ecol. Monogr. 63(4):327–366.

Milchunas, D.G., V.E. Sala, and W.K. Lauenroth. 1988. A generalized model of the effects of grazing by large herbivores on grassland community structure. Am. Natural. 132:87–106.

Morgan, J.A., D.R. LeCain, J.J. Read, H.W. Hunt, and W.G. Knight. 1998. Photosynthetic pathway and ontogeny affect water relations and the impact of CO_2 on *Bouteloua gracilis* (C_4) and *Pascopyrum smithii* (C_3). Oecologia. 114:483–493.

Mutz, J.L., and D.L. Drawe. 1983. Clipping frequency and fertilization influence herbage yield and crude protein content of four grasses in South Texas. J. Range Manage. 36:582–585.

Naeth, M.A., A.W. Bailey, D.J. Pluth, D.S. Chanasyk, and R.T. Hardon. 1991. Grazing impacts on litter and soil organic matter in mixed prairie and fescue grassland ecosystems in Alberta. J. Range Manage. 44:7–12.

National Oceanic and Atmospheric Administration. 1994. Local climatological data, Cheyenne, Wyoming. Natl. Climate Data Center, Ashville, NC.

Nelson, D.W., and L.E. Sommers. 1982. Total carbon, organic carbon, and organic matter. p. 539–581. *In* A.L. Page et al. (ed.) Methods of soil analysis. Part 1. 2nd ed. Agron. Monogr. 9. ASA and SSSA, Madison, WI.

Schuman, G.E., J.D. Reeder, J.T. Manley, R.H. Hart, and W.A. Manley. 1999. Impact of grazing management on the carbon and nitrogen balance of a mixed-grass rangeland. Ecol. Appl. 9(1):65–71.

Schuster, J.L. 1964. Root development of native plants under three grazing intensities. Ecology 45:63–70.

Shariff, A. R., M.E. Biondini, and C.E. Grygiel. 1994. Grazing intensity effects on litter decomposition and soil nitrogen mineralization.. J. Range Manage. 47:444–449.

Smith, D. 1973. Physiological consideration in forage management. p. 425–436. *In* M.E. Heath et al. (ed.) Forages. Iowa State Univ. Press, Ames, IA.

Smoliak, S., J.F. Dormaar, and A. Johnston. 1972. Long-term grazing effects on *Stipa-Bouteloua* prairie soils. J. Range Manage. 25:246–250.

Stevenson, A., R.E. Baumgartner, and G.E. Schuman. 1984. High Plains Grasslands Research Station, detailed soil survey. Publ. 1-84/1C/3.62. USDA, ARS, Cheyenne WY; USDA, SCS, Casper, WY; and Wyoming Agric. Exp. Stn., Laramie, WY.

Warren-Wilson, J. 1963. Estimation of foliage denseness and foliage angle by inclined point quadrats. Aust. J. Bot. 11:95–105.

7

Total Carbon, Aggregation, Bulk Density, and Penetration Resistance of Cropland and Nearby Grassland Soils

Paul W. Unger

USDA-ARS
Bushland, Texas

ABSTRACT

Converting native grassland (NGL) to cropland (CL) decreases soil organic matter contents (components of soil total C contents, STCC), which often leads to soil degradation. Re-establishing grass on CL generally increases soil organic matter, which improves soil conditions. This study was conducted to determine effects of land uses [CL, NGL, and Conservation Reserve Program (CRP) land (CRPL)] on STCC (mainly organic C), aggregation, bulk density, and penetration resistance. Cropland, NGL, and CRPL sites were sampled at 11 locations in the southern Great Plains. Penetration resistance was measured at the sites. Mean STCC at 0- to 2-cm depths were lowest on CL, intermediate on CRPL, and highest on NGL (2.7, 5.4, and 7.6 Mg ha^{-1}, respectively). Trends were the same at greater depths. Re-establishing grass on CL resulted in increased STCC, but achieving contents comparable to those on NGL will require more than 10 yr. Water-stable aggregate mean weight diameter and water-stability of 1- to 2-mm aggregates were positively related and percentage of <0.25-mm water-stable aggregates was negatively related to STCC. The aggregate results indicate the importance of high STCC for maintaining conditions conducive to using precipitation and irrigation water efficiently. Land use affected bulk density and penetration resistance slightly, but relationships between them were not significant. Few relationships between soil water content and penetration resistance were significant, probably because of widely diverse conditions involved. Establishing grass on CL reverses soil degradation, but crop production is lost. Hence, practices that minimize STCC losses should be used to minimize CL degradation.

Soil organic matter contents commonly decrease when grasslands are converted to cropland (Angers et al., 1992; Balesdent et al., 1988; Haas et al., 1957; Hobbs & Brown, 1965; Tiessen et al., 1982; Woods & Schuman, 1988). Such decreases often lead to deterioration of soil physical conditions such as decreased aggregate stability, reduced water infiltration, and increased runoff and erosion (Angers et al., 1992; Dalal & Henry, 1988; Moldenhauer et al., 1994). When grass is re-established on cropland (CL), soil organic matter contents generally increase, which involves increased C sequestration and improves soil physical conditions (Robles & Burke, 1998; Schumacher et al., 1994).

Much of the CL in the Texas High Plains portion of the southern Great Plains was converted from native grasslands (NGL) in the late 1800s to early 1900s, but some conversions occurred later. Grass has been re-established on some CL in the region under the Soil Bank Program, the Great Plains Conservation Program, and the CRP, with nearby land remaining in crop production or in native grass. The different land uses undoubtedly have resulted in different soil organic C contents. These, in turn, may have resulted in differences in soil texture, aggregation, bulk density, and penetration resistance. These soil properties also may have been affected directly by the different land uses independent of soil C contents. Effects of different land uses (CL, NGL, and CRP grassland) on soil C contents and related soil properties, however, are not well documented. The objective of this study was to determine the effects of long- and short-term cropping practices and differently managed grasslands on STCC (mainly organic C for the soils involved), texture, water-stable aggregation, bulk density, and penetration resistance.

MATERIALS AND METHODS

Nearby CL, NGL, and CRP land (CRPL) sites were selected at 11 locations in the Texas High Plains portion of the southern Great Plains. Comments pertaining to management of the sites are included in Table 7–1. Soil samples were obtained from each site in 1998 for determining texture, STCC, water-stable aggregation, and bulk density. Cone penetrometer resistances were determined at each site in 1999.

For determining texture, STCC, and aggregation, bulk soil was obtained with a flat-bottomed spade at three positions at each site at depths of 0 to 2, 2 to 5, 5 to 10, 10 to 20, and 20 to 30 cm. Soil from the three positions was composited into one sample separately for each depth, thoroughly mixed, passed through a screen having 12.7-mm openings, air-dried, and stored in closed containers until making the determinations.

Texture was determined by the hydrometer method (Day, 1965) on two subsamples of bulk soil. After grinding and passing a portion of the soil through a sieve with 0.18-mm openings, total C concentration of duplicate subsamples was determined by a high-temperature combustion method using a Leco CNS-2000 analyzer (Leco Corp., St. Joseph, MI[1]). Soil bulk density values for each site were used to convert the results to a STCC (Mg ha^{-1}) basis. Duplicate subsamples of bulk soil were wetted under a vacuum to determine water-stable aggregate size distribution by the Kemper and Rosenau (1986) procedure. From these determinations, aggregate mean weight diameter (MWD) and percentage of aggregates <0.25 mm in diameter were calculated. A portion of the bulk soil was sieved to obtain aggregates in the 1.0- to 2.0-mm size range. Duplicate subsamples of these aggregates were wetted under vacuum to determine their stability by Kemper's (1965) procedure. Results of all duplicate subsample determinations were averaged before analyzing the data.

---

[1] The mention of trade or manufacturer names is made for information only and does not imply an endorsement, recommendation, or exclusion by the USDA-Agricultural Research Service.

Table 7–1. Location, county of occurrence, soil series and texture†, and general comments regarding the sites.

Series‡	Depth	Soil Land use CL	NGL	CRPL	Comments§
	cm				
			*Location 1, Oldham County*		
Pullman	0–2	SiC	SiCL	SiCL	CL = dryland winter wheat and grain sorghum
	2–5	SiC	SiCL	CL	since 1920s; NGL = broken out 1920s, re-estab-
	5–10	SiC	SiCL	SiCL	ished native grass in 1960s under Great Plains
	10–20	SiC/C	SiCL	SiCL	program; CRPL = long-term use for dryland
	20–30	C	SiCL	SiC	crops, CRP grass established in 1988.
			Location 2, Deaf Smith County		
Ulysses	0–2	CL	CL	CL	CL = dryland winter wheat and grain sorghum
	2–5	CL	CL	CL	since 1940s; NGL = never plowed; CRPL =
	5–10	CL	CL	CL	long-term use for dryland crops, CRP grass
	10–20	C	CL	C	established in 1988.
	20–30	C	CL	C	
			Location 3, Swisher County		
Pullman	0–2	CL	CL	CL	CL = irrigated cotton, winter wheat, and grain
	2–5	CL	CL	CL	sorghum since 1950s; NGL = never plowed;
	5–10	CL	CL	CL	CRPL = irrigated crops until early 1980s, dry-
	10–20	CL	C	CL	land crops until 1987, CRP grass established
	20–30	C	C	C	in 1988.
			Location 4, Hutchinson County		
Gruver	0–2	CL	CL	CL	CL = dryland winter wheat and grain sorghum
	2–5	SiCL	SiCL	CL	for more than 40 yr; NGL = never plowed;
	5–10	SiCL	SiCL	CL	CRPL = long-term use for dryland crops,
	10–20	C/SiC	SiCL	SiCL	CRP grass established in 1989.
	20–30	C/SiC	SiC	C	
			Location 5, Moore County		
Sherm	0–2	C	SiCL	SiCL	CL = irrigated corn, winter wheat, and grain
	2–5	C	SiCL	SiC	sorghum for more than 40 yr; NGL = never
	5–10	SiC	SiCL	SiC	plowed; CRPL = irrigated winter wheat and
	10–20	SiC	SiC	SiC	grain sorghum until 1988, CRP grass
	20–30	SiC	C	C	established in 1989.
			Location 6, Armstrong County		
Pullman	0–2	CL	C	CL	CL = dryland winter wheat and grain sorghum
	2–5	CL	SiC	CL	for more than 50 yr; NGL = never plowed;
	5–10	CL	SiCL	CL	CRPL = dryland crops until 1987, CRP
	10–20	C	CL	CL	grass established in 1988.
	20–30	C	C	C	
			Location 7, Briscoe County		
Pullman	0–2	CL	CL	CL	CL = irrigated cotton and grain sorghum since
	2–5	CL	CL	CL	1955; NGL = under Soil Bank program 1959–
	5–10	CL	CL	CL	1969, native grasses since 1969; CRPL =
	10–20	CL	C	CL	irrigated cotton and winter wheat 1960–1987,
	20–30	C	C	C	CRP grass established in 1988.

(continued on next page)

Table 7–1. Continued.

Series†	Depth	Soil			Comments§
		Land use			
		CL	NGL	CRPL	
	cm				
		Location 8, Dallam County			
Dallam	0–2	SCL	SL	SCL	CL = dryland winter wheat and grain sorghum
	2–5	SCL	SL	SCL	for more than 40 yr; NGL = never plowed;
	5–10	SCL	SL	SCL	CRPL = dryland winter wheat and grain
	10–20	SCL	SL	SCL	sorghum for more than 40 yr, CRP grass
	20–30	SCL	SL	SCL	established in 1988.
		Location 9, Dallam County			
Sunray	0–2	SL	CL	L/CL	CL = irrigated corn, winter wheat, and grain
	2–5	SL	CL	L	sorghum since 1960s; NGL = never plowed;
	5–10	SL	CL	L	CRPL = dryland winter wheat and grain
	10–20	SL	C	CL	sorghum until 1988, CRP grass established
	20–30	SL	C	CL	in 1988.
		Location 10, Carson County			
Pullman	0–2	SiC	SiCL	SiC	CL = dryland winter wheat and grain sorghum
	2–5	SiC	SiCL	SiC	for more than 40 yr; NGL = never plowed;
	5–10	SiC	SiCL	SiC	CRPL = long-term use for dryland crops,
	10–20	SiC	SiC	C/SiC	CRP grass established in 1988.
	20–30	SiC	SiC	SiC	
		Location 11, Potter County			
Pullman	0–2	SiCL	SiCL	CL	CL = dryland and irrigated crops since 1920s;
	2–5	SiC	SiCL	C	NGL = never plowed; CRPL = long-term use
	5–10	SiC	SiCL	C	for irrigated crops, CRP grass established
	10–20	SiC	SiCL	C	in 1988.
	20–30	SiC	C	C	

† Soil textures: L = loam; SL = sandy loam; SCL = sandy clay loam; CL = clay loam; SiCL = silty clay loam; SiC = silty clay; and C = clay.

‡ Soil series classifications: Dallam = fine-loamy, mixed, mesic Aridic Paleustalf; Gruver = fine, mixed, mesic Aridic Paleustoll; Pullman = fine, mixed, superactive, thermic Torrertic Paleustoll; Sherm = fine, mixed, mesic Torrertic Paleustoll; Sunray = fine-loamy, mixed, mesic Calciorthidic Paleustoll; Ulysses = fine-silty, mixed, mesic Aridic Haplustoll.

§ Botanical names for crops are: corn (*Zea mays* L.), winter wheat (*Triticum aestivum* L.), and grain sorghum [*Sorghum bicolor* (L.) Moench].

Bulk density was determined by 15-cm increments to a 60-cm depth from two 54-mm diam. cores at each site. The core segments were weighed, oven-dried, and re-weighed before calculating bulk density. Although determined by 15-cm increments, bulk densities also were calculated for 10-cm increments (using a weighting procedure) for determining relationships between bulk density and penetration resistance, which was determined for 10-cm increments. Soil penetration resistance to a 50-cm depth was determined from 10 cone penetrometer readings at each site. The 10 readings were averaged, with resistances at 5-, 15-, 25-, 35-, and 45-cm depths used for analyzing the data. The penetrometer, as described by Morrison and Bartek (1987), had a 30° cone tip that was 12.83 mm in diameter and mounted on a 9.53-mm diam. shaft. Soil water contents were determined at depths of 0 to 10,

10 to 20, 20 to 30, 30 to 40, and 40 to 50 cm when penetration resistances were determined.

The study did not involve replicated fields devoted to CL, NGL, or CRPL sites. Rather, soil samples were obtained randomly from similar positions at several points in the field of the different sites and composited into one sample separately for each depth. As a result, statistical analyses involving replicated data were not possible. Hence, data were analyzed by a repeated measures design (Cody & Smith, 1987) using SAS (1989). When significant at the $P \leq 0.05$ level of probability, the least significant difference (LSD) was determined. Regression procedures were used to establish relationships between soil water content and penetration resistance, between bulk density and penetration resistance, and between STCC and aggregate variables.

RESULTS

Texture

Soil textures were similar at the different sites at most locations (Table 7–1). Differences that occurred are attributed to past management. For example, tillage possibly shifted the position of different-sized soil particles (sand, silt, and clay) within and between tillage layers. Also, if wind or rain storms caused erosion, some soil particles possibly were selectively removed, especially from soil layers at or near the surface. Other particles possibly were moved by water from surface to subsurface layers.

Soil Total Carbon Content

Mean STCC (across locations) were highest on NGL and lowest on CL at all, except the 20- to 30-cm depth where the differences were not significant (Table 7–2). The STCC on CRPL were intermediate at all depths. Mean STCC were 5.2 Mg ha^{-1} at 0 to 2, 5.6 at 2 to 5, 8.0 at 5 to 10, 15.0 at 10 to 20, and 13.9 at 20 to 30 cm. Higher STCC at 10 to 20 and 20 to 30 cm than at other depths are due to the larger increments involved at the lower depths. On a concentration basis (data not shown), STCC generally decreased with depth at all locations and for all sites.

Mean total STCC (across locations and summed across depths) was highest at NGL sites (57.2 Mg ha^{-1}), lowest at CL sites (37.8 Mg ha^{-1}), and intermediate at CRPL sites (48.4 Mg ha^{-1}). The total at CRPL sites was not different from the total at other sites. Based on the means, CL contained about 66% and CRPL contained about 85% as much total C as NGL. Also, CL contained about 78% as much total C as CRPL. If the STCC at CRPL and CL sites were similar when grass was re-established at CRPL sites (about 10 yr before sampling for this study), then the mean annual increase in STCC at CRPL sites averaged about 1.1 Mg ha^{-1}.

The highest mean total STCC was at Location 2 (69.9 Mg ha^{-1}), with the total at the CRPL site at that location being 87.0 Mg ha^{-1}. The high totals at that location are attributed to the soil (Ulysses series, Table 7–1), which contains some carbonate nodules. Carbonates were not detected at other locations. The lowest mean

Table 7–2. Soil total C contents at nearby cropland (CL), native grassland (NGL), and Conservation Reserve Program grassland (CRPL) sites at 11 locations in the Texas High Plains portion of the southern Great Plains.

Location	0–2				2–5				5–10				10–20				20–30			
	CL	NGL	CRPL	Mean†	CL	NGL	CRPL	Mean†	CL	NGL	CRPL	Mean†	CL	NGL	CRPL	Mean†	CL	NGL	CRPL	Mean†
									Mg ha⁻¹											
1	1.8	7.5	3.4	4.2c	2.8	7.0	4.6	4.8bc	4.6	8.4	5.7	6.3b	9.9	15.8	10.5	12.1a	8.5	16.2	10.9	11.9a
2	2.8	7.5	9.0	6.4b	3.2	9.0	9.1	7.1b	6.4	11.5	14.8	11.0b	26.9	16.5	32.5	25.3a	22.2	16.5	21.6	20.1a
3	2.8	5.1	5.3	4.4c	4.1	4.6	6.4	5.0c	6.6	7.2	7.5	7.1b	12.4	14.4	11.7	12.8a	12.5	13.6	11.3	12.5a
4	2.4	8.4	2.3	4.4b	3.7	7.6	3.4	4.9b	6.0	8.9	6.4	7.1b	12.4	16.4	13.0	13.9a	11.4	15.0	17.4	14.6a
5	3.6	8.8	7.2	6.5c	5.6	6.6	7.6	6.6c	8.8	8.8	11.8	9.8b	15.4	15.2	17.1	15.9a	11.2	13.8	17.1	14.0a
6	2.3	14.6	4.9	7.3b	3.2	11.9	5.9	7.0b	7.5	11.3	8.8	9.2b	11.5	18.6	15.9	15.3a	11.1	15.4	13.7	13.4a
7	1.6	3.1	5.9	3.5b	2.5	3.6	5.2	3.8b	4.3	5.9	5.7	5.3b	7.7	14.1	10.4	10.7a	8.4	11.9	11.1	10.5a
8	2.7	5.9	6.4	5.0b	3.9	6.4	7.5	6.0b	6.2	7.0	6.8	6.7b	11.3	9.2	15.3	11.7a	12.9	9.3	7.8	10.0a
9	3.5	3.2	5.1	3.9b	2.9	5.0	4.5	4.1b	7.5	8.5	7.0	7.7b	8.8	22.2	11.8	14.3a	8.6	33.2	9.8	17.2a
10	3.2	9.1	4.1	5.5d	5.0	8.5	5.0	6.2d	7.5	11.8	8.2	9.2c	12.9	21.4	15.6	16.6a	11.9	18.4	14.0	14.8b
11	2.7	10.4	6.1	6.4c	3.9	8.6	6.6	6.4c	6.7	10.8	8.8	8.8b	13.0	21.0	15.3	16.4a	12.9	17.8	12.5	14.4a
Mean	2.7	7.6	5.4		3.7	7.2	6.0		6.6	9.1	8.3		12.9	16.8	15.3		11.9	16.5	13.4	
LSD:	sites = 1.8, locations = NS‡;				sites = 0.9, locations = NS;				sites = 0.6, locations = NS;				sites = 0.7, locations = NS;				sites = NS; locations = NS			

† Mean values within a row followed by the same letter or letters are not different at the P = 0.05 level.

‡ NS = not significant.

total was at Location 7 (38.6 Mg ha^{-1}), with the overall lowest STCC on CL at that location (24.5 Mg ha^{-1}).

When considered on an equal depth increment basis (data not shown), the STCC at all locations usually was highest at 0 to 2 cm, either at a NGL or a CRPL site. It was several times higher than the lowest STCC, which usually occurred at CL or CRPL sites. The difference between the highest (on NGL or CRPL) and lowest (on CL) STCC indicates that a grass cover was effective for maintaining or increasing the STCC of surface soil, which is important for maintaining or improving soil conditions as shown in the following sections.

Aggregate Mean Weight Diameter

At all depths, mean aggregate weight diameters (MWD) of water-stable soil aggregates (across locations) were similar at NGL and CRPL sites and higher than at CL sites (Table 7–3). Although the MWD at individual locations was not always lowest at the CL site, the low overall mean suggests that soil conditions (low aggregate stability and filling of soil pores) would be conducive to poor water infiltration at all depths at most CL sites, which could hamper soil water storage when precipitation or irrigation rates exceed a soil's water infiltration rate.

Mean MWD for locations (across sites) did not differ at any depth, but differed among depths. The overall means for depths were: 1.96 mm (c) at 0 to 2 cm, 2.87 mm (ab) at 2 to 5 cm, 3.09 mm (a) at 5 to 10 cm, 2.88 mm (a) at 10 to 20 cm, and 2.26 mm (bc) at 20 to 30 cm. Means followed by the same letter or letters in parentheses are not different at the $P = 0.05$ level.

Mean MWD for depths (across sites) differed only at Location 4 where it was highest at 10 to 20 cm and lowest at 0 to 2 cm (Table 7–3). The MWD for sites at the different locations and depths showed no consistent trends. The lowest and highest MWD at each location are underlined. The MWD was lowest on CL at 10 locations, but the low values occurred at different depths. Highest MWD occurred at depths of 2 to 5 cm (four times), 5 to 10 cm (four times), and 10 to 20 cm (three times). Six were on NGL sites and five on CRPL sites.

The MWD is calculated from the percentages of soil aggregates in different size ranges. The percentage of fine aggregates (<0.125 or <0.25 mm) is sometimes used to indicate the potential of a soil for surface sealing and, hence, reduction in water infiltration (Loch, 1989). Although determined, percentage differences in <0.25-mm aggregates in this study were not significant (data not shown), but were related to STCC (discussed later).

Water-Stability of One- to Two-Millimeter Aggregates

Mean water-stability of 1- to 2-mm aggregates (across locations) was lowest on CL and highest on NGL at all depths (Table 7–4). The generally low percentages (54.0–57.3) at the different depths on CL suggest water infiltration and, hence, soil water storage could be hampered as compared with that on NGL and CRPL. Only one value on NGL (51.0% at the 10- to 20-cm depth at Location 8) was below the lowest mean on CL (54.0%). No values on CRPL were below the

Table 7–3. Mean weight diameter (MWD) of water-stable soil aggregates at nearby cropland (CL), native grassland (NGL), and Conservation Reserve Program grassland (CRPL) sites at 11 locations in the Texas High Plains portion of the southern Great Plains.

| | Soil depth, cm | | | | | | | | | | | | | | | | | | | |
| | 0–2 | | | | 2–5 | | | | 5–10 | | | | 10–20 | | | | 20–30 | | | |
Location	CL	NGL	CRPL	Mean	CL	NGL	CRPL	Mean	CL	NGL	CRPL	Mean	CL	NGL	CRPL	Mean	CL	NGL	CRPL	Mean
								MWD, mm												
1	0.46†	2.20	1.77	1.48a‡	0.50	3.09	3.52	2.37a	0.51	3.08	3.70	2.43a	0.96	3.40	2.73	2.36a	0.82	3.07	1.59	1.83a
2	1.38	2.69	5.17	3.08a	2.96	2.96	5.72	3.88a	2.80	2.77	4.03	3.20a	2.20	3.11	3.86	3.06a	2.13	2.40	3.94	2.83a
3	2.04	1.90	2.49	2.14a	1.54	4.84	4.36	3.58a	2.26	3.95	3.65	3.29a	2.01	3.93	2.66	2.87a	1.71	3.79	1.39	2.30a
4	0.66	1.91	0.75	1.11c	2.17	3.61	1.66	2.48ab	2.15	3.42	3.76	3.11ab	2.39	4.61	2.91	3.30a	1.34	2.96	2.54	2.28b
5	2.03	3.29	4.28	3.20a	2.16	3.41	4.52	3.36a	2.51	4.85	3.35	3.57a	2.66	3.31	2.72	2.90a	1.09	2.21	1.89	1.73a
6	1.57	3.02	2.42	2.33a	1.54	5.55	3.57	3.55a	1.26	4.65	5.25	3.72a	1.78	5.42	4.07	3.76a	1.31	2.71	2.97	2.33a
7	1.15	1.15	2.73	1.68a	0.78	2.97	3.88	2.54a	1.54	3.19	4.38	3.04a	1.82	4.16	3.36	3.11a	1.49	4.25	1.25	2.33a
8	1.09	1.23	1.10	1.14a	0.66	1.64	2.06	1.45a	0.85	1.81	3.34	2.00a	2.24	1.15	3.12	2.17a	2.18	1.33	2.16	1.89a
9	1.01	1.69	0.97	1.22a	0.86	3.35	1.86	2.02a	1.01	3.12	2.82	2.32a	1.21	3.27	3.46	2.64a	0.69	2.82	3.54	2.35a
10	2.15	2.35	1.07	1.86a	1.77	3.92	3.35	3.01a	2.80	4.36	4.62	3.93a	0.45	4.43	3.16	2.68a	0.88	3.32	2.39	2.20a
11	1.16	3.82	1.88	2.29a	0.88	5.87	3.20	3.32a	1.23	5.30	3.60	3.38a	1.74	4.22	2.83	2.93a	2.95	2.13	3.36	2.81a
Mean	1.33	2.30	2.24		1.44	3.75	3.43		1.72	3.68	3.87		1.76	3.72	3.17		1.51	2.82	2.45	
LSD:	sites = 0.75, locations = NS§				sites = 0.87, locations = NS;				sites = 0.66, locations = NS;				sites = 0.72, locations = NS;				sites = 0.79, locations = NS			

† Underlined values in a given row are the lowest and highest for that location.
‡ Mean values within a row followed by the same letter or letters are not different at the P = 0.05 level.
§ NS = not significant.

Table 7–4. Water-stability of 1- to 2-mm soil aggregates at nearby cropland (CL), native grassland (NGL), and Conservation Reserve Program grassland (CRPL) sites at 11 locations in the Texas High Plains portion of the southern Great Plains.

Location	Soil depth, cm																			
	0–2				2–5				5–10				10–20				20–30			
	CL	NGL	CRPL	Mean	CL	NGL	CRPL	Mean	CL	NGL	CRPL	Mean	CL	NGL	CRPL	Mean	CL	NGL	CRPL	Mean
									%											
1	72.6	81.4	72.2	75.4a†	74.2	80.7	79.0	78.0a	73.4	81.3	73.1	75.9a	75.3	84.2‡	75.1	78.2a	70.4	82.6	77.8	76.9a
2	56.9	90.8	80.3	76.0a	48.5	91.2	74.7	71.5a	52.1	91.1	76.5	73.2a	73.1	83.6	75.6	77.4a	73.8	81.9	74.1	76.6a
3	57.7	87.4	75.3	73.5a	55.7	93.3	76.8	68.6a	47.2	90.0	70.9	69.4a	70.7	87.2	69.5	75.8a	76.5	83.3	69.5	76.4a
4	56.1	90.4	75.6	74.0b	63.0	92.7	82.6	79.4a	52.4	91.7	79.0	74.4b	57.5	88.0	74.4	73.3b	55.6	84.4	75.7	71.9b
5	69.3	78.4	70.7	72.8a	71.2	71.9	62.2	68.4ab	72.3	77.7	64.1	71.4ab	64.8	75.6	59.0	66.5bc	55.3	68.3	57.8	60.5c
6	50.6	88.3	75.5	71.5a	55.3	89.2	76.6	73.7a	60.3	86.7	65.5	70.8a	68.9	85.3	64.9	73.0a	67.4	79.2	59.0	68.5a
7	37.8	63.3	82.0	61.0a	41.2	78.5	81.2	67.0a	34.2	76.1	72.7	61.0a	36.8	77.5	63.1	59.1a	42.2	73.9	64.0	60.0a
8	59.2	74.4	73.6	69.1a	47.5	67.0	68.1	60.9b	45.7	66.9	62.1	58.2bc	41.6	51.0	64.6	52.4c	42.4	62.9	62.5	55.9bc
9	42.8	81.1	63.4	62.4a	43.4	86.8	80.3	70.2a	43.4	83.7	91.6	72.9a	45.0	80.4	90.7	72.0a	41.1	76.4	80.8	66.1a
10	75.3	91.2	78.2	81.6a	77.6	90.6	82.1	83.4a	71.8	89.8	79.2	80.3ab	58.7	89.2	72.4	73.4bc	57.9	86.4	68.8	71.0c
11	52.7	84.0	75.0	70.6a	46.5	76.3	62.7	61.8b	41.4	77.8	60.9	60.0bc	37.3	71.3	57.1	55.2c	39.6	64.9	58.4	54.3c
Mean	57.3	82.8	74.7		56.7	81.6	75.1		54.0	82.9	72.4		57.2	79.5	69.7		56.5	76.7	68.0	
LSD:	sites = 7.2, locations = NS§;				sites = 7.8, locations = NS;				sites = 8.4, locations = NS;				sites = 8.4, locations = 16.0;				sites = 6.9, locations = 13.2			

† Mean values within a row followed by the same letter or letters are not different at the $P = 0.05$ level.
‡ Underlined values in a given row are the lowest and highest for that location.
§ NS = not significant.

lowest mean on CL. The higher water-stability of aggregates suggests that water infiltration would be greater on NGL and CRPL than on CL.

Overall means for water-stable aggregates for locations (across sites) differed only at 10 to 20 and 20 to 30 cm. The individual mean was highest at Location 1 at both depths. These means, however, were not different from those at some other locations. The mean was lowest at 10 to 20 cm at Location 8 and at 20 to 30 cm at Location 11. At both depths, means were not significantly different from some other means.

Means for depths (across sites) differed at Locations 4, 5, 8, 10, and 11. At Location 4, it was higher at 2 to 5 cm than at other depths. For other locations, the mean was among the highest at 0 to 2 cm and among the lowest at 20 to 30 cm. Water stable surface aggregates are important for minimizing their dispersion and surface seal development, which reduce water infiltration (Loch, 1989) and, thereby, hamper soil water storage.

Lowest and highest percentages of water-stable aggregates at the locations are underlined (Table 7–4). At 10 locations, percentages were highest at 0- to 2- and 2- to 5-cm depths, with 9 occurring on NGL and 1 on CRPL. The other high percentage was on NGL at 10 to 20 cm. Lowest percentages were on CL at 10 locations and on CRPL at one location, with the lows occurring at different depths. The low and high percentages occurred at different sites and depths, indicating a significant site by depth interaction.

Bulk Density

Mean soil bulk densities (across sites and locations) were 1.39 Mg m^{-3} (c) at 0- to 15-, 1.47 (b) at 15- to 30-, 1.57 (ab) at 30- to 45-, and 1.63 (a) at 45- to 60-cm depths. Values followed by the same letter or letters in parentheses are not different at the $P = 0.05$ level. At all depths, means (across locations) were similar on CL, NGL, and CRPL sites (Table 7–5). Means for locations (across sites) differed only at the 0- to 15-cm depth, where the mean was highest (1.58 Mg m^{-3}) at Location 8 and lowest (1.30 Mg m^{-3}) at Location 10. The highest and lowest densities were not different from those at some other locations. Means (across sites) increased with depth at all locations, except Locations 8 and 9.

Mean bulk densities for individual locations increased with depth and most high densities occurred at 45 to 60 cm. (The low and high densities at each location are underlined.) However, one high density occurred at 15 to 30 cm and two occurred at 30 to 45 cm. Likewise, most low densities occurred at 0 to 15 cm, but two occurred at 15 to 30 cm and another at 30 to 45 cm. High densities occurred on CL six times, on NGL three times, and on CRPL three times (the same on NGL and CRPL at Location 5). Low densities occurred on CL three times, on NGL six times, and on CRPL two times.

Penetration Resistance

Mean penetration resistances for sites (across locations) (Table 7–6) differed only at the 5-cm depth, with resistance lower on CL (1.08 MPa) than on NGL and

Table 7–5. Soil bulk density at nearby cropland (CL), native grassland (NGL), and Conservation Reserve Program grassland (CRPL) sites at 11 locations in the Texas High Plains portion of the southern Great Plains.

	Soil depth, cm															
	0–15				15–30				30–45				45–60			
Location	CL	NGL	CRPL	Mean	CL	NGL	CRPL	Mean	CL	NGL	CRPL	Mean	CL	NGL	CRPL	Mean
								Mg m^{-3}								
1	1.33	1.30†	1.32	1.32c‡	1.52	1.43	1.49	1.48b	1.53	1.57	1.58	1.56a	1.62	1.50	1.60	1.57a
2	1.35	1.23	1.38	1.32b	1.46	1.15	1.33	1.32b	1.48	1.28	1.39	1.38b	1.64	1.33	1.60	1.52a
3	1.43	1.28	1.39	1.37b	1.50	1.40	1.50	1.47b	1.59	1.62	1.60	1.60a	1.64	1.65	1.54	1.61a
4	1.30	1.33	1.32	1.32c	1.57	1.38	1.51	1.49b	1.67	1.56	1.67	1.63a	1.82	1.68	1.67	1.72a
5	1.44	1.33	1.44	1.40b	1.38	1.48	1.47	1.44b	1.63	1.59	1.57	1.60a	1.56	1.70	1.70	1.65a
6	1.22	1.46	1.55	1.41c	1.52	1.54	1.52	1.53bc	1.72	1.69	1.61	1.67ab	1.86	1.75	1.73	1.78a
7	1.34	1.32	1.40	1.35d	1.41	1.52	1.55	1.49c	1.63	1.58	1.65	1.62b	1.69	1.71	1.74	1.71a
8	1.58	1.63	1.52	1.58a	1.51	1.55	1.87	1.64a	1.41	1.58	1.65	1.57a	1.42	1.73	1.85	1.67a
9	1.66	1.53	1.44	1.54a	1.80	1.47	1.34	1.54a	1.71	1.49	1.46	1.55a	1.65	1.50	1.54	1.56a
10	1.44	1.19	1.27	1.30b	1.42	1.38	1.38	1.39b	1.55	1.51	1.47	1.51a	1.64	1.58	1.47	1.56a
11	1.42	1.31	1.27	1.33b	1.49	1.37	1.31	1.39b	1.61	1.64	1.33	1.53a	1.59	1.63	1.47	1.56a
Mean	1.41	1.36	1.39		1.50	1.43	1.48		1.60	1.56	1.54		1.65	1.62	1.62	
LSD:	sites = NS§, locations = 0.15;				sites = NS, locations = NS;				sites = NS, locations = NS;				sites = NS, locations = NS			

† Underlined values in a given row are the lowest and highest for that location.
‡ Mean values within a row followed by the same letter or letters are not different at the $P = 0.05$ level.
§ NS = not significant.

Table 7–6. Soil penetration resistance at nearby cropland (CL), native grassland (NGL), and Conservation Reserve Program grassland (CRPL) sites at 11 locations in the Texas High Plains portion of the southern Great Plains.

Location	5				15				25				35				45			
	CL	NGL	CRPL	Mean	CL	NGL	CRPL	Mean	CL	NGL	CRPL	Mean	CL	NGL	CRPL	Mean	CL	NGL	CRPL	Mean
												MPa								
1	1.17†	2.35	1.80	1.77a‡	1.59	2.30	1.93	1.94a	1.69	2.19	2.03	1.97a	1.92	2.01	2.17	2.03a	2.09	2.11	2.15	2.12a
2	0.98	2.12	2.38	1.82a	1.69	2.46	2.44	2.20a	1.49	2.41	2.18	2.03a	1.45	2.42	2.15	2.01a	1.68	2.32	2.22	2.07a
3	1.71	2.16	2.52	2.13a	2.66	1.66	1.88	2.07a	2.80	1.61	2.15	2.19a	3.06	2.64	2.18	2.63a	3.06	2.64	2.14	2.61a
4	1.24	2.08	2.39	1.90a	1.71	2.10	2.52	2.11a	1.80	2.13	2.71	2.21a	1.95	1.91	2.68	2.18a	1.97	1.17	2.77	1.97a
5	1.24	2.05	1.00	1.43b	1.71	2.35	1.08	1.71ab	1.80	2.36	1.42	1.86a	1.95	2.18	1.63	1.92a	1.97	2.18	1.69	1.95a
6	0.51	1.35	0.80	0.89b	2.11	2.46	1.48	2.02a	2.56	2.72	1.62	2.30a	2.53	2.77	1.74	2.35a	2.22	1.63	1.85	1.90a
7	0.90	2.34	1.73	1.65a	0.87	2.31	1.85	1.68a	1.50	1.85	1.49	1.61a	1.48	2.23	1.46	1.72a	1.59	2.16	1.79	1.85a
8	1.22	1.98	1.57	1.59a	2.44	2.24	1.85	2.18a	2.20	2.11	2.29	2.20a	1.85	2.05	2.04	1.98a	1.80	2.15	1.97	1.97a
9	0.45	2.17	2.36	1.66a	1.64	2.21	2.47	2.11a	2.09	1.75	1.62	1.82a	1.90	2.04	1.56	1.83a	1.25	2.19	1.79	1.74a
10	1.71	2.34	1.68	1.91a	2.66	2.50	2.11	2.42a	2.80	2.44	2.05	2.43a	2.99	2.22	1.93	2.38a	2.99	2.14	1.82	2.32a
11	0.81	2.27	0.74	1.27b	1.46	2.66	1.53	1.88a	2.02	2.74	1.49	2.08a	2.15	2.42	1.51	2.03a	2.21	2.58	1.75	2.18a
Mean	1.08	2.11	1.73		1.86	2.30	1.92		2.06	2.21	1.92		2.11	2.27	1.92		2.08	2.12	2.00	
LSD:	sites = 0.36, locations = NS§;				sites = NS, locations = NS;				sites = NS, locations = NS;				sites = NS, locations = NS;				sites = NS, locations = NS			

† Underlined values in a given row are the lowest and highest for that location.
‡ Mean values within a row followed by the same letter or letters are not different at the $P = 0.05$ level.
§ NS = not significant.

CRPL (2.11 and 1.73 MPa, respectively). Mean resistances (across sites) did not differ among locations at any depth.

For locations individually, mean penetration resistances among depths (across sites) differed only at Locations 5, 6, and 11, where resistance was lower at 5 cm than at other depths. Mean resistance (across depths) (data not shown) was lowest on CL at Locations 1, 2, 4, and 7, and on CRPL at Locations 5, 6, 10, and 11. It was not different at other locations. The different trends possibly resulted from water content (data not shown) and bulk density differences at the different locations.

Based on regression analyses involving results from all sites, relationships between soil water content and penetration resistance were significant at the 5-, 15-, and 25-cm depths. The respective equations are:

$$y = 2.706 - 4.850x \quad (r^2 = 0.267, P = 0.01) \tag{1}$$

$$y = 2.830 - 3.232x \quad (r^2 = 0.245, P = 0.01) \tag{2}$$

$$y = 2.826 - 2.846x \quad (r^2 = 0.193, P = 0.02) \tag{3}$$

where y is penetration resistance (MPa) and x is soil water content ($m^3\ m^{-3}$). At the 5-, 15-, and 25-cm depths, the relationships between soil water content and penetration resistance were significant on CRPL sites. The respective equations are:

$$y = 3.491 - 7.441x \quad (r^2 = 0.791, P = 0.001) \tag{4}$$

$$y = 3.581 - 5.768x \quad (r^2 = 0.686, P = 0.01) \tag{5}$$

$$y = 3.234 - 4.358x \quad (r^2 = 0.431, P = 0.05) \tag{6}$$

with terms the same as defined above. No other relationships for water content and penetration resistance were significant. Also, none were significant for bulk density and penetration resistance.

DISCUSSION

Soil STCC were much lower at CL than at NGL sites at most locations at the 0- to 2-cm depth. Some rather large differences occurred also at 2 to 5 cm, with lesser differences at greater depths. The reductions at 0 to 2 cm were greatest where the conversion from NGL to CL occurred over 40 yr ago (Table 7–1, Locations 1, 2, 4, 5, 6, 8, 10, and 11), especially when CL was used for dryland crops. These results are consistent with previously published results (Angers et al., 1992; Balesdent et al., 1988; Dalal & Henry, 1988; Haas et al., 1957, 1974; Hobbs & Brown, 1965; Johnson et al., 1974; Tiessen et al., 1982). The decline generally was less where CL was used for irrigated crops (Locations 3, 7, and 9), especially where only grain crops and no cotton (*Gossypium hirsutum* L.) were grown (Location 9). Cotton, which produces relatively low amounts of residue, apparently does not contribute greatly to maintaining soil STCC at levels similar to those obtained where only grain crops are produced.

The STCC at CRPL sites usually were intermediate between those at CL and NGL sites. Overall means at different depths also were intermediate at CRPL sites. The reason for the similar or higher contents at CRPL sites as compared to those at NGL sites at Locations 2, 3, 7, 8, and 9 is not known. The intermediate contents at CRPL sites indicate that re-establishing grasses on cropland results in increasing STCC. However, in the semiarid region of the southern Great Plains where these sites are located, considerably more than 10 yr will be needed in most cases for STCC at re-established grassland sites to approach the contents at NGL sites. Results from southeastern Wyoming also indicate that STCC under dryland conditions increase slowly when grasses are re-established on cropland. After 6 yr, soil C contents in CRP fields had increased only slightly as compared with contents in cultivated fields (Robles & Burke, 1998).

Soil STCC influence numerous soil properties, including aggregation (Angers et al., 1992; Schumacher et al., 1994). At the 0- to 2-cm soil depth, water-stable aggregate MWD, percentage of <0.25-mm water-stable aggregates, and water-stability of 1- to 2-mm aggregates were related to STCC. The relationship for MWD (y, mm) and STCC (x, Mg ha^{-1}) is:

$$y = 0.653 + 0.249x \quad (r^2 = 0.478, P = 0.001) \tag{7}$$

For <0.25-mm water-stable aggregates (y, %) and STCC (x, Mg ha^{-1}), the relationship is:

$$y = 53.70 - 3.205x \quad (r^2 = 0.510, P = 0.001) \tag{8}$$

The relationship for water-stability of 1- to 2-mm aggregates (y, %) and STCC (x, Mg ha^{-1}) is:

$$y = 55.14 + 3.149x \quad (r^2 = 0.462, P = 0.001) \tag{9}$$

Soil STCC affects soil aggregation, which, in turn, affects numerous other factors related to crop production (including water infiltration, soil water storage, soil aeration, and plant rooting). Therefore, the highly significant relationships between soil STCC and each aggregate variable indicate the major importance of maintaining soil STCC (mainly, soil organic C contents) at high levels for achieving satisfactory crop production. Favorable aggregate conditions at or near the soil surface are especially important for achieving favorable water infiltration rates, which are important for using precipitation and irrigation water resources efficiently, increasing vegetative growth, and increasing C input into the soil system.

Soils with large MWD aggregates are important for reducing erosion. Water-stable aggregates minimize aggregate dispersion and decrease the percentage of small aggregates (<0.25 mm). Small aggregates result in surface sealing and, in turn, reduce water infiltration (Loch, 1989). They also are easily transported by water, thus resulting in clogging of soil pores and contributing to soil erosion.

Surface soil aggregate stability is highly important regarding water infiltration and soil erosion. Aggregate stability of subsurface soil, however, is important also because stable aggregates help maintain soil structure. Differences in aggregate conditions were found at soil depths below 2 cm (Tables 7–3 and 7–4), with overall mean values paralleling the mean values found at the 0- to 2-cm depth. Because STCC at depths below 2 cm also paralleled those at 0 to 2 cm, each aggregate variable was related to STCC at the different depths, but at lower levels of significance than at the surface layer (results not shown).

Relatively few differences in soil bulk density and penetration resistance were significant. Also, penetration resistance was not related to soil bulk density at any depth and penetration resistance was related to soil water content in only a few cases. These results are contrary to expectations because penetration resistance previously was shown to be significantly related to both bulk density and water content of Pullman soil (Unger & Jones, 1998), which was the soil at 6 of the 11 locations in this study. The wide range of soil textures (loam, sandy loam, sandy clay loam, clay loam, silty clay loam, silty clay, and clay), management factors (dryland vs. irrigated), tillage methods, crops grown, etc., apparently overshadowed the direct influence of soil bulk density and water content on soil penetration resistance.

CONCLUSIONS

Results of this study involving soil at CL, NGL, and CRPL sites suggested the following conclusions for the semiarid portion of the southern Great Plains:

1. Management practices (tillage depth and type) and past erosion resulted in slight soil texture differences among sites at the different locations, especially in the surface soil layer (0- to 2-cm soil depth).
2. Long-term use of former grasslands as cropland resulted in major decreases in STCC. The decreases were greater under dryland than under irrigated conditions. Re-establishing grass on former cropland areas under the CRP resulted in increased STCC, but considerably more than 10 yr will be needed to raise the contents to levels comparable to those on NGL.
3. Soil water-stable aggregate mean weight diameter and water-stability of 1- to 2-mm diam. aggregates were positively related to soil STCC. The percentage of <0.25-mm water-stable aggregates was negatively related to the STCC. The relationships were highly significant at the 0- to 2-cm depth and less significant at greater depths. The aggregate stability results indicated the importance of maintaining STCC at high levels for maintaining favorable soil conditions for using precipitation and irrigation water resources efficiently.
4. Land use had little effect on soil bulk density and penetration resistance. No relationships between soil bulk density and penetration resistance and few relationships between soil water content and penetration resistance were significant. This general lack of significance was attributed to the wide diversity of conditions at the different locations [soil textures, management factors (dryland vs. irrigated), tillage methods, crops grown, etc.].

ACKNOWLEDGMENTS

The assistance of F. B. Pringle, USDA-NRCS, Soil Scientist (retired), Amarillo, Texas, in selecting the locations and sites for sampling; L. J. Fulton, USDA-

ARS, Biological Technician, Bushland, Texas, in conducting the research and summarizing and analyzing the data; and R. C. Schwartz, USDA-ARS, Soil Scientist, in assisting with the data analyses is gratefully acknowledged.

REFERENCES

Angers, D.A., A. Pesant, and J. Vigneux. 1992. Early cropping-induced changes in soil aggregation, organic matter, and microbial biomass. Soil Sci. Soc. Am. J. 56:115–119.

Balesdent, J., G.H. Wagner, and A. Mariotti. 1988. Soil organic matter turnover in long-term field experiments as revealed by carbon-13 natural abundance. Soil Sci. Soc. Am. J. 52:118–124.

Cody, R.P., and J.K. Smith. 1987. Applied statistics and the SAS programming language. 2nd ed. North-Holland, New York, Amsterdam, London.

Dalal, R.C., and R.J. Henry. 1988. Cultivation effects on carbohydrate contents of soil and soil fractions. Soil Sci. Soc. Am. J. 52:1361–1365.

Day, P.R. 1965. Particle fractionation and particle-size analysis. p. 545–567. *In* C.A. Black et al. (ed.) Methods of soil analysis. Part 1. Agron. Monogr. 9. ASA, Madison, WI.

Haas, H.J., C.E. Evans, and E.F. Miles. 1957. Nitrogen and carbon changes in Great Plains soils as influenced by cropping and soil treatments. USDA Tech. Bull. 1164. U.S. Gov. Print. Office, Washington, DC.

Haas, H.J., W.O. Willis, and J.J. Bond. 1974. Summer fallow in the Northern Great Plains (spring wheat). p. 12–35. *In* Summer fallow in the western United States. USDA-ARS Conserv. Res. Rep. 17. U.S. Gov. Print. Off., Washington, DC.

Hobbs, J.A., and P.L. Brown. 1965. Effects of cropping and management on nitrogen and organic carbon contents of a western Kansas soil. Bull. 144. Kansas Agric. Exp. Stn., Manhattan, KS.

Johnson, W.C., C.E. Van Doren, and E. Burnett. 1974. Summer fallow in the southern Great Plains. p. 86–109. *In* Summer fallow in the western United States. USDA-ARS Conserv. Res. Rep. 17. U.S. Gov. Print. Office, Washington, DC.

Kemper, W.D. 1965. Aggregate stability. p. 511–519. *In* C.A. Black et al. (ed.) Methods of soil analysis. Part 1. Agron. Monogr. 9. ASA, Madison, WI.

Kemper, W.D., and R.C. Rosenau. 1986. Aggregate stability and size distribution. p. 425–442. *In* A. Klute (ed.) Methods of soil analysis. Part 1. 2nd ed. Physical and mineralogical methods. Agron. Monogr. 9. ASA and SSSA, Madison, WI.

Loch, R.J. 1989. Aggregate breakdown under rain: Its measurement and interpretation. Ph.D. diss. Univ. New England, Queensland, Australia.

Moldenhauer, W.C., W.D. Kemper, and B.A. Stewart. 1994. Long-term effects of tillage and crop residue management. p. 55–66. *In* B.A. Stewart and W.C. Moldenhauer (ed.) Crop residue management to reduce erosion and improve soil quality—Southern Great Plains. USDA-ARS Conserv. Res. Rep. 37. U.S. Gov. Print. Office, Washington, DC.

Morrison, J.E., Jr., and L.A. Bartek. 1987. Design and field evaluation of a hand-pushed digital soil penetrometer with two cone materials. Trans. ASAE 30:646–651.

Robles, M.D., and I.C. Burke. 1998. Soil organic matter recovery on Conservation Reserve Program fields in southeastern Wyoming. Soil Sci. Soc. Am. J. 62:725–730.

SAS Institute. 1989. SAS/STAT user's guide. Vers. 6. 4th ed. Vol. 2. SAS Inst., Cary, NC.

Schumacher, T.E., M.J. Lindstrom, M.L. Blecha, and B.A. Stewart. 1994. Management options after leaving the Conservation Reserve Program. p. 67–70. *In* B.A. Stewart and W.C. Moldenhauer (ed.) Crop residue management to reduce erosion and improve soil quality–Southern Great Plains. USDA-ARS Conserv. Res. Rep. 37. U.S. Gov. Print. Office, Washington, DC.

Tiessen, H., J.W.B. Stewart, and J.R. Bettany. 1982. Cultivation effects on the amounts and concentrations of carbon, nitrogen, and phosphorus in grassland soils. Agron. J. 74:831–835.

Unger, P.W., and O.R. Jones. 1998. Long-term tillage and cropping systems affect bulk density and penetration resistance of soil cropped to dryland wheat and grain sorghum. Soil Till. Res. 45:39–57.

Woods, L.E., and G.E. Schuman. 1988. Cultivation and slope position effects on soil organic matter. Soil Sci. Soc. Am. J. 52:1371–1376.

8   Carbon Sequestration on the Canadian Prairies— Quantification of Short-Term Dynamics

C. A. Campbell, E. G. Gregorich, and W. Smith

Agriculture and Agri-Food Canada, ECORC
Ottawa, Ontario, Canada

R. P. Zentner, B.-C. Liang, and B. G. McConkey

Agriculture and Agri-Food Canada, SPARC
Swift Current, Saskatchewan, Canada

G. Roloff

Universidade Federal do Paraná
Brazil

H. H. Janzen

Agriculture and Agri-Food Canada
Lethbridge, Alberta, Canada

K. Paustian

Colorado State University
Fort Collins, Colorado

ABSTRACT

Carbon sequestration in soils is being promoted as a partial solution to the increasing levels of atmospheric CO_2 and its implications for global warming. There is much interest among the international community regarding the concept of "C credits trading" for C sequestered in agricultural soils, although this concept was not approved at the 1997 International Conference on Greenhouse gases in Kyoto, Japan. For it to be approved there must be credible evidence that C sequestration in soils can be significant, and can be reasonably and accurately quantified. We can directly measure C changes in soils, but increases are small relative to C in soil, changes are slow, and soil C is so variable, that this approach is not practical for assessing C credits. A more reasonable approach may be to use simulation models to estimate C changes. In this study we tested two process-based models, CENTURY and EPIC, but neither was effective in estimating soil C changes in a 30-yr crop rotation experiment being conducted in the semiarid prairie of southwestern Saskatchewan. CENTURY failed partly because it was unable to accurately estimate yields (and thus residue inputs), of stubble-seeded crops adequately, and both models were inadequate in estimating C dynamics even when they made reasonable estimates of yields. We developed a simpler empirical model, which uses two first-order kinetic expressions to describe residue decomposition and soil C mineralization simultaneously, and used this to estimate soil C dynamics effectively. Input data to this model included C inputs from crop residues, (either measured or estimated from grain yields) that were used to estimate soil C formation based on a residue decomposition equation that we obtained from the scientific literature. We also used estimates of the size of the actively cycling C and more resistant C pools, and appropriate decomposition rate constants. These we obtained from C-dating results of humic fractions for prairie soils, and by assuming the active C was that present in the light fraction organic matter and microbial biomass (about 20% of total C in the soil at Swift Current). This empirical model provided useful first approximation of the direction and amount of C change that occurred in the Chernozemic soil. We found that it was important to have soil management history to allow proper estimation of the direction of C change. Although our empirical model provided useful information, we will need to use process-based models if we are to more accurately estimate and extrapolate C changes over time and space. However, current versions of these models require improvement, especially regarding the magnitude of the influence of crop residue inputs compared to soil decomposition processes, on net changes in soil C.

Soil organic matter is derived from tissues of plants, animals, and microorganisms. It is closely linked to soil productivity, and its preservation is a key element of soil conservation (Campbell, 1978). The amount and composition of organic matter in a soil is subject to change; mismanagement of soils can result in substantial losses, while adoption of appropriate conservation practices can maintain or increase organic matter content (Janzen et al., 1997).

Disregarding soil erosion, the amount of organic matter in an agricultural soil is the net result of two processes: the input of C through photosynthesis (primary production), and the loss of C via decomposition (respiration) (Janzen et al., 1997, 1998). In photosynthesis, atmospheric CO_2 is reduced to organic C forms that, upon the death of the plant, enter the soil and become the materials from which soil organic matter is formed. At the same time, organic matter already in the soil is gradually being decomposed and oxidized by soil microorganisms, resulting in the loss of organic matter to the atmosphere as CO_2. These two processes therefore form a complete cycle. Any factor that favors greater C input relative to decomposition will increase organic matter; conversely, stimulation of decomposition relative to C input will induce net organic matter loss (Janzen et al., 1997).

The organic matter present in North American soils was formed over thousands of years under the native vegetation. During that time, the counteracting processes of C input and decomposition eventually equilibrated, resulting in the stabilization of organic matter content at a level reflecting the environmental conditions at the given site (Janzen et al., 1997). Conversion of these lands to arable agriculture, however, disrupted this equilibrium, resulting in a loss of soil organic matter. Although the magnitude of this loss varies, agricultural soils have typically lost from 15 to 60% of the organic matter present at the time of initial cultivation. A significant fraction of this loss was due to erosion; losses due to decomposition were probably no more than 30% (Acton & Gregorich, 1995).

The loss of organic matter upon adoption of arable agriculture can be attributed to several factors. First, the amount of C returned to the soil in plant litter is typically much lower in agricultural systems than in native systems. This is partly because a portion of the organic material is harvested and removed. For example, about one-third of the C content of the wheat (*Triticum aestivum* L.) plant is exported from the land as grain. Second, the change in plant species upon conversion to agriculture typically affects the distribution of C in the soil and conditions for decomposition. For example, native grasses usually have more root C than cereal crops and decomposition rates tend to be lower under native grass than under wheat. Finally, conversion to arable agriculture typically involves intensive tillage, which often accelerates organic matter decomposition and losses by erosion.

The prairies of western Canada account for about 80% of the arable agricultural land in Canada and contain large reserves of C as soil organic matter (Curtin et al., 1994; Dumanski et al., 1998). However, the size of this pool may change in response to management practices. In the past, such changes were a concern because of the potential negative impact on soil productivity (Campbell, 1978). In more recent times the possible impact of such changes on the concentration of CO_2 in the atmosphere and the latter's potential as a contributor to global warming through the "greenhouse gas effect" is also a concern to society (Janzen et al., 1998; Bruce et al., 1999). Nonetheless, soil organic matter can also act as a sink for C and may therefore be an asset in helping to mitigate the CO_2 component of the greenhouse gas effect (Janzen et al., 1998; Bruce et al., 1999). Although the concept of trading "C credits" for C sequestered in agricultural soils was discussed at the International Conference to debate global warming at Kyoto in 1997, this concept was not yet approved. Further, it is unlikely that it will be approved unless scientists can provide credible evidence indicating that soils can indeed be managed to result in significant amounts of sequestered C, and that such C gains can be reasonably and accurately quantified (Bruce et al., 1999).

Detailed discussion on how crop management may influence soil organic C (SOC) storage on the Canadian prairies has been presented in recent review articles (Janzen et al., 1997, 1998). In these reviews it was concluded that: (i) the loss of SOC upon conversion of soils to arable agriculture has abated; (ii) significant gains in SOC (typically about 3 Mg C ha^{-1} or less within a decade) can be achieved in some soils by adoption of improved practices such as more intensive cropping, reducing tillage intensity, improved crop nutrition, addition of organic amendments, and reversion to perennial vegetation; (iii) changes in SOC occur predominantly in labile organic matter fractions (e.g., light fraction C and microbial biomass C); (iv) the change in SOC, either gain or loss, is of finite duration and magnitude; (v) estimates of SOC change from individual studies are subject to limitations and are best viewed as part of a multisite network; and (vi) the energy inputs into agroecosystems (e.g., fertilizers, fuel, pesticides) need to be included in the calculation of the net C balance.

The objectives of this chapter were to quantify the trends in soil C that have been measured in two long-term crop rotation experiments conducted on the Canadian prairies, and to discuss some of the challenges we face regarding quantifying C sequestration in prairie soils.

QUANTIFYING CARBON CHANGES IN SOIL

If we are to gain international acceptance that C can be sequestered in agricultural soils scientists must develop "an agreed upon methodology for measuring variable changes in C stocks" (Bruce et al., 1999). This is important because in the USA and Canada, the major industrial contributors to CO_2 emissions such as the public utility companies, and some policy-makers, are very interested in "carbon credit trading" (Bruce et al., 1999). We know that changes in soil C can be measured directly, but these changes are small compared to the size of stocks present in the soil, they occur slowly and, often spatial variability of soil C is large, thus a direct measurement approach becomes complicated and may be impractical for this purpose (Bruce et al., 1998). Simulation modeling of soil C changes in agricultural soils is well developed for assessing long-term changes (i.e., >50–100 yr changes) (Paustian et al., 1997; Smith et al., 1998). But, quantifying short-term changes (<5–20 yr), which is necessary to determine the value of C sequestered for C trading purposes, is presently much less feasible. It is imperative that the latter be achieved soon if the concept of "C credits trading" of C sequestered in agricultural soils is to be accepted by the international community and the Intergovernmental Panel on Climate Change (IPCC).

The CENTURY Model

Of several simulation models developed to describe C dynamics in soil (Smith et al., 1998), CENTURY, developed in Colorado by Parton et al. (1987), has been the most widely used by North American scientists. However, it was developed to assess C changes over the long-term (>30 yr) and it is more reliable when used in this manner (Smith et al., 1998). Recently, some workers have suggested that CENTURY can be used effectively for short-term assessment of C dynamics at specific sites (Voroney & Angers, 1995; Liang et al., 1996; Monreal et al., 1997). If this claim is valid, then this model could prove to be a useful tool for satisfying a significant part of our requirements for the Kyoto accords. Nonetheless, despite the these claims, it appears that current versions of CENTURY may not yet be a satisfactory tool for estimating short-term changes in soil C, as was concluded during a special workshop held at University of Saskatchewan, Saskatoon (25–27 Aug. 1998), to investigate this problem. For example, the CENTURY model Version 4 performed poorly when simulating tillage effects for most workshop participants (data not shown).

One of the main deficiencies of CENTURY with respect to its use for simulating short-term dynamics of soil C may be related to its inability to accurately estimate grain (and thus residue) yields (Kelly et al., 1997). This may be particularly true for crops grown in semiarid prairie climates. When we attempted to estimate grain yields of wheat grown on fallow (F-W) and continuous wheat (Cont W), for a 30-yr crop rotation experiment conducted at Swift Current, Saskatchewan, Canada, the results though very good for F-W, were poor for stubble-crop wheat (Fig. 8–1) . Since residue input is directly related to grain yields (Campbell et al., 1997) and since soil C is directly related to residue input (Rasmussen et al., 1980; Campbell et al., 1997; Huggins et al., 1998), then any model failing to estimate yields accurately is likely to be deficient in simulating soil C dynamics over the short term. Thus, it was not surprising that CENTURY did not accurately estimate soil C (Fig. 8–1).

We believe that the poor yield simulation by CENTURY for stubble crops in this experiment may be related to its use of monthly precipitation as one of the main determinants of crop production. Monthly precipitation may not be very effective for estimating grain yields in cold, semiarid climates where stored soil water (overwinter precipitation) plays a major role in determining yields (Campbell et al., 1997), and where seasonal distribution of growing season precipitation is so critical to crop production (Campbell et al., 1988). The developers of the CENTURY model recognize this weakness (Kelly et al., 1997) because they are presently involved in a project to tie CENTURY to the more robust plant growth model, CERES (J. Ritchie,1997, personal communication).

We hypothesize that in our situation, CENTURY greatly underestimated available water for stubble crops (Fig. 8–1b,c). This would not only influence estimates of yields directly, but probably influenced net N mineralized. This apparent shortcoming in estimating yields, would negatively influence CENTURY's efficiency for making accurate short-term C simulations in semiarid climates because estimates of residue C inputs would be erroneous. But there are other apparent inaccuracies in the estimate of soil C by CENTURY. Two obvious discrepancies are the failure of CENTURY to reflect the impact of 7 yr (1989–1996) of above-average residue inputs for F-W (Fig. 8–1a), and that the soil C estimates were either showing a downward trend (Fig. 8–1a,b), or constancy (Fig. 8–1c) even though measured soil C was either constant or increasing over time. This may be an indication that CENTURY requires some modifications regarding the relative weighting currently given to the influence of crop residue inputs on soil C changes. The trajectory of trend lines in the CENTURY soil C estimates may also be related to the high initial values estimated by the model (4 Mg ha^{-1} higher than our estimate for 1967). Perhaps if the model was re-initialized so that the estimated starting C values were similar to the measured values, then the trajectory of the estimated and measured trends in C may be closer.

Despite the apparent shortcomings of CENTURY, we believe the soil C dynamics parameters of this model are generally sound. Consequently, once the production (C inputs) aspects are satisfactorily addressed, this model should provide a reasonably accurate simulation of even short-term, site-specific conditions. Of course, further modifications will need to be made to improve its ability to "scale-up" results to a landscape basis and thereby account for erosion and deposition effects on soil C levels. (This latter aspect is currently being researched on the Canadian prairies by Drs. D. Anderson, D. Pennock, and A. Frick in the Dep. of Soil Science at the University of Saskatchewan.)

Environmental Policy Integrated Climate Model

We used the Environmental Policy Integrated Climate (EPIC) model (Williams, 1995) to estimate spring wheat grain yield, soil erosion loss, and soil organic C in the 0- to 15-cm depth for continuous wheat (Cont W) and fallow-wheat (F-W), both fertilized adequately with N and P, in a 30-yr crop rotation experiment

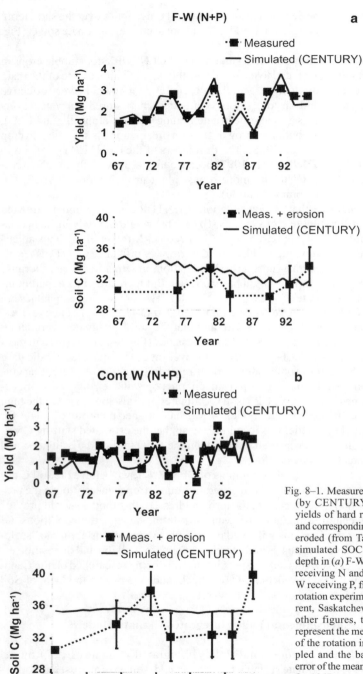

Fig. 8–1. Measured and simulated (by CENTURY model) grain yields of hard red spring wheat and corresponding measured plus eroded (from Table 8–3) C and simulated SOC in 0- to 15-cm depth in (*a*) F-W and (*b*) Cont W receiving N and P, and (*c*) Cont W receiving P, from a 30-yr crop rotation experiment at Swift Current, Saskatchewan. In this and other figures, the SOC values represent the mean for all phases of the rotation in the years sampled and the bars are standard error of the mean (S_x). SOC value for 1967 was assumed to be the same as the 1976 value for F-W (N+P).

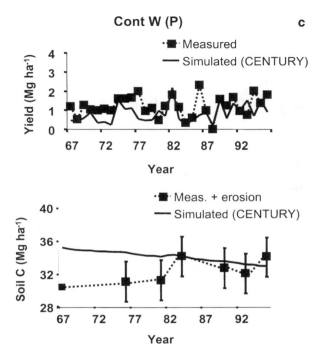

Fig. 8–1. Continued.

at Swift Current. We used EPIC Version 5300 with the Baier-Robertson potential evapotranspiration approach to estimate available water and thus yield and C dynamics. Roloff et al. (1998a,b) provides a detailed discussion on how this model was used.

In general, EPIC provided reasonable accuracy and precision in estimating yield trends in Cont W and F-W over the 30-yr period of study (Fig. 8–2a). Although these yield simulations were superior to those obtained for stubble-wheat with the CENTURY model, there is still room for improvement in estimating annual yield variability (Roloff, 1998b). Like CENTURY, EPIC was not effective in estimating trends in SOC (Fig. 8–2b). The model suggested C would decrease steadily with time in both F-W and Cont W rotations, this despite correctly simulating the above-average yield trends from 1989 to 1996 which likely caused the upturn in measured SOC (Fig. 8–2b). This suggests that EPIC also does not give sufficient weighting to the impact of crop residue inputs on build-up of SOC. As well, in EPIC, SOC may also be decomposing too rapidly. In any event, we believe that the C dynamics aspects of EPIC need further improvements, such as explicitly accounting for the various C pools and crop-specific residue decomposition rates (results of simulation with EPIC of soil and C loss by erosion are discussed later).

Other Models

Since CENTURY and EPIC, in their present form, do not adequately simulate short-term C dynamics in soil, will other ecosystem-type models, such as

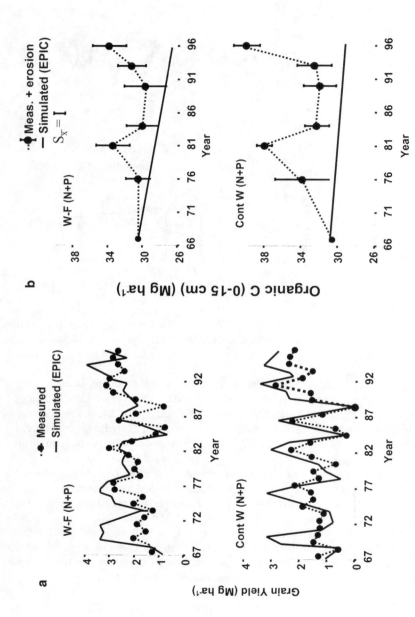

Fig. 8–2. (a) Measured and simulated (by EPIC model) grain yields for the same rotations assessed in Fig. 8–1(a) and 1(b). (b) Corresponding measured plus eroded (from Table 8–3) C and simulated (EPIC) SOC for rotations assessed in Fig. 8–2a.

Rothamsted C model (Roth-C) (Jenkinson & Rayner 1977), or denitrification-de-composition (DNDC) (Smith et al., 1998), do better? We doubt it because these models generally operate in a similar manner differing only in degree.

While scientists are working to modify and improve models such as CEN-TURY and EPIC so that they can perform more effectively for short- and long-term situations, what can we do? We propose, that a workable approximation may be to use a simpler method of estimating soil C dynamics by treating recent residue additions and pre-existing soil C separately, similar to the method proposed 50 yr ago by Woodruff (1949). The proposed equation is:

$$C_t = C_1 e^{-k_1 t} + C_2 e^{-k_2 t} + \sum_{n=0}^{t} [A_n (p_1 e^{r_1(t-n)} + p_2 e^{r_2(t-n)})] \qquad [1]$$

where C_t is the amount of soil organic C per unit mass of soil remaining in soil after t yr, C is the amount of C in the soil on a mass basis initially ($t = 0$), k is the annual rate of soil C decomposition, A_n is the C addition as plant residue in yr n, p is a proportion of residue C (note $p_1 + p_2 = 1$), and r is the annual rate of residue decomposition. The subscripts 1 and 2 refer to differing degrees of susceptibility to decomposition with 1 representing the more active and 2 representing the slower decomposing pool of the plant residue and the soil humus. Although in theory, the output from A would feed into C in Eq. [1], we overlooked this factor to keep calculations simple. We solved the A portion and C portions of Eq. [1] separately then added the results (Campbell et al., 2000).

The advantage of using Eq. [1] is that values for A (related to residue inputs) are usually available, or can be readily estimated from grain yields, harvest index and straw/root ratios. These yield data integrate climatic conditions thus helping to make this approach robust.

Equations based on a 10-yr ^{14}C-labeled ryegrass (*Lolium perenne* L.) residue decomposition experiment conducted in Britain (Jenkinson, 1977), and a similar experiment conducted with wheat straw in Saskatchewan, Canada, by Voroney et al. (1989) suggest that cereal and grass residues decompose at similar rates in temperate climates. Thus, for hard red spring wheat straw decomposing in a fallow-wheat-wheat (F-W-W) rotation in southwestern Saskatchewan, Voroney et al. (1989) obtained the following two-component first-order decomposition equation in an Aridic Haploboroll, Sceptre clay soil:

$$y = 0.72 e^{-1.4t} + 0.28 e^{-0.081t} \qquad [2]$$

where y is proportion of residue C remaining in the soil after t yr since residue addition. In an Alfisol, Waitville loam soil in northeastern Saskatchewan, they found the rate of decomposition was slightly faster than for the clay. Equation [2] is similar to one obtained by Jenkinson (1977) at Rothamsted. He reported:

$$y = 0.71 e^{-2.83t} + 0.29 e^{-0.087t} \qquad [3]$$

These results suggest that we may be able to represent the two pool proportions for residue additions (p_1 and p_2) of Eq. [1] with those from Eq. [2]. Further, this suggests that the latter may also provide good representation of plant residue

Table 8–1. Some typical soil fractions and their decomposition rate-constants estimated from C dating for Saskatchewan soils.

Soil	Soil zone	Humic fraction†	Proportion of organic C (%)	^{1}/mrt (k) (yr^{-1})‡	Reference
Melfort sic	Black chernozem	HA and humin	71	0.000 85	Campbell (1965)
		FA and acid extract	29	0.002 1	
Oxbow cl	Black chernozem	Unhydrolyzed	50	0.000 68	Martel (1972)
		Acid hydrolyzed	50	0.050	
Sceptre c	Brown chernozem	Unhydrolyzed	40	0.003 6	Martel (1972)
		Acid hydrolyzed	60	0.000 63	
Waitville l	Gray luvisol	HA and humin	83	0.014	Campbell (1965)
		FA	17	0.004	

† HA = humic acids, FA = fulvic acids.
‡ mrt = mean residence time.

decomposition for large areas of arable land growing cereals and grasses throughout temperate climates. The parameters for the active and slow decomposing pools of soil C (C_1 and C_2, respectively in Eq. [1]) can be estimated from information in the scientific literature, as can values of k_2 (Table 8–1). Thus, we might estimate the rate constant for the slow decomposing fraction of soil humus from the mean residence time (mrt) of the humic fractions ($\approx$0.000 66 yr^{-1} for chernozemic soils). We could also assume that the active fraction (C_1) is represented by the light fraction and microbial biomass C which, though variable through the year, averages about 20% in Swift Current soil (Campbell et al., 2000). Thus C_2 would be 80% of soil C in this soil. We could then estimate k_1 by solving Eq. [1] iteratively to match the measured C for F-W system. However, because in Eq. [1], the value of k must embody differences not only in inherent susceptibility to decomposition of soil C fractions, but also differences in the effect of soil moisture, temperature, etc., on rate of decomposition, estimates of k are empirical in nature. This is a disadvantage compared to process-based models where k and r are functions of moisture and temperature. Our rate constants are therefore only pseudo rate constants. We reasoned that k_1 for F-W-W must be less than for F-W (less frequent fallow and fallow favors decomposition). Similarly, k_1 for F-W-W will be greater than for Cont W.

Process-based models such as CENTURY, because they provide more rigorous quantitative representation of soil C processes than the empirical Woodruff-type equations, will be the method of choice for more accurate estimates of SOC dynamics, but, as we showed earlier, they require some refinement. The Woodruff-type equation has been shown to provide a good first approximation of the direction and magnitude of relative effects of various agronomic treatments on soil C dynamics in Sanborn plots and some treatments of the Morrow plots (Huggins et al., 1998). Therefore we decided to test Eq. [1] on two data sets from the Canadian prairies.

Testing the Woodruff-Type Equation

We tested Eq. [1] using data from two long-term crop rotation experiments, one from an Orthic Brown Chernozem (Aridic Haploboroll) at Swift Current,

Saskatchewan, and the other from Orthic Dark Brown Chernozem (Typic Haploboroll) at Lethbridge, Alberta. The 30-yr study at Swift Current was initiated in 1967 on land that had been cropped to F-W since 1922 with primarily P fertilization (Campbell et al., 1983, 1992; Campbell & Zentner, 1993). Thus, this soil (Swinton loam–silt loam) would have been somewhat degraded at start of the experiment. The 42-yr experiment at Lethbridge was established in 1951 on land that had been broken from tall and short grass prairie in 1910 that was then seeded to mixed crop rotations that included legumes and to which light manure applications were made every 6 yr (Bremer et al., 1994; Johnston et al., 1995). Thus, at the start of the latter experiment, SOC in the 0- to 15-cm depth at this site was about 34.5 Mg ha^{-1} while at Swift Current it was estimated to be about 30.5 Mg ha^{-1}. Both of these studies were conducted with stubble-mulch shallow tillage after the 1940s. There are 13 treatments at Lethbridge (we discuss only two, unfertilized F-W and Cont W), and nine treatments at Swift Current (we discuss eight).

Swift Current Rotations

The rotation treatments and their fertilizer regimes are described in Table 8–2. We collected grain yields from all seeded plots and straw yields from specified plots (Table 8–2) annually, and we measured soil C and N (only C discussed) in all rotation phases periodically (viz., 1976, 1981, 1984, 1990, 1993 and 1996). We did not measure soil C at the initiation of the study in 1967, but estimated it to be about 30.5 Mg C ha^{-1} in the 0- to 15-cm depth (Campbell et al., 2000). The latter was the value for F-W in 1976 and we reasoned this system was at steady state because it was cropped to F-W for over 50 yr prior to our experiment. We only discuss C in the 0- to 15-cm depth because at no time was there a significant change of C in 15- to 30-cm depth (Campbell et al., 2000). Bulk density was only measured at start of the study. It was assumed to be constant throughout the years, being 1.22 Mg m^{-3} for the 0- to 15-cm depth.

The values used in Eq. [1] were derived as follows: where straw was not measured, we estimated straw values from grain-to-straw equations that we developed over the life of the study (Campbell et al., 2000). We converted straw yields to total residue C inputs by assuming root-to-straw ratio = 0.59 (Campbell et al., 1977) and C content of residues = 45% (Millar et al., 1936). We used the constants derived by Voroney et al. (1989) for the Sceptre clay to calculate the dynamics of residue C decomposition because the location of this experiment was only 60 km from Swift Current, while the Waitville experiment, though similar in texture to the Swift Current study, was located several hundred kilometers from Swift Current. Thus, the p_1, p_2, r_1, r_2 values are those for Eq. [2] and they were used to estimate residue C dynamics for all rotation phases separately then values were averaged over rotation phases (Campbell et al., 2000).

As stated earlier, C_1 and C_2 for this soil were taken as 20 and 80% of the C, respectively, and k_2 assumed to be 0.00066 yr^{-1}. These values were used for all treatments. Because we observed a close similarity between soil C trends over the years for the degrading treatments (F-W and three F-W-W systems) (Campbell et al., 2000), we averaged soil C measurements over these four treatments (Fig. 8–3). Similarly, we averaged values for three aggrading systems [Cont W (P), W-Lent (N +

Table 8–2. Crop rotations and fertilizer applied to treatments at Swift Current, Saskatchewan (adapted from Campbell et al., 2000).

Rotation #	Rotation†	Fertilizer criteria	Cropped phase	Average N and P applied (kg ha^{-1} yr^{-1})					
				1967–1976		1977–1986		1987–1996	
				N	P	N	P	N	P
11	F-(W)	N and P applied as required	Fallow	6.1	9.4	6.8	10.2	21.3	9.2
2	F-W-(W)	N and P applied as required	Fallow	7.4	9.4	6.4	10.2	22.3	9.2
			Stubble	19.5	9.4	32.1	10.2	43.6	9.2
1	(F)-W-(W)	P applied as required but no N applied except that in P fertilizer	Fallow	4.9	9.4	4.9	10.2	4.6	9.2
			Stubble	4.9	9.4	10.0	10.2	4.6	9.2
5	F-W-W	N applied as required no P applied	Fallow	3.4	0	2.0	0	21.0	0
			Stubble	10.1	0	22.6	0	31.4	0
4a	F-(Rye)-W‡	N and P applied as required	Fallow	3.5	6.7	4.9	10.3	25.0	9.5
			Stubble	20.9	9.6	31.0	10.3	53.6	9.8
4b	CF-WW-WW	N and P applied as required	Fallow	--	--	4.9	10.0	11.0	6.6
			Stubble	--	--	4.9	10.0	34.1	7.5
3	F-Flx-(W)	N and P applied as required	Fallow	5.2	3.8	7.4	10.2	8.8	9.2
			Stubble	16.4	9.4	26.3	10.2	39.3	9.2
8	Cont (W)	N and P applied as required	Stubble	23.5	9.4	35.7	10.2	42.8	9.2
9	Cont (W)§	(Fallow if less than 60 cm moist soil exists at seeding time, N and P applied as required)	Stubble	22	9.4	55.5	10.8	--	--
10	Cont (W)§	(Fallow if grassy weeds become a problem, N and P applied as required)	Stubble	24.6	9.4	52.7	10.8	--	--
19	(W)-Lent	N and P applied as required	Lentil	--	--	26.0	10.1	10.7	7.4
			Wheat	--	--	24.6	10.1	31.6	9.2
12	Cont (W)¶	P applied as required but no N applied except that in P fertilizer	Stubble	4.9	9.4	15	10.2	4.6	9.2

† Selected plots, indicated in parentheses, were sampled for straw weight at harvest: F = fallow, CF = chemical fallow, W = spring wheat, WW = winter wheat, Rye = fall rye, Flx = flax, Lent = grain lentil, Cont = continuous.

‡ After 1984, rotation 4a was changed to chemical fallow–winter wheat–winter wheat (spring wheat whenever winter wheat failed to survive the winter). But in 1993 it was changed again to CF-Rye-W.

§ During the first 12 yr, the criteria necessary for summer fallowing in these two rotations were met on several occasions but the action was not implemented. In 1979, these two rotations were changed to the spring wheat-lentil rotation.

¶ In 1980 and 1982 N was inadvertently applied to this rotation at rates of 70 and 40 kg N ha^{-1}, respectively.

P) and F-Rye-W (N + P)] (Fig. 8–3). We then used iterative calculations with Eq. [1] and these two curves to estimate approximate values for k_1 (the only remaining unknown) in degrading and aggrading systems. This process suggested the k_1 value for F-W would be about 0.02 yr^{-1}. We then assumed the value for F-W-W would be 0.01 yr^{-1} and for aggrading systems 0.001 yr^{-1}.

Soil erosion redistributes surface soil, which is richest in C, removing it from some sites and depositing it in others. There is little research to indicate that a significant proportion of C is lost during the translocation of eroded material within the landscape. Some labile C (e.g., light fraction) may be lost, during, or shortly after translocation within the landscape; but the majority of C, which is associated with clay minerals, probably is not lost by decomposition (Gregorich et al., 1998). Thus we used EPIC to estimate C lost by erosion over the 30-yr period (Table 8–3) and added these values to the measured C values before comparing these sums to the estimated soil C values derived with Eq. [1] (Fig. 8–4).

Equation [1] estimated the trends in SOC consistently well, though it tended to underestimate the measured + erosion values for W-Lent (N + P). This may be an indication that the narrower C/N ratio crop residues in the latter system supports a more efficient conversion of residue C to SOC than does wider C/N ratio material such as for monoculture wheat. Unlike results for CENTURY and EPIC, the trends in SOC estimated by Eq. [1] showed a flat response in the 1980s when

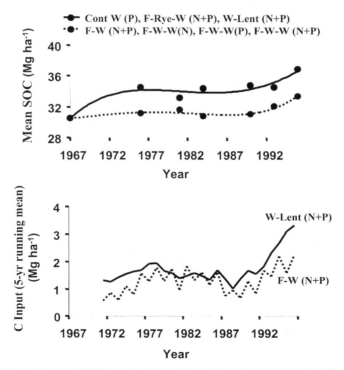

Fig. 8–3. Pattern of change in SOC for selected aggrading and degrading systems at Swift Current compared to the corresponding typical pattern of change in residue C inputs (from Campbell et al., 2000).

Table 8–3. Organic C lost by erosion† from selected rotations, estimated by EPIC model at Swift Current, Saskatchewan (adapted from Campbell et al., 2000).

Rotation	Erosion losses	Period sampled					
		1967–1976	1977–1981	1982–1984	1985–1990	1991–1993	1994–1996
F-W (N+P)‡	Soil lost during period (Mg ha⁻¹)	12.14	9.69	9.35	18.75	11.88	11.88
	C lost up to end of period (Mg ha⁻¹)§	0.20	0.38	0.53	0.83	1.03	1.25
F-W-W (N+P)‡	Soil lost during period (Mg ha⁻¹)	10.51	9.02	7.49	15.60	8.99	8.99
	C lost up to end of period (Mg ha⁻¹)§	0.18	0.33	0.45	0.73	0.89	1.06
F-Rye-W (N+P)‡	Soil lost during period (Mg ha⁻¹)	4.51	2.49	1.94	5.29	3.18	3.18
	C lost up to end of period (Mg ha⁻¹)§	0.09	0.13	0.17	0.27	0.33	0.40
Cont W (N+P)‡	Soil lost during period (Mg ha⁻¹)	2.82	1.28	1.68	4.30	2.55	2.55
	C lost up to end of period (Mg ha⁻¹)§	0.05	0.08	0.11	0.18	0.23	0.29

† This includes both wind and water erosion. Wind erosion was usually up to 10% higher than water erosion for fallow-containing systems except those with fall rye where water erosion was twice that of wind erosion. Continuous cropping systems had water erosion four times that of wind erosion.

‡ The weighted mean annual soil loss was 2.3, 1.9, 0.6, and 0.5 Mg ha⁻¹ yr⁻¹ for F-W, F-W-W, F-Rye-W and Cont W, respectively.

§ These are cumulative values. Carbon loss obtained by multiplying estimated soil loss and measured %C.

droughts in 1984, 1985 and 1988 resulted in very low yields, and then a consistent trend to higher SOC values in the wet 1990s. In a few instances the measured data appeared to be unreasonably high or low (see "?" on Fig. 8–4). We suspect that the (?) points reflect texture differences at this site as reported by O'Halloran (1986). Campbell et al. (1996) have shown a direct relationship between C gains in response to adoption of no-tillage management and clay content of soil in a study conducted in southwestern Saskatchewan. As stated earlier, soil texture varies from loam to silty clay loam at this site.

Note that if we estimate C gains using Eq. [2], and assume that no soil C decomposition occurred, then the estimates fit the measured data for summerfallow-fall rye-wheat (F-Rye-W) (N + P) and wheat-lentil (W-Lent) (N + P) almost exactly (Fig. 8–5). This implies that soil humus decomposition is very low in these

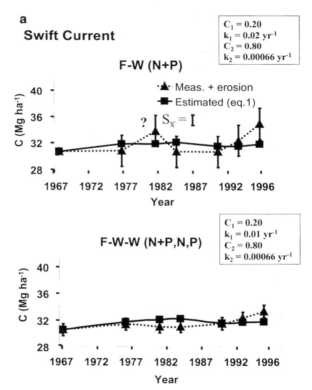

Fig. 8–4. Comparison of measured SOC and estimates made using a Woodruff-type equation (Eq. [1]) for various rotations from the Swift Current 30-yr crop rotation experiment. In using the Woodruff-type equation we employed the Voroney et al. (1989) constants for straw decomposition in the Sceptre clay as the basis for the A components, and constants derived from C-dating (k_2) and proportion of soil C in light fraction plus biomass C for C_1 and remaining C as C_2 (see Eq. [1]). We estimated k_1 by iteration against measured data. $S_{\bar{x}}$ represents the standard error of the mean for the measured values. Eroded C, estimated by EPIC model (Table 8–3) were added to the measured C values before comparing to the simulated values. [(a) = F-W (N+P) and mean of F-W-W (N, P, and N+P); (b) = Cont W (P) and Cont W (N+P); (c) = F-Rye-W (N+P) and W-Lent (N+P)] (from Campbell et al., 2000). (Continued on next page)

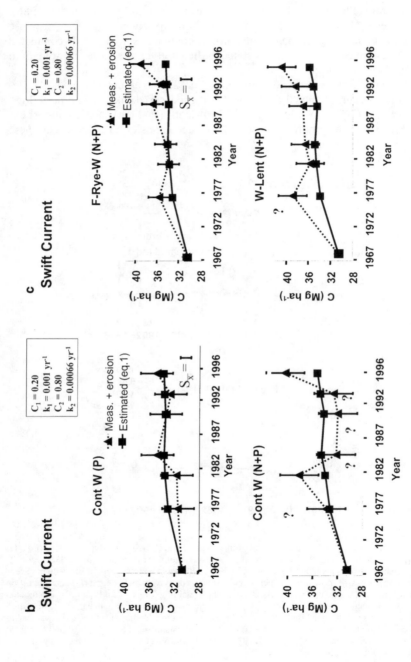

Fig. 8–4. Continued.

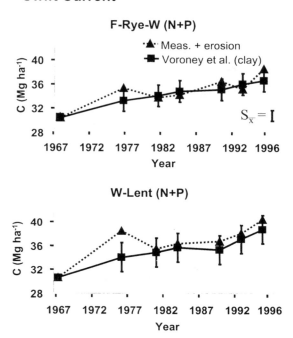

Swift Current

Fig. 8–5. Comparison of measured plus estimated eroded C vs. estimates of C gains calculated from Voroney et al. (1989) residue C decomposition model (Eq. [2]) with no allowance made for C lost via soil decomposition, for the F-Rye-W (N+P) and W-Lent (N+P) rotations at Swift Current (from Campbell et al., 2000).

two systems or, conversely that some of our assumptions for lentil and for fall rye using spring wheat based model (Eq. [2]) are not appropriate.

Generally, the measured soil C was of two patterns (Fig. 8–3). In the degrading (fallow-spring seeded crop) systems, SOC was generally constant until 1990, then increased sharply in response to a 7-yr period of above-average yields (residue input). For the aggrading systems there was an early gradual increase, due to change from over 60 yr of F-W (poor fertility) to continuous cropping (with good fertility); then values leveled off in the drought-prone 1980s, then increased sharply due to high yields in the 1990s. These results strongly suggest that crop residue inputs play a very significant role in influencing short-term soil C dynamics, perhaps an even more significant role than soil decomposition mechanisms. These results further underscore the need for the more process-related models, such as CENTURY and EPIC, to improve their estimates of crop production and to re-examine the parameter values they use in weighting the relative contribution of crop residue inputs to soil C changes as compared to the contribution of soil processes such as mineralization, immobilization and denitrification. The W-Lent and F-Rye-W systems suggest that the mineralization may not be great in the frigid semiarid prairie conditions of western Canada.

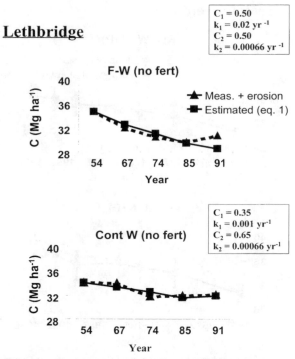

<figcaption>Fig. 8–6. Comparison of measured plus estimated eroded C (0–15 cm depth) vs. estimated C (Eq. [1]) for unfertilized F-W and Cont W rotations at Lethbridge, Alberta. Calculations made as in Fig. 8–4 but with different C_1 and C_2 values to account for more fertile soil at Lethbridge than at Swift Current. Erosion estimated by assuming rates of soil loss similar to those at Swift Current (Table 8–3) and multiplying by C concentrations on sampling dates × number of years between sampling dates. Carbon concentrations for F-W were 1.76, 1.60, 1.51, 1.45, and 1.50% in 1954, 1967, 1974, 1985 and 1991, respectively; for Cont W the values were 1.83, 1.82, 1.70, 1.70, and 1.74%, respectively.</figcaption>

Lethbridge Rotations

When we used the constants developed to assess Swift Current's Cont W and F-W systems to simulate C trends in these two systems at Lethbridge, our simulation results far overestimated the measured data (data not shown). We believe this was partly because no fertilizer has been applied to these two treatments since 1951. Second, at initiation of this experiment this soil was not in as degraded a state as was the soil at Swift Current. The long use of mixed rotations that included legumes, plus regular additions of manure, from breaking in 1910 to 1951, likely maintained this soil in a very fertile state. Evidence of this was the much higher organic C level at Lethbridge at start of the study (Fig. 8–6) than at Swift Current (Fig. 8–4) even though both soils are in the same ecozone. Consequently, even the switch to continuous cropping (albeit no fertilization) was not sufficient to maintain SOC levels in this system and measurements showed a steady decline (Fig. 8–6). This un-

Table 8–4. Soil organic C gained (or lost) from start of experiment to last measurement (measured and estimated†).

Rotation	Measured	Measured plus eroded	Equation estimate	Mean annual change based on measured data	Mean annual change based on equation estimates
		MgC ha⁻¹		MgC ha⁻¹ yr⁻¹	
Swift Current (1967–1996)					
Fallow-wheat (N+P)	3.4	4.6	1.8	0.11	0.06
Fallow-wheat-wheat (avg. N+P, +P, +N)	2.8	3.9	3.1	0.09	0.10
Fallow-rye-wheat (N+P)	6.9	7.3	4.9	0.23	0.16
Continuous wheat (N+P)	9.5	9.7	6.5	0.32	0.22
Continuous wheat (+P)	3.6	3.9	4.1	0.12	0.14
Wheat-lentil (N+P)	8.5	8.8	6.6	0.28	0.22
Lethbridge (1954–1991)					
F-W (unfert.)	−5.3	−6.6‡	−6.1	−0.14	−0.16
Continuous wheat (unfert.)	−2.3	−2.6‡	−2.2	−0.06	−0.06

† Carbon change estimated by Eq. [1] and erosion estimated by EPIC model (Campbell et al., 2000).
‡ We assumed the rate of soil erosion at Lethbridge was same as at Swift Current and estimated period C losses by multiplying by C concentration in soil at each sampling time and then accumulated C lost to erosion as done for Swift Current (see Table 8–3).

derscores an important point about which Janzen et al. (1998) have previously hypothesized. They suggest that our ability to increase C in soils will depend not only on the amount of residue C inputs, but also on the C content of the soil at the time the residue was added. It is difficult to increase the C content of a soil that is already high in C, but much easier to increase the C content of a degraded soil (Janzen et al., 1998).

We used Voroney et al. (1989) model for residue decomposition (Eq. [2]), using the same constants as at Swift Current. Mean residence times for the fractions of the Lethbridge soil were not available, but we assumed that k_1 and k_2 were the same as for the respective Swift Current rotations. Thus k_2 was 0.0006 yr⁻¹ for F-W and Cont W (unfertilized), and k_1 was 0.02 yr⁻¹ for F-W and 0.001 yr⁻¹ for Cont W. To account for the more fertile soil conditions initially present in the Lethbridge than in Swift Current soil, we estimated (by iteration analysis to match the measured data) that for F-W, C_1 was 50% and C_2, 50% of the soil C, while for Cont W, C_1 was 35% and C_2 was 65% (Fig. 8–6).

We calculated the change in SOC between the start of the experiment and the last date sampled, for the rotations at Swift Current and Lethbridge, comparing values derived by measurements and those estimated by Eq. 8–1 (Table 8–4). We also calculated a mean annual rate of increase in SOC over the life of each experiment. The measured and estimated SOC changes and rates of change were generally similar and generally confirmed our expectations regarding the influence of management on SOC changes. Based on our estimates at Swift Current, SOC changes range between 0.06 Mg ha⁻¹ yr⁻¹ in F-W to 0.22 Mg ha⁻¹ yr⁻¹ in well-fertilized continuously cropped systems (Table 8–4). At Lethbridge where net losses have occurred, losses were almost three times greater in F-W than in Cont W.

CHALLENGES REMAINING TO QUANTIFYING
SOIL CARBON ON PRAIRIES

If the concept of "C credits trading" is to gain credibility as part of the solution in mitigating the "greenhouse gas" phenomenon for which CO_2 is partly responsible, scientists must be able to provide reasonable evidence that soil C changes in the short term (<20 yr) can be measured accurately over time and space. For this to be done effectively, we will need to use models that are less empirical than Woodruff-type models; models that are more process oriented, like CENTURY. However, our results suggest that CENTURY may have a shortcoming with respect to its ability to estimate C inputs in stubble crops under Canadian prairie conditions. The EPIC model may do a better job of estimating residue C inputs, but its submodel for estimating soil C appears to be weak. Both models were unable to simulate the increases and decreases in soil C due to periods of above- or below-average production, even when they were able to estimate yields acceptably. We hypothesize that they may not be giving sufficient weighting to the contribution of input crop residues to soil C changes (as compared to the weighting being given to soil C decomposition processes). Certainly, the Woodruff-type equation was much more effective in simulating these weather (yield)-related trends, probably because accurate estimates of C inputs were used in the model. Perhaps the process-based models have over-emphasized the impact of soil processes such as mineralization, immobilization, and denitrification. It is apparent that improved versions of the process-based models are required if we are to make accurate regional or national extrapolations of C changes in agricultural soils in the short term.

The Woodruff-type equation performed well when it was provided with adequate data to characterize a site. However, it suffers from shortcomings related to our limitations in modifying the rate of soil and residue C decomposition as a function of weather factors. Second, unless we know the k-values for a specific soil, it is difficult to decide what this value should be. The assumed value of k_2 seems reasonable enough (0.00066 yr^{-1} for chernozems). The rate constants for the active fraction (k_1) may vary with soil texture, inherent fertility of the soil, tillage and weather factors, and we have no method of predicting to what extent these rate parameters should be modified to satisfy these variables. Nonetheless, we can certainly use data from the scientific literature and intuitive reasoning to estimate appropriate coefficients for use in this equation. Then, using known values or estimates of residue C inputs we could at least make a credible estimate of relative changes in soil C. This would then guide us in deciding whether measured values, with their propensity for large variability, are reasonable. This is important because we often use the measured values to calibrate process-based models and, if the calibration is faulty, later extrapolations made with these models will be even more faulty.

Equation [1] could be refined to account for differences in decomposition of residues in different management systems, such as no-tillage and moldboard plow, by data obtained from the literature. For example, there are data available that quantify the total amount of C decomposed in surface-placed and buried residues (Curtin et al., 1998). Douglas et al. (1980) provide data that show consistent differences in the decomposition rate constants of surface-placed and buried wheat straw. This information can be used to adjust the rate constants in Eq. [1] for use in specific management systems.

Finally, we suggest that the Woodruff-type of calculation can be used to estimate effects of agronomic treatments on soil C changes over the short term (<20 yr). This can be a useful tool in providing an initial estimate of the influence of agronomic practices on soil C sequestration.

REFERENCES

Acton, D.F., and L.J. Gregorich. 1995. The health of our soils—toward sustainable agriculture in Canada. Publ. 1906/E. Centre Land Biol. Resour. Res., Res. Branch, Agric. Agri-Food Canada, Ottawa.

Bremer, E., H.H. Janzen, and A.M. Johnston. 1994. Sensitivity of total, light fraction, and mineralizable organic matter to management practices in a Lethbridge soil. Can. J. Soil Sci. 74:131–138.

Bruce, J.P., M. Frome, E. Haites, H. Janzen, R. Lal, and K. Paustian. 1999. Carbon sequestration in Soils. J. Soil Water Conserv. 54:382–389.

Campbell, C.A. 1965. The use of naturally occurring ^{14}C to measure the persistence of organic components in soil. Ph.D. diss. Dep. Soil Sci., Univ. Saskatchewan, Saskatoon, Canada.

Campbell, C.A. 1978. Soil organic carbon, nitrogen and fertility. p. 173–271. In M. Schnitzer and S.U. Khan (ed.) Soil organic matter. Developments in soil science. Elsevier Sci. Publ. Co., Amsterdam, The Netherlands.

Campbell, C.A., D.R. Cameron, W. Nicholaichuk, and H.R. Davidson. 1977. Effects of fertilizer N and soil moisture on growth, N content and moisture use by spring wheat. Can. J. Soil Sci. 57:289–310.

Campbell, C.A., H. Janzen, and N.G. Juma. 1997. Case studies of soil quality in the Canadian prairies: Long-term field experiments. p. 351–397. In E.G. Gregorich and M.R. Carter (ed.) Soil quality for crop production. Elsevier Sci. Publ., Amsterdam, The Netherlands.

Campbell, C.A., B.G. McConkey, R.P. Zentner, F. Selles, and D. Curtin. 1996. Long-term effects of tillage and crop rotations on soil organic C and total N in a clay soil in southwestern Saskatchewan. Can. J. Soil Sci. 76:395–401.

Campbell, C.A., D.W.L. Read, R.P. Zentner, A.J. Leyshon, and W.S. Ferguson. 1983. The first 12 years of a long-term crop rotation study in southwestern Saskatchewan—yields and quality of grain Can. J. Plant Sci. 63:91–108.

Campbell, C.A., F. Selles, R.P. Zentner, B.G. McConkey, S.A. Brandt, and R.C. McKenzie. 1997. Regression model for predicting yield of hard red spring wheat grown on stubble in the semiarid prairie. Can. J. Plant Sci. 77:43–52.

Campbell, C.A., and R.P. Zentner. 1993. Soil organic matter as influenced by crop rotations and fertilization. Soil Sci. Soc. Am. J. 57:1034–1040.

Campbell, C.A., R.P. Zentner, B.C. Liang, G. Roloff, E. Gregorich, and B. Blomert. 2000. Organic C accumulation in soil over 30 years in semiarid southwestern Saskatchewan—effect of crop rotations and fertilizers Can. J. Soil Sci. 80:179–192.

Campbell, C.A., R.P. Zentner, and P.J. Johnson. 1988. Effect of crop rotation and fertilization on the quantitative relationship between spring wheat yield and moisture use in southwestern Saskatchewan. Can. J. Soil Sci. 68:1–6.

Campbell, C.A., R.P. Zentner, F. Selles, V.O. Biederbeck, and A.J. Leyshon. 1992. Comparative effects of grain lentil-wheat and monoculture wheat on crop production, N economy and N fertility in a brown Chernozem. Can. J Plant Sci. 72:1091–1107.

Curtin, D., F. Selles, C.A. Campbell, and V.O. Biederbeck. 1994. Canadian prairie agriculture as a source and sink of the greenhouse gases, carbon dioxide and nitrous oxide. Publ. 379M0082. Res. Stn., Res. Branch, Agric. Canada, Swift Current, SK, Canada.

Curtin, D., F. Selles, H. Wang, C.A. Campbell and V.O. Biderbeck. 1998. Carbon dioxide emissions and transformation of soil carbon and nitrogen during wheat straw decomposition. Soil Sci. Soc. Am. J. 62:1035–1041.

Douglas, C.L., Jr., R.R. Allmaras, P.E. Rasmussen, R.E. Ramig, and N.C. Roager, Jr., 1980. Wheat straw decomposition and placement effects on decomposition in dryland agriculture of the Pacific northwest. Soil Sci. Soc. Am. J. 44:833–837.

Dumanski, J., R.L. Desjardins, C. Tarnocai, C. Monreal, E.G. Gregorich, C.A. Campbell, and V. Kirkwood. 1998. Possibilities for future carbon sequestration in Canadian agriculture in relation to land use changes. Climate Change 40:81–103.

Gregorich, E.G., K.J. Greer, D.W. Anderson, and B.C. Liang 1998. Carbon distribution and losses: erosion and deposition effects. Soil Till. Res. 47:291–302.

Huggins, D.R., G.A. Buyanovsky, G.H. Wagner, J.R. Brown, R.G. Darmody, T.R. Peck, G.W. Lesoing, M.B. Vanotti, and L.G. Bundy. 1998. Soil organic C in the tallgrass prairie-derived region of the corn belt: Effects of long-term crop management. Soil Till. Res. 47:219–234.

Janzen, H.H., C.A. Campbell, E.G. Gregorich, and B.H. Ellert. 1997. Soil carbon dynamics in Canadian agroecosystems. p. 57–80. In R. Lal et al. (ed.) Soil processes and the carbon cycle. CRC Press, Boca Raton, FL.

Janzen, H.H., C.A. Campbell, R.C. Izaurralde, B.H. Ellert, N. Juma, W.B. McGill, and R.P. Zentner. 1998. Management effects on soil C storage in the Canadian prairies. Soil Till. Res. 47:181–195.

Jenkinson, D.S. 1977. Studies on the decomposition of plant material in soil. V. The effects of plant cover and soil type on the loss of carbon from ^{14}C labelled ryegrass decomposing under field conditions. J. Soil Sci. 28:424–434.

Jenkinson, D.S., and J.H. Rayner. 1977. The turnover of soil organic matter in some Rothamsted classical experiments. Soil Sci. 123:298–305.

Johnston, A.M., H.H. Janzen, and E.G. Smith. 1995. Long-term spring wheat response to summerfallow frequency and organic amendment in southern Alberta. Can. J. Plant Sci. 74:327–330.

Kelly, R.H., W.J. Parton, G.J. Crocker, P.R. Grace, J. Klir, M. Körschens, P.R. Poulton, and D.D. Richter. 1997. Simulating trends in soil organic carbon in long-term experiments using the century model. Geoderma 81:75–90.

Liang, B.C., E.G. Gregorich, and A.F. MacKenzie. 1996. Modeling the effects of inorganic and organic amendments on organic matter in a Quebec soil. Soil Sci. 161:109–114.

Martel, Y. 1972. The use of radiocarbon dating for investigating the dynamics of soil organic matter. Ph.D. diss. Dep. Soil Sci., Univ. Saskatchewan, Saskatoon, SK, Canada.

Millar, H.C., F.B. Smith, and P.E. Brown. 1936. The rate of decomposition of various plant materials in soils. Am. Soc. Agron. J. 28: 914–923.

Monreal, C.M., R.P. Zentner, and J.A. Robertson. 1997. An analysis of soil organic matter dynamics in relation to management, erosion and yield of wheat in long-term crop rotation plots. Can. J. Soil Sci. 77: 553–563.

O'Halloran, I.P. 1986. Phosphorus transformations in soils as affected by management. Ph.D. diss. Dep. Soil Sci., Univ. Saskatchewan, Saskatoon, SK, Canada.

Parton, W.J., D.S. Schimel, C.V. Cole, and D.S. Ojima. 1987. Analysis of factors controlling soil organic matter levels in Great Plains grasslands. Soil Sci. Soc. Am. J. 51:1173–1179.

Paustian, K., E. Levine, W.M. Post, and I.M. Ryzhova. 1997. The use of models to integrate information and understanding of soil C at the regional scale. Geoderma 79:227–260.

Rasmussen, P.E., R.R. Allmaras, C.R. Rohde, and N.C. Roager, Jr. 1980. Crop residue influences on soil carbon and nitrogen in a wheat-fallow system. Soil Sci. Soc. Am. J. 44:596–600.

Roloff, G., R. de Jong, C.A. Campbell, and V.W. Benson. 1998a. EPIC estimates of soil water, nitrogen and carbon under semi-arid temperate conditions. Can. J. Soil Sci. 78:539–550.

Roloff, G., R. de Jong, R.P. Zentner, C.A. Campbell, and V.W. Benson. 1998. Estimating spring wheat yield variability with EPIC. Can. J. Soil Sci. 78:541–549.

Smith, P., D.S. Powlson, J.U. Smith, and E.T. Elliott. 1998. Evaluation and comparison of soil organic matter models using data sets from seven long-term experiments. Geoderma (Spec. Issue) 81:1–225.

Voroney, R.P., and D.A. Angers. 1995. Analysis of the short-term effects of management on soil organic matter using the CENTURY model. p. 113–120. In R. Lal et al. (ed.) Soil management and greenhouse effect. CRC Press, Inc., Boca Raton, FL.

Voroney, R.P., E.A. Paul, and D.W. Anderson. 1989. Decomposition of wheat straw and stabilization of microbial products. Can. J. Soil Sci. 69:63–77.

Williams, J.R. 1995. The EPIC model. p. 909–1000. In V.P. Singh (ed.) Computer models of watershed hydrology. Water Resour. Publ., Littleton, CO.

Woodruff, C.M. 1949. Estimating the nitrogen delivery of soil from the organic matter determination as reflected by Sanborn field. Soil Sci. Soc. Proc. 14:208–212.

9

Possibilities for Changes in the Greenhouse Gas Balance of Agroecosystems in Canada

R. L. Desjardins, W. Smith, B. Grant, C. Tarnocai, and J. Dumanski

Research Branch
Agriculture and Agri-Food Canada
Ottawa, Ontario, Canada

ABSTRACT

Agroecosystems emit about 10 to 13% of the greenhouse gases in Canada. This percentage could be reduced, without producing undue hardship to farmers, by either selecting management practices that reduce emissions of CO_2, CH_4 and N_2O or by selecting measures that increase the capacity of soils to sequester C. Examples of possibilities to increase C sequestration in agricultural soils are presented. It is shown that, based on the Century model, a shift from wheat (*Triticum aestivum* L.) wheat fallow (WWF) in 1990 to continuous wheat (Cont W) on 3.1 million hectares (Mha) in the prairies would sequester 8 Tg C by 2010. The Century model is also used to show that the gradual adoption of no-till across Canada is converting agricultural soils from a source of CO_2 to a sink. A study of the amount of C sequestered in agricultural soils across Canada due to a shift from conventional tillage to no-till demonstrated that on the average the observed impact of no-till on C sequestration was 1.6 times larger than predicted by the Century model.

The increase in concentration of greenhouse gases such as CO_2, CH_4 and N_2O is of great environmental concern, both nationally and globally because of the potential impact of these gases on climate. As a result, 174 countries recently signed an agreement in Kyoto to set specific targets for reducing greenhouse gas emissions. Canada promised to reduce its greenhouse gas emissions by 6% below the 1990 level by the years 2008 to 2012. However, with business as usual it is generally accepted that, by 2010, Canada's greenhouse gas emissions will be about 20% higher than projected, that is the emissions will be about 180 Mt CO_2 equivalent higher than the 1990 level.

Most of the surplus is due to increases in CO_2, CH_4 and N_2O emissions. The potential effect of these gases is summed up using the respective global warming potential of each gas. The global warming potentials are a function of the absorption characteristics of the gas and the average lifetime of the gas. For a 100-yr time horizon, CH_4 and N_2O have global warming potentials 21 and 310 times greater than CO_2 (IPCC, 1996).

Large quantities of CO_2 are absorbed by vegetation. Some rough estimates indicate that approximately 500 Tg CO_2 is absorbed each year by agricultural crops in Canada. If a fraction of the CO_2 that plants absorb could be immobilized in the soil, then C sequestration would help reduce the greenhouse gas buildup in the atmosphere. The soil C is a function of the difference between inputs to soil of photosynthetically fixed C and losses of soil C via decomposition and respiration. The balance between these two processes is affected by the interaction of climate, soil type and land use (Cole et al., 1996). In this chapter, we present the greenhouse gas emissions estimates for Canada due to agricultural sources and we discuss potential changes in the greenhouse gas balance. A long list of possibilities for changing soil C is presented, but because of space limitation we focus on the impact of reduced summer fallow and reduced tillage.

Greenhouse Gas Emissions from Agroecosystems

The greenhouse gas emissions from agroecosystems are given in Table 9–1. A more comprehensive explanation of these numbers is available in recent publications (Smith et al., 1997; Desjardins & Riznek, 2000; Monteverde et al., 1997; Janzen et al., 1999). In the cases without inputs, CO_2 emissions are primarily from soils, CH_4 from livestock and manure and N_2O from the N associated with manure, fertilizers, biological fixation and crop residues. In the cases with inputs, the CO_2 produced by fossil fuel used on farm and CO_2 emitted during the production of inputs such as fertilizers and agrochemicals, and the CO_2 emitted during the production of machinery is also included. The first set of numbers represents about 10% of the anthropogenic annual emissions in Canada, while the second set represents about 13% of the emissions. These two sets of numbers show the relative magnitude of the emissions of CO_2, CH_4, and N_2O.

In contrast to other biomes, agroecosystems are "interventionist" in the sense that human decisions can change an agroecosystem from a source to a sink or vice versa. This fact makes agroecosystems very important, even though they are not a large component of the terrestrial biosphere. Because this chapter examines the pos-

Table 9–1. Greenhouse gas emissions from Canadian agroecosystems in teragrams CO_2 equivalent.†

	1981	1986	1991	1996
		Without inputs		
CO_2	8	7	5	2
CH_4	22	19	20	23
N_2O	31	30	31	37
Total	61	56	56	62
		With inputs		
CO_2	30	28	26	26
CH_4	22	19	20	23
N_2O	31	30	31	37
Total	83	77	77	86

† Emissions due to food processing are not included.

sibilities for changing the greenhouse gas balance due to changes in soil C stock, it is also very important to consider the magnitude of N_2O and CH_4 emissions— particularly N_2O emissions, which can increase substantially as a result of efforts to increase C sequestration (Janzen et al., 1999).

Soil Carbon Change in Agricultural Soils

Agriculture has transformed large areas from natural ecosystems to agroecosystems under extensive and intensive management. This has had major impacts on C pools and on the increase in atmospheric CO_2. In Canada, based on pair sampling, Dumanski et al. (1998) have reported that agricultural soils have lost 0.77 Pg C in the first 30 cm since cultivation began. This is equivalent to a loss ranging from 13 to 35% depending on the soil types. It is also equivalent to 2820 Mt CO_2. This loss has caused an increase of 0.4 ppmv in the global atmosphere CO_2 concentration. The C loss estimate, which was obtained by measuring the soil C at thousands of locations, is still based on a relatively small sample. In order to obtain more representative estimates of the change in soil C, it is essential to use models.

The site specific Century model that makes use of simplified relationships of the soil-plant-climate interactions to describe the dynamics of soil, C has extensively been tested in recent years (Parton et al., 1987; Smith et al., 1997). Figure 9–1 shows the estimated cumulative loss of soil C from agroecosystems in Canada in the first 30 cm from 1910 to 2010. This diagram represents the aggregated results of thousands of model runs for the major cropping systems and soil management practices in Canada. It shows that agricultural soils have lost about 1050 Tg C since 1910, this is equivalent to about 3850 Mt CO_2. The rate of change of C has slowed considerably since the 1970s. This means that land degradation has also slowed considerably. Figure 9–2 shows the numerous factors that can lead

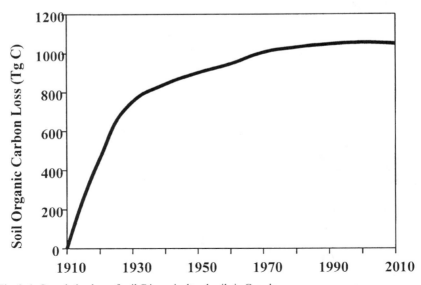

Fig. 9–1. Cumulative loss of soil C in agricultural soils in Canada.

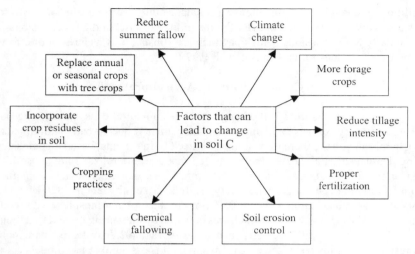

Fig. 9–2. Factors that can influence the C content in agricultural soils.

to a change in soil C, such as planting more forage crops, implementing soil conservation practices, climate change, etc. (Lal et al., 1998). Soil erosion control measures, which minimizes the impact of either wind, water or tillage practices, is another method of conserving soil organic matter (Smith et al., 2000).

Many of the agricultural measures that can be used to control soil erosion are listed on Fig. 9–3. These either slow down the depletion of soil organic matter or contribute to an increase in soil C storage. Any of these measures could have been selected for further study. We focus on the impact of reduced summer fallow and

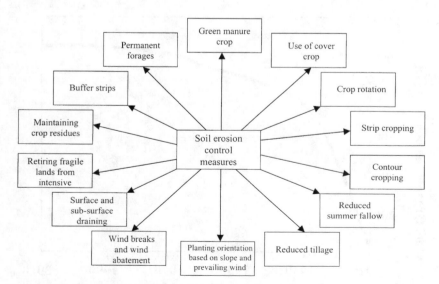

Fig. 9–3. Soil erosion measures that can reduce CO_2 emissions.

Table 9–2. Most probable distribution of land uses where reductions in summer fallow is possible (50-yr scenario).

	Current summer fallow area	Most probable future summer fallow area	Additional land for continuous crops due to summer fallow reduction
		Mha	
Chernozemic (brown–dark brown)	4.4	2.7	1.7
Chernozemic (black)	1.9	0.8	1.1
Luvisolic (gray–dark gray)	0.5	0.2	0.3

reduced tillage on the soil C stock and how a change in these practices can affect the greenhouse gas balance of agroecosystems in Canada.

Impact of a Reduction in Summer Fallow on the Greenhouse Gas Balance

Leaving land unplanted for a growing season (summer fallowing) is frequently done in the brown, dark brown, and black soils in the prairies to either replenish soil moisture or control weeds and diseases. This is still a common practice but the area of land under summer fallow has steadily decreased in recent years. Between 1981 and 1991, the total agricultural area in the prairies under summer fallow decreased from 31 to 24%. Since 1991, an additional 6% reduction has been reported in the 1996 Census. The summer fallow area is expected to stabilize at 3.7 Mha in the future, that is 3.1 Mha less than in 1990 (Dumanski et al., 1998).

The Century model was used to simulate the effects of a probable shift from WWF to Cont W on the major soil types identified in Table 9–2. These data represent the area as a function of soil type where summer fallow reduction could be safely implemented. Three representative polygons of each soil type were selected to conduct simulations using the Century model. A wheat fallow rotation was assumed from 1910 to 1990 and then we introduced either WWF or Cont W until 2050 (Fig. 9–4).

The cumulative gain in soil C from black chemozemic, brown and dark brown chemozemic and dark gray luvisolic soils are shown in Fig. 9–5. A gain of 8 Tg C is predicted by 2010. This graph is derived from the total area for the various soil types and the soil C change estimated for the various soil types. It should be understood that by selecting Cont W we are simply assuming continuous cropping. In the Brown soils, Cont W is a reasonable cropping system but extra efforts are required to enhance available water, such as leaving long stubble to trap the snow, etc.

In the black soils, a monoculture system like Cont W would result in undesirable diseases and weed pressures. Producers use mixed rotations that include oilseeds [such as canola (*Brassica napus* L.)] and pulse crops together with cereals [wheat, barley (*Hordeum vulgare* L.), etc.). Such mixed rotations are likely to result in even greater soil C sequestration than predicted for Cont W.

Fallowing on gray luvisols is not frequently done. In 1990 it was practiced on 0.5 Mha and it is assumed that about 0.3 Mha could be planted with continuous

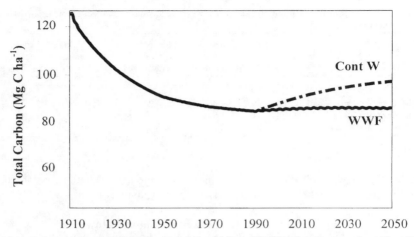

Fig. 9–4. Estimated change in soil C for wheat wheat fallow (WWF) and continuous wheat (Cont W) for a black chernozenic soil.

crop. These soils are inherently low in organic matter but can support high production and therefore they have a great potential to increase organic matter. In such soils, farmers usually use rotations that have a high proportion of forage crops that result in higher increases in soil organic matter than Cont W. Hence the estimated 8 Tg C is probably an underestimation of the potential.

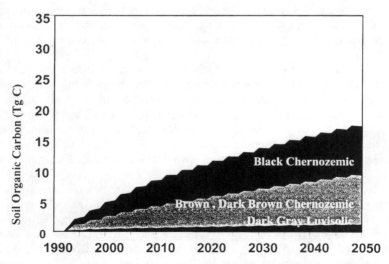

Fig. 9–5. Cumulative gain in soil C due to a shift in cropping practice in 1990 from wheat wheat fallow (WWF) to continuous wheat (Cont W).

Table 9–3. Average no-till agriculture as a percentage of total area within the province or country.

Province/national	1991	1995	2000	2005	2010
			No-till (%)		
Atlantic	2	2	2	2	2
Quebec	3	4	7	9	11
Ontario	4	15	20	20	20
Manitoba	5	8	12	15	20
Saskatchewan	10	20	30	35	38
Alberta	3	9	17	23	28
B.C.	3	9	13	16	20
Canada	7	14	22	26	30

Impact of a Reduction in Tillage Intensity
on the Greenhouse Gas Balance

Tillage was once necessary to control weeds and prepare soil for planting. With technical advances this is no longer necessary and a growing number of farmers are using no-till or direct-seeding practices. Table 9–3 presents actual percentage of the adoption of no-till in 1991 and 1996 based on the Census data. Projected increases in the use of no-till for up to 2010 are also given at the provincial level.

The predicted rate of C gain using the Century model is given for the gradual adoption of no-till (Fig. 9–6). These data show that on the average C losses from agricultural soils have slowed considerably and they are in the process of reversing. Soil C pools have either stabilized or are beginning to increase.

Assuming a full adoption of no-till in 1990, Fig. 9–7 shows that the predicted cumulative increase of C by 2010 would have been about 30 Tg C. This rate is equivalent to about 0.07 Mg C ha^{-1} yr^{-1} for about 20 yr. This is an upper limit that is independent of climate and cultural practices. The 0.07 value compares favorably with

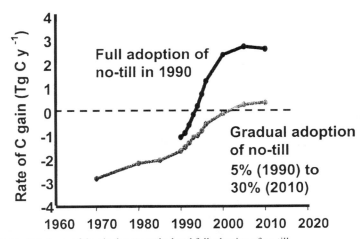

Fig. 9–6. Predicted rate of C gain due to gradual and full adoption of no-till.

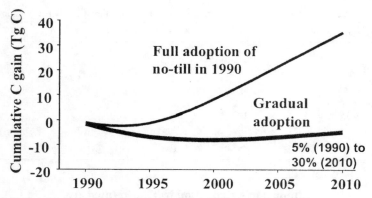

Fig. 9–7. Predicted cumulative C gain due to gradual and full adoption of no-till.

Bruce et al. (1998) estimates of 0.2 Mg C ha^{-1} yr^{-1} using best management practices. It shows that it is feasible to sequester soil C but that substantial change in
management practices will be required if C sequestration is to significantly change
the greenhouse gas balance.

In order to evaluate the accuracy of the predicted results of no-till by the Century model, we have calculated the predicted C sequestration for a wide range of
soils across Canada for a 20-yr period. We compare these estimates to actual soil
C change observed under no-till management. The duration of the no-till experiments ranged from 4 to 18 yr (Janzen et al., 1997). The duration of these trials were
all shorter than the 20 yr in the simulation. Omitting the two extreme data points
in the trials, the predicted mean value from the Century model was 1.2 Mg C ha^{-1}
yr^{-1} while the observed mean value was 1.7 Mg C ha^{-1} yr^{-1} (Fig. 9–8). This means
that the predicted cumulative gains in Fig. 9–7 might be low and should be increased
by a factor of about 1.6.

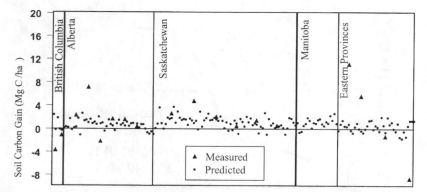

Fig. 9–8. Comparison of measured and predicted change in soil C across Canada due to no-till. Note
that the locations of the measured data points do not correspond with the location of the predicted
values.

SUMMARY

In this chapter we presented a long list of management practices that could influence the greenhouse gas balance of agroecosystems by changing the C sequestration potential of soils. Before recommending any of these practices many factors must be considered such as the feasibility, cost, influences on soil quality and the whole environment. We limited our analysis to the impact of reduced summer fallow and reduced tillage on C sequestration. We showed that if in 1991 3.1 Mha under WWF were converted to Cont. W, the C stock in these soils would increase by 8 Tg C by 2010. Because of the close interactions between C and N cycles, any study dealing with soil C sequestration must also consider the potential increases in N_2O emissions. We have not examined this in this chapter. We also recognize that the quantity of C that can be sequestered in soils has an upper limit that is dependent on climate and cultural practices.

REFERENCES

Bruce, J.P., M. Frome, E. Haites, H. Janzen, R. Lal, and K. Paustian. 1998. Carbon sequestration in soils. J. Soil Water Conserv. 54:382–389.

Campbell, C.A., R.P. Zentner, J.G. Dormaar, and R.P. Voroney. 1986. Land quality, trends and wheat production in western Canada. p. 318–353. In A.E. Slenkard and D.B. Fowler (ed.)Wheat production in Canada: A review. Div. Ext. Commun. Relat., Univ. Saskatchewan, Saskatoon, SK.

Cole, V., C. Cerri, K. Minami, A. Mosier, N. Rosenberg, and D. Sauerbeck. 1996. Agricultural options for mitigation of greenhouse gas emissions. p. 745–772. In Climate change 1995. Contrib. Work. Group II, 2nd Assessment Rep. Intergovern. Panel Climate Change.

Desjardins, R.L., and R. Riznek. 2000. Agricultural greenhouse gas budget. p. 133–142. In T. McRae et al. (ed.) Environmental sustainability of Canadian agriculture: Report of the Agri-Environmental Indicator Project. Catalogue no. A22-201/2000E. Agriculture and Agri-Food Canada, Ottawa, ON, Canada.

Dumanski, J., R.L. Desjardins, C. Tarnocai, C. Monreal, E. Gregorich, C.A. Campbell, and V. Kirkwood. 1998. Possibilities for future carbon sequestration in Canadian agriculture in relation to land use change. Climate Change 40:81–103.

Janzen, H.H., C.A. Campbell, E.G. Gregorich, and B.H. Ellert. 1997. Soil carbon dynamics in Canadian agroecosystems. p. 57–80. In R. Lal et al. (ed.) Soil carbon dynamics in Canadian agroecosystems. Soil processes and carbon cycles. CRC Press, Boca Raton, FL.

Janzen, H.H., R.L. Desjardins, R. Asselin, and B. Grace. 1999. The health of our air: Towards sustainable agriculture in Canada. Catalog. no. A53-1981/1998E. Res. Branch, Agric. Agri-Food Canada, Ottawa.

Lal, R., J.M. Kimble, R.F. Follett, and C.V. Cole. 1998. The potential of U.S. cropland to sequester carbon and mitigate the greenhouse effect. Ann Arbor Press, Ann Arbor, MI.

Intergovernmental Panel on Climate Change. 1996. Climate change 1995: The science of climate change. Tech. Summ. Work. Group I. Cambridge Univ. Press, Cambridge, UK.

Monteverde, C.A., R.L. Desjardins, and E. Pattey. 1997. Agroecosystem greenhouse gas balance indicator: Nitrous oxide component. Rep. 20. Agri-Environ. Indicat. Proj. Agric. Agri-Food Canada, Ottawa, Canada.

Parton, W.J., D.S. Schimel, C.V. Cole, and D.S. Ojima. 1987. Analysis of factors controlling soil organic matter levels in Great Plains grasslands. Soil Sci. Soc. Am. J. 51:1173–1179.

Smith, W.N., P. Rochette, C. Monreal, R.L. Desjardins, E. Pattey, and A. Jaques. 1997. The rate of carbon change in agricultural soils in Canada at the landscape level. Can. J. Soil Sci. 77:219–229.

Smith, W.N., G. Wall, R.L. Desjardins, and B. Grant. 2000. Soil organic carbon. p. 85–92. In T. McRae et al. (ed.) Environmental sustainability in Canadian agriculture: Report of the Agri-Environmental Indicator Project. Catalogue no. A22-201/2000E. Agriculture and Agri-Food Canada, Ottawa, ON, Canada.

10 Forest Inventory Data, Models, and Assumptions for Monitoring Carbon Flux

R. A. Birdsey and L. S. Heath

USDA Forest Service
Newtown Square,, Pennsylvania

ABSTRACT

Estimates from forest inventories indicate that U.S. forest ecosystems are a net sink for C, currently removing 0.2 to 0.3 Pg C yr^{-1} from the atmosphere. Estimates of C storage in forest biomass are based on comprehensive forest inventory data collected periodically by the USDA, Forest Service. Estimates of C storage in forest soils (including mineral soil, litter and coarse woody debris) are based on intensive ecosystem studies and models relating soil C to climate variables, forest type, and land use history. Estimates of C storage in wood products are based on comprehensive models of biomass utilization, recycling, and disposal. Opportunities to improve the estimation process include better estimates of soil C responses to land use and environmental changes, and the need for more comprehensive data to estimate C flux on remote forest lands that are not significantly influenced by human activities.

Estimates of C flux based on periodic national-scale forest inventory data show that increases in biomass and organic matter on U.S. forest lands added an average of 0.3 Pg yr^{-1} of stored C to forest ecosystems from 1952 to 1992, enough to offset 25% of U.S. emissions for the period (Birdsey & Heath, 1995). Projections show additional increases of approximately 0.2 Pg yr^{-1} through 2040. About 60% of the projected increase is estimated to be soil C. The projected baseline includes forest policies in effect at the time the projections were made; in particular, reduced harvest levels on National Forest lands, decreased clearcutting and increased partial cutting, and continuation of federal tree-planting programs at recent historical levels. In the 1980s, U.S. forest trees (excluding other forest ecosystem components) were accumulating C at an annual rate of 0.461 Pg yr^{-1}. Removals of C from the forest due to timber harvesting and land clearing totaled 0.355 Pg yr^{-1}, for an annual net gain of 0.106 Pg yr^{-1} in U.S. forest trees (Birdsey, 1992). Turner et al. (1995) used a similar methodology to estimate an annual accumulation of C in forest trees of 0.331 Pg yr^{-1}, and removals of 0.266 Pg yr^{-1}, for a net annual gain of 0.79 Pg yr^{-1}. Schroeder et al. (1997) presented yet another variation on the methodology for using forest inventory data to estimate C in forest biomass. For the two broadleaf forest types studied, they developed estimates that were about 30% higher than those reported by Birdsey (1992) and in the pioneering work of Johnson and Sharpe (1983).

There are significant regional differences in past and projected C storage. These differences reflect long-term changes in land use and harvesting. Millions of acres of forests in the Northeast have regrown on abandoned agricultural land, causing a steep historical increase in C including a substantial buildup on C-depleted soils. The regrowing forests have restored much of the soil C lost when the land was used for crops or pasture. These regrowing forests are now maturing, and there-fore the rate of C buildup is expected to slow substantially. The historical pattern is similar in the South, but the intensive utilization of southern forests for wood prod-ucts over the last few decades has effectively halted regional gains in C, as growth and removals have come into rough balance. In the Pacific Coast states, C stocks are expected to increase after a recent decline, mainly due to reduced harvest on public lands as more forest land has been reserved from timber production.

Forest soils comprise a huge reservoir of C. An average of 61% of the C in forest ecosystems of the USA is found in soils, with substantial regional variabil-ity (Fig. 10–1). These estimates are consistent with other analyses that show the high-est percentages of soil C in high-latitude forests, and the lowest percentages in low latitude forests (Dixon et al., 1994). About 8% of forest ecosystem C is in litter and coarse woody debris on the soil surface. The total amount of soil C in U.S. forests was estimated to be 33 Pg in 1992. On timberland, which is two-thirds of all for-est land, soil C has increased from 14.4 Pg in 1952 to 20.8 Pg in 1992, an average annual increase of 0.16 Pg yr^{-1}. This rate of increase is sufficient to have offset about 14% of U.S. CO_2 emissions for the period.

An important feature of the national forest inventory is that the sampling in-tensity and frequency of re-measurement is greatest in areas that are most affected

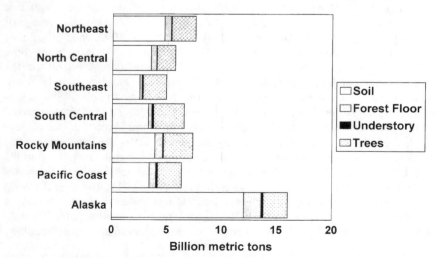

Fig. 10–1. Average carbon storage in forests of the U.S. by region and ecosystem component (from Bird-sey & Heath 1995).

by human activities such as harvesting and land use change. Consequently, the forest inventory represents anthropogenic influences on forests better than it represents forests with little direct human influences. Because field measurements are focused on above-ground biomass, national forest inventories cannot provide all needed statistics about C in forests. Nevertheless the high spatial resolution and detailed tree measurements form a good statistical foundation that can be enhanced with other sources of data. For example, current C flux estimation procedures rely heavily on data from intensive-site ecosystem studies for estimating relationships between below-ground C and above-ground observations. A potential future enhancement to the estimation procedure is to use data from CO_2 flux measurement towers to enhance the temporal resolution of the C flux estimates.

Many other countries have comprehensive national forest inventories and have used these inventories as a basis for estimating changes in carbon storage. For example, Kauppi (1992) estimated the magnitude of the European C sink, Kurz and Apps (1993) did the same for Canada, and Kolchugina and Vinson (1995) estimated C storage and flux for Russia.

OVERVIEW OF FORCARB AND ASSOCIATED MODELS

The C budget of forest ecosystems of the USA is estimated using a core model, FORCARB, and several subroutines that calculate additional information, including C in wood products (Plantinga & Birdsey, 1993; Birdsey et al., 1993; Birdsey & Heath, 1995; Heath et al., 1996). A conceptual model of the relationships among the various components of the forest C budget is presented in Fig. 10–2. FORCARB is part of an integrated model system consisting of an area change model (Alig, 1985), a timber market model (TAMM, Adams & Haynes, 1980), a pulp and paper model (NAPAP-Ince, 1994) and an inventory projection model (ATLAS, Mills & Kincaid, 1992). Through linkage with these models, FORCARB projects changes in C storage in private forests as a function of management intensity and land use change. A spreadsheet version of FORCARB, unlinked with economic models, is used for public forest lands managed primarily through a policy and planning process, and for forest land not meeting the minimum productivity and land use criteria for timberland (formerly called "productive" forest).

The current version of FORCARB partitions C storage in the forest into four separate components: trees, soil, forest floor, and understory vegetation. A new version under construction adds additional ecosystem partitions for coarse woody debris and standing dead wood. The definitions of these components are broad enough to include all sources of organic C in the forest ecosystem. The tree portion includes all above-ground and below-ground portions of all live and dead trees, including the merchantable stem, limbs, tops, cull sections, stump, foliage, bark and rootbark, and coarse tree roots (>2 mm). The soil component includes all organic C in mineral horizons to a depth of 1 m, excluding coarse tree roots. The forest floor includes all dead organic matter above the mineral soil horizons except standing dead trees: litter, humus, and other woody debris. Understory vegetation includes all live vegetation other than live trees.

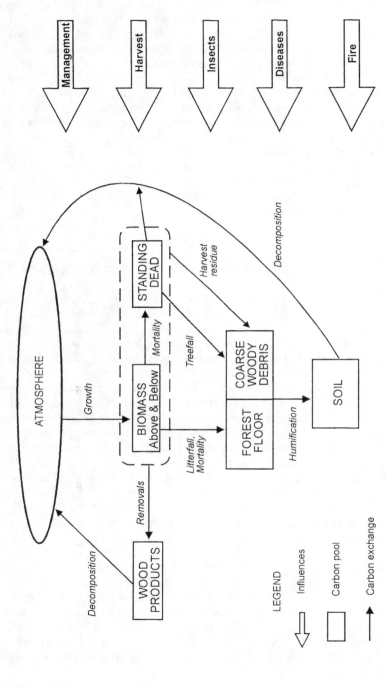

Fig. 10–2. Conceptual model of the net carbon budget of U.S. forest ecosystems.

Using data from forest inventories and intensive-site ecosystem studies, estimates of average C storage by age or volume classes of forest stands (analogous to a forest yield table) are made for each ecosystem component for forest classes defined by region, forest type, productivity class, and land use history. Equations are derived to estimate C storage in the forest floor, soil, and understory vegetation for each forest class. Additional details about estimating C storage for different regions, forest types, site productivity class, and past land use are provided in Birdsey (1996). These equations are then applied to projections of growing stock inventory and increment, harvested area and volumes, and timberland area obtained from ATLAS.

The C pools of wood from timber harvests on both private and public lands are estimated with a model based on the work of Row and Phelps (1991). Carbon pools from forest harvests before 1980 are estimated using a similar method (Heath et al., 1996). There are four disposition categories: products, landfills, energy, and emissions. Products are goods manufactured or processed from wood, including lumber and plywood for housing and furniture, and paper for packaging and newsprint. Landfills store C as discarded products that eventually decompose, releasing C as emissions. Emissions also include C from wood burned without generation of usable energy, and from decomposing wood. Energy is a separate category from emissions because wood used for energy may be a substitute for fossil fuels.

Potential changes in forest productivity as a consequence of environmental change, as estimated with an ecosystem process model, have been used as input to the integrated FORCARB/economic models to estimate the impacts of climate change on timber production and the national C budget (Joyce et al., 1995). The key to successful integration of the different classes of models has been an attempt to use consistent assumptions about how changes in productivity affect volume and biomass.

FOREST ECOSYSTEM DATABASES

An extensive and comprehensive forestry data collection, management, and reporting system underlies FORCARB and related models and analyses (Powell et al., 1993). The comprehensive national inventory of forest lands began in the USA in the 1930s. By the early 1950s, all states had been inventoried at least once, with the exception of interior Alaska. Currently, states are inventoried on a cycle of about 10 yr, with national statistics compiled every 5 yr. Recent compilations of national statistics are for the years 1987, 1992, and 1997. For long-term trend monitoring and projections, 5-yr intervals are sufficient and are consistent with proposals to report greenhouse gas emissions in 5-yr periods. Ongoing changes in the way national forest inventories are implemented will facilitate annual reporting of basic statistics, which in turn will facilitate reporting of C flux on an annual basis.

Since World War II, U.S. forest inventories have used multiphase sampling designs involving remote sensing and ground measurements (Schreuder et al., 1995). The phase one sample typically consists of interpretation of high-altitude color infrared photography, widely available and highly accurate for estimating changes in forest area and locating field sample plots. Interpreters classify more than 3 000 000 sample points nationally to monitor activities such as timber harvest and land use that may change the photo classification from forest to nonforest cover. Current research involves techniques to switch from high-altitude photography to satellite imagery for the first sample phase.

The phase two sample consists of more than 150 000 permanent field sample locations that are re-measured periodically to provide statistics on disturbance (harvest and mortality), growth, species composition change, and a host of observed and calculated site descriptors such as ownership and forest type. At each sample location, a rigorous protocol is followed to select and measure a representative sample of trees. These measurements are then expanded to the population level using the statistics from the phase one sample.

A subsample of the phase two plots (known as "Forest Health Monitoring" plots) is the basis for more intensive ecosystem measurements. Soils, coarse woody debris, understory vegetation, and other ecological variables may be collected on this subsample, which is linked statistically to the phase 1 and 2 samples. The subsample of phase 2 consists of approximately 5000 sample plots. Successive measurements have been initiated on about one-half of this subsample of field plots.

The most comprehensive ecosystem measurements available are from intensive, long-term ecosystem studies such as those comprising the Long Term Ecological Research (LTER) site network. The LTER and similar sites typically have a rather long history of repeated measurements of a common and comprehensive suite of ecological variables, including those such as soil and litter C that are unavailable from extensive statistical sampling networks. Unlike national forest inventories, intensive studies are concentrated on relatively undisturbed sites.

ESTIMATING CARBON IN TREES

The quantity of C in live and dead trees is derived from volume and biomass estimates of the national forest inventory. Methods to estimate volume, biomass, and the components of change (growth, removals, and mortality) are reviewed in Birdsey and Schreuder (1992). Estimates of growing-stock volume are converted to tree C in a two-stage process. First, total tree volume is estimated from growing-stock volume (the merchantable part of trees) using a ratio to account for the additional tree parts excluded from the definition of growing-stock: tops and branches, rough and rotten trees, small trees (less than 5.0 in. dbh, where dbh = diameter at breast height), standing dead trees, stump sections, roots, and bark. A factor is added to account for C in foliage. Separate ratios are computed for softwoods and hardwoods to account for differences in the average ratio of total volume to growing-stock volume. Ratios are derived from two principal sources: a nationwide biomass study prepared by the USDA, Forest Service containing estimates of above-ground biomass by tree component (Cost et al., 1990), and a special report prepared by Koch (1989, personal communication) containing estimates of the proportion of below-ground tree volume. Separate ratios are derived for each region of the USA to account for differences in tree form, and to be consistent with regional data used to develop yield tables for timber projection models. The validity of this method rests on the assumption that the ratio of total above-ground biomass to merchantable biomass (estimated in dry weight units) is equivalent to the ratio of total above-ground volume to merchantable volume. There is considerable variation in the ratios of total to merchantable volumes among regions and species groups (Birdsey, 1992). For the USA as a whole, the average ratio of total to merchantable volume for softwoods is 1.91 and for hardwoods is 2.44.

The second step involves converting total tree volume in cubic feet to C in pounds. Separate factors are used for major forest types and for softwoods and hardwoods within each forest type, and for broad geographical regions. The volume-to-C conversion factor is computed in two steps. First, volume in cubic feet is converted to biomass in dry pounds by multiplying the number of cubic feet times the mean specific gravity times the weight of a cubic foot of water (28.3 kg or 62.4 lb.). A weighted mean specific gravity for softwoods or hardwoods is estimated from the relative frequency of the three predominant hardwood or softwood species in each forest type and region. The second step is to multiply the biomass in dry pounds by a factor to account for the average C content of the tree. Estimates of the C content of trees used in past studies have generally ranged from 45 to 50% (Houghton et al., 1985); however, Koch (1989, personal communication) found that, for the USA as a whole, the average percentage C for softwoods was 52.1 and for hardwoods was 49.1, with some slight regional variations. The final factors used to convert volume to C for U.S. forest types range from 11.41 to 17.76 for softwoods, and from 11.76 to 19.82 for hardwoods (Birdsey, 1992). A separate set of conversion factors for pure stands of plantation species was also developed (Birdsey, 1996).

ESTIMATING CARBON IN SOILS

Carbon in soil and litter is estimated with models that relate the quantity of organic matter to temperature, precipitation, age class, and land use history, using data from ecosystem studies compiled by various authors. The approach follows that of Burke et al. (1989), who used a multiple regression procedure to find the best predictive equations for soil organic C in cropland and grassland in the Central Plains and adjacent areas. The data in Post et al. (1982) were used to estimate regression coefficients for a similar, compatible model for forest lands. The methodology is explained in more detail in Plantinga and Birdsey (1993) and Birdsey (1992).

The estimates of soil C developed by Post et al. (1982) for temperate forests and used to derive estimates for the USA represent relatively undisturbed, secondary forests. These estimates are considered reference points and are used to generate simple functions to describe the changes in soil C associated with harvesting and land use change. Diagrams depicting the different cases of harvesting and land use change are presented in Plantinga and Birdsey (1993).

Because we lack comprehensive statistical data bases of soil C linked with above-ground measurements, which could be used to derive empirical estimates of soil C changes from harvesting and land use, we develop a series of assumptions based on continuing literature reviews. The most recent compilation of our assumptions is presented in Heath and Smith (2000). In general, we use assumptions about: (i) soil C at initial conditions, (ii) age associated with the reference estimates for mature secondary forests, (iii) rate of transition from initial conditions to reference conditions, and (iv) changes after reference conditions are attained. The literature is inconclusive about many aspect of soil C dynamics. For example, Johnson (1992) found a variety of soil C responses to harvesting, with some studies showing a soil C decrease and other studies a soil C increase.

We based our assumptions on those made by Houghton et al. (1983, 1985), modified to include findings reported in recent literature. For the South, we assumed that clearcut harvest was followed by intensive site preparation, producing a 20% loss of the soil C by age 10. For less intensive harvesting such as partial cutting, or regeneration methods that exclude soil disturbance, no soil C loss is estimated. Changes between reference points are assumed to be linear, and the rate of change after reaching reference levels (assumed to be 50 yr) is reduced linearly to zero. For other regions besides the South, loss of soil C after clearcut harvest is assumed to be zero, resulting in a constant level throughout the yield period.

Tree plantations or natural vegetation established on agricultural land with depleted organic matter can cause a substantial accumulation of soil organic matter, depending on species, soil characteristics, and climate (Johnson, 1992). For example, *Populus* spp. established on sandy soils showed large increases in soil and forest floor C due to high litter production (Dewar & Cannell, 1992).

For replanted pasture in all regions, soil C at age 0 was the higher of either: (i) the level estimated with the equation from Burke (1989), or (ii) two-thirds of the average for secondary forests at the reference age. For replanted cropland in all regions, soil C at age 0 was the higher of either: (i) the level estimated by Burke et al. (1989), or (ii) one-half of the average for secondary forests at the reference age. It was assumed that soil C would increase linearly from the lowest level to the reference age. In all cases after reaching the reference age, the rate of accumulation of soil C was assumed to decline as the forest matured.

ACCOUNTING FOR LAND USE CHANGE

Because land use leaves a long-term legacy of changes in soil C, accurate accounting for historical land use, even on land that has been in forest for decades, is essential for estimating the rate of current soil C change using the assumptions described above. Unfortunately, available data that can be tied directly to the national forest inventory is sparse and only goes back, at best, a few decades. The USDA "Census of Agriculture" reports, and periodic statewide forest inventory reports compiled since the 1930s, can be used to approximate the percentage of current forest land that is derived from other land uses, by owner, for different regions of the USA. These percentages are used in FORCARB to estimate weighted average soil C changes using the assumptions described in the previous section for forest land with different land use histories.

By explicitly tracking the changes to and from forest use, estimates of the changes in C due to land use change relative to estimates of changes in C due to harvesting and management can be compiled. In one simulation for the USA, the assumed accretion of C in soils on forest land that was previously in agricultural use accounted for approximately 0.03 Pg yr^{-1} of the average projected C increase of 0.2 Pg yr^{-1} (Heath & Alig, 1998, unpublished data).

ESTIMATION ERRORS AND DATA GAPS

It is a long and complicated process to compile large inventory datasets and review copious literature to contruct a comprehensive C budget for a large area such as the USA. Even with the large amount of data available, many assumptions and simplified models are needed to present a complete estimate for all components of forest ecosystems. Yet it is instructive to make the attempt, not only because of the widespread interest in such estimates, but also because the process facilitates identification of the data gaps and model development needed to make truly verifiable estimates of C flux for monitoring policy outcomes.

The most comprehensive and accurate regional estimates of C flux using inventory data are for above-ground biomass. However, there are significant gaps in data for areas that are not inventoried frequently, such as Interior Alaska. There are also estimation errors of the regression models used to estimate tree biomass from field measurements, in addition to sampling and measurement errors (which are typically very small).

Data on soil and litter C are from ecosystem studies reported in the literature that are not usually part of a regional statistical sample. Therefore, regional estimates include unknown estimation errors when such data is extrapolated using empirical models. In addition, for many long-term but suspected significant changes in quantities of soil C, we use assumptions that are logical and peer-reviewed, but that are still untested hypotheses.

Some important progress has been made in applying the principles of uncertainty analysis (Smith & Heath, 2000) and error analysis (Phillips et al., 2000) to overcome some of the difficulty in evaluating the results of our estimation process, and determining where resources should be allocated to make significant improvements. Another way to validate C estimates using our inventory approach for large regions is to compare the estimates with those of completely different estimation procedures, for example, methods based on atmospheric sampling networks (e.g., Fan et al., 1998). However, there are significant errors in alternate estimation approaches as well, so that it is impossible to determine the relative validity of different estimates.

Finally, the inventory approach does not include all factors of environmental change. For example, atmospheric deposition of nitrogen compounds to forest soils affects soil C dynamics and perhaps the allometric relationships used to estimate tree biomass, yet deposition effects are not considered. In the future, such factors could be addressed by linking the current integrated modeling system with process models that model key dynamic factors.

REFERENCES

Adams, D.M., and R.W. Haynes 1980. The timber assessment market model: Structure, projections, and policy simulations. For. Sci. Monogr. 22. Soc. Am. For., Bethesda, MD.

Alig, R.J. 1985. Modeling area changes in forest ownerships and cover types in the Southeast. Res. Pap. RM-260. USDA, For. Serv., Rocky Mount. For. Range Exp. Stn., Ft. Collins, CO.

Birdsey, R.A. 1992. Carbon storage and accumulation in United States forest ecosystems. Gen. Tech. Rep. WO59. USDA, For. Serv., Washington, DC.

Birdsey, R.A., and H.T. Schreuder. 1992. An overview of forest inventory and analysis estimation procedures in the Eastern United States. USDA, For. Serv., Rocky Mount. For. Range Exp. Stn., Ft. Collins, CO.

Birdsey, R.A., A.J. Plantinga, and L.S. Heath. 1993. Past and prospective carbon storage in United States forests. For. Ecol. Manage. 58:33–44.

Birdsey, R.A., and L.S. Heath. 1995. Carbon changes in U.S. forests. p. 56–70. In L.A. Joyce (ed.) Productivity of America's forests and climate change. Gen. Tech. Rep. RM-271. USDA, For. Serv., Fort Collins, CO.

Birdsey, R.A. 1996. Carbon storage for major forest types and regions in the conterminous United States. p. 23–39 and 261–308. In R.N. Sampson and D. Hair (ed.) Forests and global change. Vol. 2. Am. For., Washington, DC.

Burke, I.C., C.M. Yonker, W.J. Parton, C.V. Cole, K. Flach, and D.S. Schimel. 1989. Texture, climate, and cultivation effects on soil organic matter content in U.S. grassland soils. Soil Sci. Soc. Am. J. 53:800–805.

Cost, N.D., J. Howard, B. Mead, W.H. McWilliams, W.B. Smith, D.D. Van Hooser, and E.H. Wharton. 1990. The biomass resource of the United States. Gen. Tech. Rep. WO-57. USDA, For. Serv., Washington, DC.

Dewar, R.C., and M.G.R. Cannell. 1992. Carbon sequestration in the trees, products and soils of forest plantations: An analysis using UK examples. Tree Physiol. 11:49–71.

Dixon, R.K., S. Brown, R.A. Houghton, A.M. Solomon, M.C. Trexler, and J. Wisniewski. 1994. Carbon pools and flux of global forest ecosystems. Science (Washington, DC) 263:185–190.

Fan, S., M. Gloor, J. Mahlman, S. Pacala, J. Sarmiento, T. Takahashi, and P. Tans. 1998. A large terrestrial carbon sink in North America implied by atmospheric and oceanic carbon dioxide data and models. Science (Washington, DC) 282:442–446.

Heath, L.S., R.A. Birdsey, C. Row, and A.J. Plantinga. 1996. Carbon pools and flux in U.S. forest products. p. 271–278. In M.J. Apps and D.T. Price (ed.) NATO ASI Ser. Vol. I 40. Springer-Verlag, Berlin.

Heath, L.S. and J.E. Smith. 2000. Soil carbon accounting and assumptions for forestry and forest-related land use change. p. 88–101. In L. Joyce and R. Birdsey (ed.) The impact of climate change on America's forests. Gen. Tech. Rep. RMRS-GTR-59. USDA, For. Serv., Pacific Northwest Res. Stn., Portland, OR.

Houghton, R.A., J.E. Hobbie, J.M. Melillo, B. Moore, B.J. Peterson, G.R. Shaver, and G.M. Woodwell. 1983. Changes in the carbon content of terrestrial biota and soils between 1860 and 1980: A net release of CO_2 to the atmosphere. Biol. Monogr. 53:235–262.

Houghton, R.A., W.H. Schlesinger, S. Brown, and J.F. Richards. 1985. Carbon dioxide exchange between the atmosphere and terrestrial ecosystems. p. 114–140. In J.R. Trabalka (ed.) Atmospheric carbon dioxide and the global carbon cycle. DOE/ER-0239. U.S. Dep. Energy, Washington, DC.

Ince, P.J. 1994. Recycling and long-range timber outlook. Gen. Tech. Rep. RM242. USDA, For. Serv., Rocky Mount. For. Range Exp. Stn., Fort Collins, CO.

Joyce, L.A., J. Mills, L. Heath, A.D. McGuire, R.W. Haynes, and R.A. Birdsey. 1995. Forest sector impacts from changes in forest productivity under climate change. J. Biogeogr. 22:703–713.

Johnson, W.C., and D.M. Sharpe, 1983. The ratio of total to merchantable forest biomass and its application to the global carbon budget. Can. J. For. Res. 13:372–383.

Johnson, D.W. 1992. Effects of forest management on soil carbon storage. Water Air Soil Pollut. 64:83–120.

Kauppi, P.E., K. Mielikainen, and K. Kuusela. 1992. Biomass and carbon budget of European forests, 1971-1990. Science (Washington, DC) 256:70–74.

Kolchugina, T.P., and T.S. Vinson, 1995. Role of Russian forests in the global carbon balance. Ambio 24:258–264.

Kurz, W.A. and M.J. Apps. 1993. Contribution of northern forests to the global C cycle: Canada as a case study. Water Air Soil Pollut. 70:163–176.

Mills, J. R., and J.C. Kincaid. 1992. The aggregate timberland assessment system atlas: A comprehensive timber projection model. Gen. Tech. Rep. PNW281. USDA, For. Serv., Pacific Northwest Res. Stn., Portland, OR.

Phillips, D.L., S.L. Brown, P.E. Schroeder, and R.A. Birdsey. 2000. Toward error analysis of large-scale forest carbon budgets. Glob. Ecol. Biogeogr. 9:305–313.

Plantinga, A.J., and R.A. Birdsey. 1993. Carbon fluxes resulting from U.S. private timberland management. Climat. Change 23:37–53.

Post, W.M., W.R. Emanuel, P.J. Zinke, and A.G. Stangenberger 1982. Soil carbon pools and world life zones. Nature (London) 298:156–159.

Powell, D. S., J.L. Faulkner, D.R. Darr, Z. Zhu, and D. MacCleery. 1993. Forest resources of the United States, 1992. Gen. Tech. Rep. RM234. USDA, For. Serv., Rocky Mount. For. Range Exp. Stn., Fort Collins, CO.

Row, C., and R.B. Phelps. 1991. Carbon cycle impacts of future forest products utilization and recycling trends. *In* Agriculture in a world of change. Proc. Outlook 1991, Washington, DC. 27–29 Nov. 1990. USDA, Washington, DC.

Schreuder, H.T., V.T. LaBau, and J.W. Hazard. 1995. The Alaska four-phase forest inventory sampling design using remote sensing and ground sampling. Photogram. Eng. Remote Sensing 61:291–297.

Schroeder, P., S. Brown, M. Jiangming, R. Birdsey, and C. Cieszewski. 1997. Biomass estimation for temperate broadleaf forests of the United States using inventory data. For. Sci. 43:424–434.

Smith, J.E., and L.S. Heath. 2000. Considerations for interpreting estimates of uncertainty in a forest carbon budget model. p. 102–111. *In* L. Joyce and R. Birdsey (ed.) The impact of climate change on America's forests. Gen. Tech. Rep. RMRS-GTR-59. USDA, For. Serv., Pacific Northwest Res. Stn., Portland, OR.

Turner, D.P., G.J. Koerper, M.E. Harmon, and J.J. Lee. 1995. A carbon budget for forests of the conterminous United States. Ecol. Applicat. 5:421–436.

11 The Potential of Soil Carbon Sequestration in Forest Ecosystems to Mitigate the Greenhouse Effect

Rattan Lal

The Ohio State University
Columbus, Ohio

ABSTRACT

The importance of soil C sequestration in managed and natural forest ecosystems is not recognized because of the lack of readily available research information on magnitude and rate of soil C sequestration. The perception that soil C sequestration is not important vis-à-vis the amount of C sequestered in the above-ground biomass is not supported by the published experimental data from different ecoregions. The soil organic carbon (SOC) pool in forest ecosystems can be equal to or more than that in the above-ground biomass. The rate of soil C sequestration due to change in land use from agricultural to natural or managed forest ecosystems depends on antecedent soil quality and ecoregional characteristics. Similar to cropland, the rate of C sequestration in forest soils also can be limited by nutrient deficiency and elemental toxicity. Lack of N availability can curtail the biomass productivity and conversion of litter and root biomass to humus. Alleviation of drought and optimization of soil-water availability can enhance C sequestration in forest soil also. Depending on soil and ecoregional factors, the magnitude of soil C sequestration can be 20 to 50 Mg C ha^{-1}. The rate of C sequestration is either low or even negative in the first 3 to 5 yr, rising to the maximum level within 10 to 15 yr, and eventually declining to a zero when soil is in equilibrium with the vegetation. The mean rate of soil C sequestration over 40- to 50-yr period may be 0.5 to 1 Mg C ha^{-1} yr^{-1}. The potential of soil C sequestration through afforestation is more in degraded soils where the needed inputs are applied to achieve the desired level of productivity. Restoration of degraded soils at the global scale has a potential to sequester up to 3 Pg C yr^{-1} in soil and biomass, and arrest the rate of increase in atmospheric concentration of CO_2.

Increasing atmospheric concentration of CO_2 and other greenhouse gases (Post et al., 1990; Houghton et al., 1999) has raised serious concerns about identifying mitigation options (U.S. DOE, 1999), and ascertaining the role of soil and biota of the terrestrial ecosystems in the global C cycle. The annual CO_2 flux from soil to the atmosphere (68 Pg yr^{-1}, Pg = petagram = 10^{15} g = 1 billion tons) is 11.3 times the emissions from fossil fuel combustion (6 Pg yr^{-1}) (Raich & Schlesinger, 1992). Furthermore, the net efflux of CO_2 from soil between 1990 and 2050 may increase by an additional 60 Pg if the global temperature were to increase at the rate of 0.03 °C yr^{-1} (Jenkinson et al., 1991). The projected additional emission of 60 Pg is equivalent to 20% increase in fossil fuel combustion between 1990 and 2050. Therefore, soil C pool and its dynamic play a major role in the global C cycle and in accentuation or mitigation of the accelerated greenhouse effect. It is with this background that several attempts have been made to assess the potential of cropland (Lal et al., 1999; Lal & Bruce, 1999), grazing systems (Follett et al., 2000), and forest ecosystems (Birdsey et al., 1993) to sequester C as possible strategies to curtail the rate of increase of atmospheric concentration of CO_2.

Carbon uptake by forest ecosystems is a key factor in assessing the potential of managed terrestrial ecosystems for mitigating the greenhouse effect. However, opinions differ with regard to the relative significance of C sequestration as SOC pool vs. the magnitude of C sequestration in the above-ground biomass. Some argue that forest soils have a small potential of C sequestration vis-à-vis the magnitude of C sequestration in the biomass (Lugo & Brown, 1992). Alban and Perala (1992) reported that soil C stayed relatively constant for 0- to 80-yr chronosequence of aspen (*Populus* spp.) stand in Minnesota. Others believe that potential of C sequestration in soils of forest ecosystems is high and needs to be seriously considered in identifying the strategies for mitigating the greenhouse effect (Freedman et al., 1992; Birdsey et al., 1993). Therefore, the objective of this review is to collate and synthesize the available research information on the rate of C sequestration in forest soils, and objectively assess its potential at global scale. Rather than preparing a comprehensive literature review on the topic, this manuscript provides specific examples that indicate the potential of C sequestration in soils of the forest ecosystem, and highlights technological options to achieve it.

SOIL CARBON VERSUS BIOMASS CARBON POOLS IN FOREST ECOSYSTEMS

Soil Organic Carbon Pool

The global SOC pool in world soils to 1-m depth estimated at 1550 Pg (Eswaran et al., 1993; Batjes, 1996) is about 2.8 times the world biotic pool estimated at 560 Pg (Schlesinger, 1995). Therefore, SOC is an important factor in the global C cycle, and it is a highly variable and dynamic entity. It is variable over space because its density differs widely among soils and ecoregions. It is variable over time because it changes with change in land use and management. The SOC pool is a function of soil characteristics and climatic factors. Even under similar rainfall conditions, changes in soil properties can have a major impact on the pool. Kimble et al. (1990) reported SOC density of principal soil orders of the world. The mean SOC density is 9.7 kg m^{-2} with coefficient of variation (CV) of 42% for tropical Oxisols, 8.3 kg m^{-2} with CV of 70% for tropical Ultisols, 9.1 kg m^{-2} with CV of 46% for temperate Mollisols, 5.5 kg m^{-2} with a CV of 62% for temperate Alfisols and 4.2 kg m^{-2} with a CV of 60% for Andisols. The mean SOC density of soils of midwestern USA to 1-m depth ranges from 8 to 70 kg m^{-2} (Waltman & Bliss, 1996). The SOC density of soils of arid regions may be as low as 0.5 to 5 kg m^{-2}, while that of organic soils may be as high as 100 to 200 kg m^{-2}. Similar estimates of SOC density are available for soils of North America (Lacelle et al., 1997). The mean SOC density of the Amazon Basin soils to 1-m depth is 10.3 kg C m^{-2} with an average C/N of 10.7 (Moraes et al., 1995). Similar to SOC density, the total pool also differs among soils and ecoregions. Fölster and Khanna (1997) assessed the SOC pool to 50-cm depth in forest soils of the tropics. It was 58 Mg ha^{-1} for atypical soil developed on alluvial deposit in West Llanos of Venezuela, 63 Mg ha^{-1} for typical subhumid soil developed on metamorphic rock in Western Nigeria, 81 Mg ha^{-1} for typical humid soil developed on an older terrace in Magdalena, and 47 Mg ha^{-1} for atypical poor-ecosystem soil developed on pleistocene sediments in Middle Orinoco piedmont in Venezuela. The SOC pool for the Brazilian Amazon, covering an area of 5×10^6 km^2 is estimated at 136 Pg to 8-m depth (Fearnside & Barbosa, 1998), 47 Pg to 1-m depth, and 21 Pg to 20-cm depth (Moraes et al., 1995), and 72 Pg in the root zone (Schroeder & Winjum, 1995). The SOC pool in tropical forest soils of the world is estimated at 206 Pg (Kimble et al., 1990).

Biotic Carbon Pool

The SOC pool and its dynamics in a forest ecosystem is as much influenced by temperature and moisture regimes as is the case in agricultural ecosystems. Simmons et al. (1996) related the SOC pool in forest floor and soil of a transect across different temperature and moisture regimes in Maine, USA. The SOC pool of the O_e/O_a layer decreased with increase in temperature, and was 41.1 Mg ha^{-1} in the northern region, 31.6 Mg ha^{-1} in central and southern regions, and 25.7 Mg ha^{-1} in the coastal region. Simons and his colleagues observed that soil respiration was related exponentially to temperature. It implies that an incremental increase in temperature caused a relatively larger increase in soil respiration. The litter fall did not change with change in temperature, but was strongly correlated with precipitation. Thus if temperature increases without change in precipitation, litter input remains the same, soil respiration increases, and C storage in the forest floor decreases. Important among soil factors affecting the SOC pool are texture, clay minerals, soil fertility, and plant available water capacity of the root zone. The SOC pool in mineral soil can differ significantly among forest types, depending on the nature of vegetation and root systems, the SOC pool increases with increase in clay content. For the same clay content, however, soils with high activity clays have more SOC content than those with low activity clays. Climate, aspect of the landscape, texture, and clay minerals are among important determinants of soil and biomass C pool. However, these are exogenous factors and cannot be readily altered. Soil factors that can be manipulated and have an important impact on the SOC pool of forest soils are soil fertility, available soil water capacity, and the interaction among them.

Within terrestrial ecosystems, forests constitute an important biome with regard to C pool and fluxes, because of the large and dynamic nature of the pool in above-ground (tree trunk, litter) and below-ground (root and soil) components. The distribution of C pool in vegetation, litter/coarse debris and soil respectively is 11.1 Pg, 6.1 Pg, and 18.3 Pg for the USA; 118 ± 28 Pg, 18 ± 4 Pg and 404 ± 38 Pg for the former USSR; and 58 to 81 Pg, 6 to 9 Pg and 72 Pg for Brazil (Schroeder & Winjum, 1995). In comparison to global biotic C pool of 560 Pg, that of the Brazilian Amazon Basin is estimated at 58 to 81 Pg in vegetation and 6 to 9 Pg in coarse woody debris (Schroeder & Winjum, 1995). Similar to soil, the C density in biomass also is highly variable among diverse forest ecosystems of the world. The biotic C pool in different vegetation types of the Amazon Basin is estimated at 137 to 200 Mg ha^{-1} for tropical moist rainforest, 60 to 172 Mg ha^{-1} for other closed forest, 25 to 40 Mg ha^{-1} for secondary and degraded forest, 20 to 30 Mg ha^{-1} for Cerrado woodlands, 15 to 20 Mg ha^{-1} for degraded woodlands, and 8 to 10 Mg ha^{-1} for savannah and grasslands (Schroeder & Winjum, 1995). Laurance et al. (1999) reported that biomass estimates varied more than twofold in the forest ecosystem of Central Amazonia, from 231 to 492 Mg ha^{-1} (mean of 356 ± 47 Mg ha^{-1}). The biomass C in temperate forests may be relatively less with above-ground plus below-ground C density of 45 to 61 Mg ha^{-1} for conifer forest, 51 to 69 Mg ha^{-1} for hardwood forest, and 42 to 57 Mg ha^{-1} for mixed wood forest (Freedman et al., 1992). The above-ground biomass in eastern hardwood forest of the USA is estimated at 197 to 330 Mg ha^{-1}, of which only 9 to 30% is in large trees (Brown et al., 1997).

Similar to the density of biomass C, the rate of growth also is variable among ecosystems. In the Brazilian Amazon, Silva et al. (1995) reported that annual increase in volume of trees 20 cm of dbh (diameter at breast height) was 1.8 m^3 ha^{-1} yr^{-1} 13 yr after logging. Logging stimulated growth but the effect was short-lived. Lugo and Brown (1992) synthesized the available information on the rate of biomass C accumulation in tropical ecosystems of Puerto Rico. The rate of biomass C accumulation was 1 to 2 Mg C ha^{-1} yr^{-1} in the 60- to 80-yr-old forest, 2 to 3.5 Mg C ha^{-1} yr^{-1} in the secondary forest fallow, and 1.4 to 4.8 Mg C ha^{-1} yr^{-1} in the plantations. Liu and Muller (1993) assessed the above-ground net primary productivity of a forest in Kentucky. The net primary productivity (NPP) was 6.3 Mg ha^{-1} yr^{-1} of which 56% was woody material and 44% was litter biomass. In Australia, Smethurst and Nambiar (1995) observed the productivity of 4.6 to 8.5 Mg C ha^{-1} yr^{-1} with a total productivity in 40- to 50-yr period ranging from 230 to 340 Mg ha^{-1}.

The Ratio of Soil/Biotic Carbon Pools

The knowledge about the ratio of SOC to the biotic C pools is useful in understanding C dynamics within an ecosystem, the impact of land use change on C dynamics, and on the magnitude of fluxes among soil, biotic and atmospheric pools. The actual C pool in soil vs. plant biomass for a specific ecosystem depends on several control factors including climate (rainfall, temperature), predominant species, drainage pattern, land aspect, and soil type (Jenny, 1941; 1980). Anderson (1991) synthesized the available information and assessed the magnitude of SOC and biotic pools in major ecosystem types of the world. The ratio of C pool in soil/biotic pools is about 0.41 in temperate deciduous forest, 0.65 in tropical forest, 1.67

Table 11–1. Carbon density in plant biomass and soil organic matter (including litter) in major ecosystems of the world (modified from Anderson, 1991; Fölster & Khanna, 1997; Neary et al., 1999).

| Ecosystem | C density | | Relative soil/ plant biomass |
| | Plant biomass | Soil organic matter | |
	Mg ha^{-1}		
Tropical forest	155	100	0.65
Boreal forest	75	125	1.67
Temperate deciduous forest	135	55	0.41
Temperate grassland	15	95	6.33
Savannahs	5	30	6.0
Tundra	5	155	31

in boreal forest, 6.0 in dryland savannas, 6.33 in temperate grasslands, and 31 in Tundra (Table 11–1). These data are approximate and merely indicative of relative trends showing that a considerable part of the productivity is transferred below ground. As much as 40 to 75% of the NPP in a forest ecosystem may be allocated below-ground.

SOIL CARBON SEQUESTRATION BY REFORESTATION AND AFORESTATION

Soil C sequestration in forest soil depends on two factors: (i) the quantity and quality of the biomass returned to the soil, and (ii) and soil factors that determine its humification efficiency. The quantity of biomass returned to the soil depends on soil productivity and species grown. Tree species affect litter quality in terms of N and lignin contents. Soils capacity to retain fraction of the C added as litter and root biomass depends on its clay content, clay minerals, and their ability to form organomineral complexes. Soils with presence of both physical and chemical protection mechanisms to protect the SOC pool can sequester C during the decomposition process.

Conversion from natural to agricultural ecosystems can lead to depletion of SOC pool, often by as much as 50% (Lal et al., 1999). Similarly, the forest harvesting practices and site preparation for afforestation and reforestation also can lead to SOC losses of 40 to 50% in the surface layer (Johnson, 1993; Khanna et al., 1998). The SOC pool remains highly dynamic in the surface layers for a number of years after deforestation (Lal, 1996), by clear cutting or slash-and-burn method. The magnitude and rate of SOC depletion is accentuated by soil degradation by erosion, salinization, compaction, etc. Similarly, the forest harvesting practices and site preparation for afforestation and reforestation also can lead to SOC losses of 40 to 50% in the surface layer (Johnson, 1993; Khanna et al., 1998). The SOC pool remains highly dynamic in the surface layers for a number of years after deforestation (Lal, 1996) by clearcutting or slash-and-burn method. The magnitude of loss of SOC may be as much as 30 to 50 Mg C ha^{-1}. Conversion of such degraded soils under agricultural and other land uses (e.g., mining) to forest ecosystems can reverse the declining trends in SOC pool, and sequester C both in above- and below-ground vegetation. Enhancement of soil quality with attendant increase in SOC pool has been the basis of the age-old system of shifting cultivation (Nye & Greenland, 1960), practiced even now in several parts of the tropics.

Reforestation

Reforestation of a soil, previously under natural or managed forest, may have less potential of soil C sequestration than afforestation of land under arable or pastoral land uses. The rate and magnitude of soil C sequestration may be high if there was loss of SOC pool due to harvesting and site preparation. The rate and magnitude of soil C sequestration also depend on the nature of tree species. Nitrogen-fixing trees usually increase soil C pool by as much as 20 to 50% more than non-N fixers (Boring et al., 1988; Johnson & Henderson, 1995). In Alabama, Wood et al. (1992) reported significant increases in soil C under stands of loblolly pine (*Pinus taeda* L.) grown in association with some herbaceous species. The positive effect was attributed to symbiotic N fixation. The increase in N availability may lead to increase in NPP and stabilization of humus (Paul & Clark, 1989; Johnson & Henderson, 1995). Improvements in SOC pool due to N-fixers raises management options in terms of establishing mixed plantation, and applying fertilizer N to intensively managed plantations. Stabilization of humus may depend on the formation of organomineral complexes through the residue returned to the soil.

Bernhard-Reversat (1987) studied the impact of seven tree species on SOC pool in the Sahel region of Senegal. Litter from native *Acacia* species had low levels of readily soluble compounds in fresh and partly decomposed residues, high respiratory output from fresh litter, and contributed a large fraction of organic light fractions into soil to form organomineral complexes. Consequently, planting *Acacia* species increased the SOC pool but *Eucalyptus* did not (Bernhard-Reversat, 1993). In Puerto Rico, Wang et al. (1991) also observed little or no effect of *Eucalyptus* plantation on SOC pool.

Afforestation

Several experiments throughout the world have shown strong increase in SOC pool by conversion of agricultural land to forest ecosystem. The magnitude and rate of C sequestration in such a system depends on nutrient availability. Changes in C storage over 88 ha of farmland in east-central Minnesota encroached by forest for a 40-yr period from 1938 to 1977 were evaluated by Johnston et al. (1996). The total C storage of 6600 Mg in the ecosystem included 3700 Mg in vegetation, 750 Mg in O (organic horizon or litter layer) and 2700 Mg in the mineral soil. Therefore, the rate of C sequestration was 1 Mg ha^{-1} yr^{-1} in vegetation biomass, 0.2 Mg ha^{-1} yr^{-1} in O horizon and 0.8 Mg ha^{-1} yr^{-1} in the mineral soil. In Denmark, Nielsen et al. (1999) observed that conversion of heathland to sitka spruce [*Picea sitchensis* (Bong.) Carrière] increased SOC content of O_i/O_e layer from 29.3 to 42.1% but decreased that of O_a layer from 9.3 to 1.8% in 60 yr. The source of plant nutrients in such unmanaged systems is either nutrient cycling from the subsoil, runoff from adjacent lands, or atmospheric deposition. The latter can be substantial in industrial areas, and also may be detrimental to vegetation. The rate of N deposition may be as much as 10 kg ha^{-1} yr^{-1} (Mäkipää, 1995).

Encroachment of agricultural land by unmanaged forests also can lead to depletion of SOC pool, especially in a nutrient-deficient system. Lack of N is a major factor responsible for SOC depletion. In a model study, Morris et al. (1997) observed that productivity of unmanaged sites dropped significantly over time, probably due to nutrient deficiency. When nutrients are limiting, a rapid decline in SOC pool also can occur. Soil C dynamics for conversion of grassland to Douglas fir at Craigieburn site in New Zealand was studied by Alfredsson et al. (1998). The site was grazed for at least 100 yr prior to conversion to fir in 1979. The SOC content measured in 0- to 30-cm depth 17 yr after conversion decreased from grassland to fir from 11.9 to 7.0% in 0- to 5-cm depth, and 6.6 to 5.4% in 5- to 15-cm depth. The SOC content of 15- to 30-cm depth was not changed and was 3.5% in grassland compared with 3.7% in fir. Similar trends were observed at the Southland site where grassland was converted to radiata pine (*Pinus radiata*). In Denmark, Nielsen et al. (1999) observed decline in SOC pool in heathland soil planted to oak (*Quercus* spp.). Heathland soils are coarse textured, low in inherent fertility, and have low N reserve. Conversion to oak decreased SOC content of O_i/O_e layer from 29.3 to 9.4% and of O_a layer from 9.3 to 4.8% in 60 yr. In a *Eucalyptus* plantation in Hawaii, Binkley and Resh (1999) observed that SOC content in 0- to 30-cm layer decreased by 0.6 Mg C ha^{-1} over a 32-yr period.

Therefore, the trends in SOC pool by conversion from agricultural to forest ecosystem depend on soil fertility, and soil moisture regime. Some of these controlling factors can be managed, especially in intensively managed forest plantations.

NUTRIENT AVAILABILITY AND CARBON POOL IN SOILS OF FOREST ECOSYSTEMS

Similar to the effects of nutrient availability on productivity of agricultural ecosystems, nutrient balance is important to biomass production, leaf litter and root turnover rate, and SOC pool in forest ecosystems. Availability of plant nutrients at adequate level and in the form available at the critical periods of growth is essential to optimum biomass production. In a natural forest ecosystem, the nutrient dynamics is primarily driven by transfer processes within the ecosystem through recycling mechanism, and also with input (atmospheric deposition, runoff and sedimentation, biological N fixation) and output (soil erosion, leaching, volatilization, and harvest of timber and other by-products). In managed plantations, however, nutrient flow can be regulated, and the input of nutrients must balance the output at the desired level of production. Intensively managed plantations of fast-growing trees are rapidly replacing degraded croplands and native forests in the tropics (Nambiar & Brown, 1997). The quantity of nutrients harvested in forest products can be substantial, and must be replaced for sustaining productivity. The amount of nutrients harvested in a 20-yr plantation in stem wood plus bark from a plantation of *Acacia* and *Eucalyptus* in Kalimantan, Indonesia, ranged from 234 to 478 kg ha^{-1} of K, 310 to 460 kg ha^{-1} of Ca, and 33 to 46 kg ha^{-1} of Mg. Biomass productivity of intensively managed plantations depends on the soil nutrient supply, which changes rapidly in such a dynamic tree-soil system. Lack of availability of sufficient nutrients to meet the requirements of rapidly growing plantation may lead to depletion of SOC pool through mineralization.

The nutrient status in soil has a major impact on the SOC pool. Gerding (1991) and Fölster and Khanna (1997) reported a strong correlation between SOC pool in the surface layer (30-cm depth) and the N and Ca^{2+} contents of soils under 20-yr-old *Pinus radiata* D. Don plantations in the Sierra Occidental region of Chile (Fig. 11–1). In central Amazonia, Brazil, Laurance et al. (1999) also observed that principal soil factors affecting the biomass were clay content, soil organic C, total soil N, and exchangeable cations. Therefore, soil quality is an important determinant of biomass production and SOC pool within an ecoregion. Principal elements are N, Ca and P, and their availability is essential to both biomass production and humification of leaf litter and root biomass.

SOIL FERTILITY MANAGEMENT AND SOIL ORGANIC CARBON POOL IN SOILS UNDER PLANTATION FORESTRY

Biomass productivity of depleted and degraded soils when converted to natural and managed forestry ecosystem is low, and constrained by lack of or low level of availability of essential nutrients, and nutrient imbalance caused by toxic concentrations (e.g., Al, Mn) of some and deficit of others (e.g., N, P, Ca). High concentrations of soluble salts in the root zone of sodic or saline soil also can restrict growth of susceptible tree species. Strong enhancement of SOC pool with application of fertilizer can thus happen only in soils and ecosystem deficient in specific plant nutrients. Soils that can meet the requirement of forest growth through inherent fertility will not respond to fertilizer application. Alleviation of soil-related constraints in nutrient deficient forest soils, therefore, can enhance production of above- and below-ground biomass, leaf litter fall, root turnover and also the SOC pool. There are several examples in the literature that support these hypotheses. In addition to change in total pool, conversion to forestry land use also can lead to stratification of SOC with higher concentration in the surface than in the subsoil (Jug et al., 1999).

For slash pine (*Pinus elliottii* Engelm.) plantation in Florida with soils of high inherent fertility, fertilizer application resulted in only a modest (and insignificant) increase in SOC content by 5 to 8% (Harding & Jakola, 1994). Harding and Jakola reported that SOC pool in a 25-yr old plantation was 111 Mg C ha^{-1} in control, 115 Mg C ha^{-1} in superphosphate + NK and 123 Mg C ha^{-1} in rock phosphate + NK treatment. In contrast, fertilizer experiments conducted on nutrient-deficient soils in Australia (Turner & Lambert, 1985) and Portugal (Käterer et al., 1995) demonstrated strong positive effects of fertilizer application on SOC pool. Thirty years after fertilizer regime was initiated on a radiata pine plantation in Australia, Turner and Lambert (1985) observed 22% increase in SOC pool.

An experiment conducted on a 59-yr-old spruce (*Picea* spp.) plantation in Norway showed that application of N, P, K, Ca, Mg and S increased the annual rate of biomass production from 6200 to 8500 kg ha^{-1} yr^{-1} (Ingerslev & Hallbacken, 1999). Mäkipää (1995) studied the effect of fertilizer use on SOC dynamics under

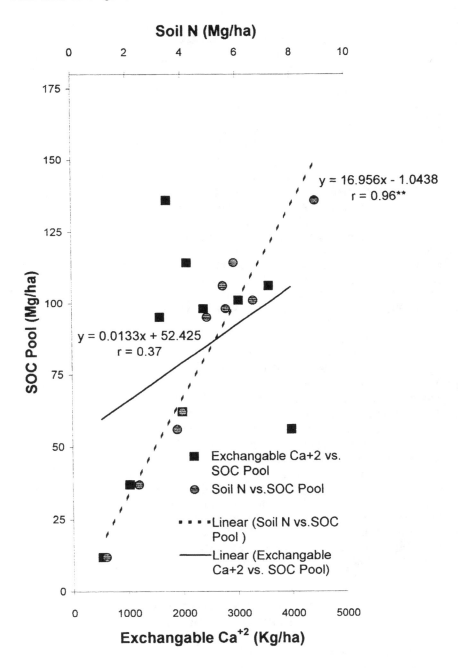

Fig. 11–1. Relationship between SOC pool and soil N and Ca^{2+} content for 10 Chilean sites under *Pinus radiata* plantation (redrawn from Gerding, 1991; Folster & Khanna, 1997).

Table 11–2. Effects of fertilizer and irrigation on the SOC profile in soil under 54-mo old plantation of *Eucalyptus globulus* in central Portugal (adapted from Kätterer et al., 1995).

Depth (cm)	Control	Fertilizer	Irrigation	Fertilizer + irrigation
		g kg^{-1}		
0–10	2.5 ± 0.6	3.4 ± 1.5	7.7 ± 2.9	7.6 ± 2.5
10–20	2.7 ± 1.1	2.5 ± 1.2	3.5 ± 0.8	5.1 ± 1.6
20–30	2.8 ± 1.1	3.0 ± 1.5	3.8 ± 1.3	5.0 ± 1.6
30–40	2.6 ± 1.5	2.5 ± 0.8	2.5 ± 1.1	6.0 ± 1.6
Mean	2.65 ± 0.13	2.85 ± 0.44	4.37 ± 2.29	5.45 ± 1.93

36- to 63-yr old Scots pine (*Pinus sylvestris* L.) and Norway spruce [*Picea abies* (L.) Karsten] plantations. The amount of C in the humus layer was more in fertilized than unfertilized treatments. The SOC pool in 0- to 30-cm layer of the mineral soil was 8 400 to 20 300 kg ha^{-1} on control plots and 17 400 to 27 000 kg ha^{-1} on plots that received N fertilizer. The increase of C due to fertilization ranged from 1900 kg ha^{-1} (14%) to 5530 kg ha^{-1} (87%). Therefore, management of soil fertility is essential to enhancing both the above-ground productivity and soil C sequestration in managed forest and plantation ecosystems.

WATER MANAGEMENT AND SOIL CARBON SEQUESTRATION

Similar to plant nutrients, water imbalance (both excess leading to anaerobiosis and deficit causing drought) also affect biomass productivity and SOC pool and fluxes in forest soils. Water deficit is a major factor in arid and semiarid regions. Application of N and other nutrients may have either no positive effect or even adverse effect on biomass productivity and SOC pool in case of severe water deficit. An experiment conducted in central Portugal on SOC dynamics under 54-mo old plantation of *Eucalyptus globulus* is a relevant example. Kätterer et al. (1995) observed significant increase in SOC pool to 30-cm depth with combined rather than separate use of irrigation and fertilizer (Table 11–2). In comparison with the control, the increase in mean SOC content was 7.5% with fertilizer, 64.9% with irrigation, and 105.7% with fertilizer plus irrigation. The mean rate of SOC sequestration was 3.8 Mg C ha^{-1} yr^{-1}. Because of the stratification of SOC pool in the surface layer, the relative increase due to irrigation and fertilizer use were more in the surface 0- to 10-cm layer than in the subsoil. Such synergistic effects leading to increase in SOC pool also may occur in soil-water conservation and use of water harvesting techniques.

SOIL RESTORATION AND SOIL ORGANIC
CARBON POOL IN FOREST SOILS

Restoration of degraded soils and ecosystems through establishment of appropriate tree species is a feasible option of sequestering C in soil by enhancing the SOC pool, improving soil quality, and increasing biomass productivity. A relevant example in support of such a hypothesis is an experiment conducted in Haryana, India, to restore a sodic soil through establishment of a plantation of *Prosopis*

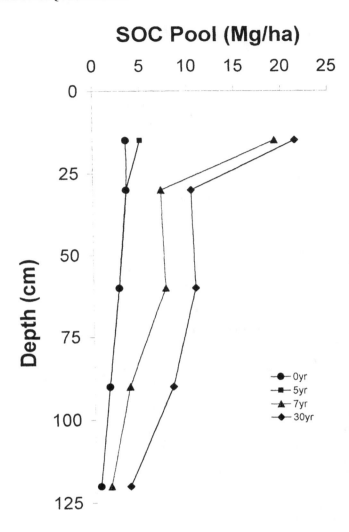

Fig. 11–2. Increase in soil C pool by establishing plantation of *Proscopis Juliflora* in sodic soil in Haryana, India (recalculated and redrawn from Bhojvaid & Timmer, 1998).

juliflora (Bhojvaid & Timmer, 1998). Temporal changes in SOC pool were studied at 0, 5, 7, and 30 yr after establishing the plantation. The total SOC pool to 120-cm depth was 11.8 Mg C ha^{-1} in the unreclaimed soil, 13.3 Mg C ha^{-1} 5 yr after, 34.2 Mg C ha^{-1} 7 yr after, and 54.3 Mg C ha^{-1} 30 yr after establishing the plantation (Fig. 11–2). The rate of soil C sequestration was 0.3 Mg C ha^{-1} yr^{-1} in the first 5 yr, 3.2 Mg C ha^{-1} yr^{-1} in the first 7 yr, and 1.42 Mg C ha^{-1} yr^{-1} for the 30-yr period (Fig. 11–3). Akala and Lal (199) assessed SOC sequestration in reclaimed mineland chronosequence in Ohio. The data in Table 11–3 show that the maximum rate of SOC sequestration under forest was 2.6 Mg C ha^{-1} yr^{-1} for 0- to 15-cm depth

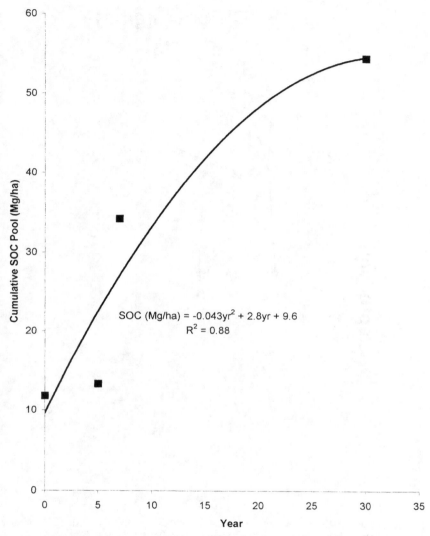

Fig. 11–3. Cumulative SOC pool in a sodic soil planted to *Proscopis Juliflora* (redrawn from data in Fig. 11–2).

measured in 14th yr and 0.9 Mg C ha^{-1} yr^{-1} measured in 12th yr of reclamation. These rates were equivalent to those observed under pastures. With large areas of degraded soils in the world, the potential of C sequestration is enormous. If 2 billion ha of degraded soils (Oldeman, 1994) can be restored and accumulate C in soil at the rate of 1.5 Mg ha^{-1} (and additional 1.5 Mg C ha^{-1} in the biomass), the potential accelerated greenhouse effect due to enrichment of atmospheric concentration of CO_2 at 3.2 Pg C yr^{-1} can be effectively nullified.

Table 11–3. The SOC sequestration rate in reclaimed minelands in Ohio sown to pasture and forest for 25 yr (Akala & Lal, 1999).

Treatment	Initial rate		Maximum rate		Year of the maximum rate	
	0–15 cm	15–30 cm	0–15 cm	15–30 cm	0–15 cm	15–30 cm
Pasture	0.13	0.02	3.1	0.68	11	13
Forest	0.26	0.001	2.6	0.87	14	12

THE POTENTIAL OF SOIL CARBON SEQUESTRATION IN FOREST ECOSYSTEMS

The importance of forest in C sequestration in biomass is well recognized (Sedjo, 1992; Lugo & Brown, 1992; Freedman et al., 1992; Birdsey et al., 1993; Fan et al., 1998). Despite its obvious potential, the importance of C sequestration in soils of forest ecosystem is neither adequately understood nor widely recognized. There are several reasons regarding the lack of this recognition.

1. Temporal changes in soil C content and bulk density to at least 1-m depth are usually not measured in forest soils. While the above-ground biomass is easily assessed by empirical relationships relating biomass to tree growth parameters, few attempts are made to measure or model temporal changes in SOC pool.

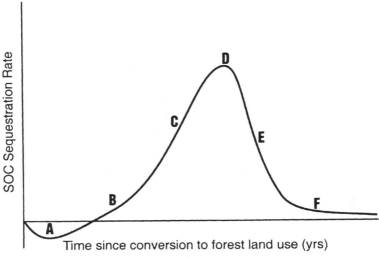

Fig. 11–4. A schematics of the rate of SOC sequestration in a degraded soil converted to managed forest. The initial decline (*A*, 0–3 yr), due to oxidation of organic matter or soil erosion caused by low vegetation cover, is followed by slow increase (*B*, 3–7 yr) and a very rapid increase (*C*, 10–15 yr). The maximum rate (*D*) may be maintained for 2 to 3 yr, and is followed by a declining rate (*E*), and finally the equilibrium rate in a mature forest of 30 to 50 yr. The segment *B–D* is a slow rise but the segment *D–E* is a rapid decline. This schematic represents the data of Bhojvaid and Timmer (1988) and Akala and Lal (1999).

2. Significant changes in SOC pool under forest ecosystem occur after relatively long periods of time, often after 30 to 50 yr. There are few, if any, experiments where SOC pool has been studied regularly over long periods of time. The rate of SOC sequestration needs to be studied over at least 30-yr period to establish definite trends. Figure 11–4 is a schematic of SOC dynamics by conversion of cropland to managed forest ecosystem. The SOC sequestration rate may follow five distinct stages. The duration and rate of each stage may differ among soils and ecosystems. It is important to develop such response functions for different soils and tree species, and to identify strategies that will enhance the duration and rate of the stage with maximum sequestration potential. The initial rate of SOC sequestration may be either negative (C loss from soil during the initial stages of tree establishment) or negligible.

3. Once the SOC pool has reached equilibrium level, as determined by inherent soil characteristics and the ecoregional parameters, the rate of SOC sequestration under mature forest may be zero. The fact that soil C sink capacity is finite and can be filled over a 25- to 50-yr period can be misleading if the observations are made only in a mature forest.

4. Similar to the soils of agricultural ecosystems, the rate of SOC sequestration in soils under forests and plantations also is limited by lack of essential nutrients, and nutrient imbalance. The humification of biomass returned to the soil (through litter and roots) requires nutrients for efficient conversion of biomass C into humus C. The sequestration of C in humus not only requires the availability of N, S, and P compounds in soils to combine with C in the biomass to produce humus, but additional quantities of these elements to support the desired NPP in trees. The ratio of C/N/P/S in humus is about 10, 50 and 70 compared to 50, 100 and 150 in leaf litter and root biomass. Himes (1998) estimated that sequestering 10 000 kg of C into humus requires 833 kg of N, 200 kg of P, and 143 kg of S. Therefore, lack of nutrients can lead to low efficiency of humification in a crop ecosystem. Such calculations, although not available, are needed for forest ecosystems. Application of fertilizers is extremely rare even to plantation forestry, especially in the developing countries of the tropics. Thus the vast potential of SOC sequestration in managed forests is not harnessed. Similar to nutrients, lack of water and other biotic and abiotic stresses can cause serious constraints to NPP and SOC sequestration.

CONCLUSIONS

1. Forest soils, similar to those under cropland and grazing systems, have a high potential of C sequestration.

2. The magnitude of potential depends on soil quality and ecoregional characteristics. The SOC content may stay relatively constant throughout the duration of stand development in reforestation of soil with high antecedent SOC content. In such cases, the rate may be zero or even negative, especially if the antecedent SOC pool is high. In contrast, degraded soils con-

verted to natural forests and managed plantations have a relatively high potential of C sequestration providing that soil-related constraints are alleviated.

3. The rate of SOC sequestration in plantations is enhanced by application of fertilizers in nutrient-deficient soils and of water in drought-prone environments. Judicious use of N, P and bases (Ca^{2+}, Mg^{2+}) to managed plantations can enhance biomass production and increase SOC sequestration.

4. The initial rate of SOC sequestration is often low for up to 5 yr after establishing the plantation. Similarly, the equilibrium rate under mature forests also may be low. The high rate of C sequestration in forest ecosystems is usually observed between 10 to 15 yr after conversion to forest land use.

ACKNOWLEDGMENT

Help received from the graduate students and staff of the soil physics laboratory at the Ohio State University in data analyses is gratefully acknowledged.

REFERENCES

Akala, V., and R. Lal. 1999. Potential of mineland reclamation for soil organic carbon sequestration in Ohio. Land Degrad. Develop. (In press.)

Alban, D.H., and D.A. Perala. 1992. Carbon storage in Lake States aspen ecosystem. Can. J. For. Res. 22:1107–1110.

Alfredsson, H., L.M. Condron, M. Clarholm, and M.R. Davis. 1998. Changes in soil acidity and organic matter following the establishment of conifers on former grassland in New Zealand. For. Ecol. Manage. 112:245–252.

Aluko, A.P. 1993. Soil properties and nutrient distribution in *Terminalia superba* stands of different age series grown in two soil types of southwestern Nigeria. For. Ecol. Manage. 58:153–161.

Anderson, J.M. 1991. The effects of climate change on decomposition process in grassland and coniferous forests. Ecol. Appl. 1:326–347.

Batjes, N.H. 1996. Total C and N in soils of the world. Eur. J. Soil Sci. 47:151–163.

Bernhard-Reversat, F. 1987. Litter incorporation to soil organic matter in natural and planted tree stands in Senegal. Pedobiologia 30:401–417.

Bernhard-Reversat, F. 1993. Dynamics of litter and organic matter at the soil-litter interface in fast-growing tree plantations on sandy ferrallitic soil (congo). Acta Ecol. 14:179–195.

Bhojvaid, P.P., and V.R. Timmer. 1998. Soil dynamics in an age sequence of *Prosopis juliflora* planted for sodic soil restoration in India. For. Ecol. Manage. 106:181–193.

Binkley, D., and S.C. Resh. 1999. Rapid changes in soils following Eucalyptus afforestation in Hawaii. Soil Sci. Soc. Am. J. 63:222–225.

Birdsey, R.A., A.J. Plantinga, and L.S. Heath. 1993. Past and prospective carbon storage in United States forests. For. Ecol. Manage. 58:33–40.

Boring, L.R., W.T. Swank, J.B. Waide, and G.S. Henderson. 1988. Sources, fates and impacts of nitrogen inputs to terrestrial ecosystems: Review and synthesis. Biogeochemistry 6:119–159.

Brown, S., P. Schroeder, and R. Birdsey. 1997. Above-ground biomass distribution of US eastern hardwood forests and the use of large trees as an indicator of forest development. For. Ecol. Manage. 96:37–47.

Eswaran, H., E. Van den Berg, and P. Reich. 1993. Organic carbon in soils of the world. Soil Sci. Soc. Am. J. 57:192–194.

Fan, S., M. Gloor, J. Mahlman, S. Pacala, J. Sarmiento, T. Takahashi, and P. Tans. 1998. A large terrestrial carbon sink in North America implied by atmospheric and oceanic carbon dioxide data and models. Science (Washington, DC) 282:442–446.

Fearnside, P.M., and R.I. Barbosa. 1998. Soil carbon changes from conversion of forest to pasture in Brazilian Amazonia. For. Ecol. Manage. 108:147–166.

Fisher, R.F. 1995. Amelioration of degraded rainforest soils by plantation of native trees. Soil Sci. Soc. Am. J. 59:544–549.

Follett, R.F., J.M. Kimble, and R. Lal. 2000. The potential of U.S. grazing land for sequestering C. CRC Press, Boca Raton, FL. (In press.)

Fölster, H., and P.K. Khanna. 1997. Dynamics of nutrient supply in plantation soils. p. 339–377. In E.K.S Nambiar and A.G. Brown (ed.) Management of soil, nutrients and water in tropical plantation forests. ACIAR/CSIRO/CIFOR Publ. ACIAR, Canberra, Australia.

Freedman, B., F. Meth, and C. Hickman. 1992. Temperate forest as carbon-storage reservoir for carbon dioxide emitted by coal-fired generating stations: A case study for New Brunswick, Canada. For. Ecol. Manage. 55:15–29.

Galloway, J.N., H. Levy II, and P.S. Kasibhatla. 1994. Year 2020: Consequences of population growth and development on deposition of oxidized nitrogen. Ambio 23:120–123.

Gerding, V.R. 1991. Pinus radiata Plantagen in Zentral-Chile. Standortfaktoren der producktivität und Nährstoffverteilung in Beständen. Ph.D diss. Univ. Göttingen, Germany.

Hall, G.M.J., and D.Y. Hollinger. 1997. Do the indigenous forests affect the net CO_2 emission policy of New Zealand. N.Z. For. 41:24–31.

Harding, R.B., and E.J. Jokela. 1994. Long-term effects of forest fertilization on site organic matter and nutrients. Soil Sci. Soc. Am. J. 58:216–221.

Harmon, M.E., W.K. Ferrell, and J.F. Franklin. 1990. Effects on carbon storage of conversion of old-growth forests to young forests. Science (Washington, DC) 247:699–702.

Himes, F.L. 1998. Nitrogen, sulfur and phosphorus and the sequestering of carbon. p. 315–320. In R. Lal et al. (ed.) Soil processes and the carbon cycle. CRC Press, Boca Raton, FL.

Houghton, R.A., J.L. Hackler, and K.T. Lawrence. 1999. The U.S. carbon budget: Contributions from land use change. Science (Washington, DC) 285:574–578.

Ingerslev, M., and L. Hallbacken. 1999. Above-ground biomass and nutrient distribution in a limed and fertilized Norway spruce (Picea abies) plantation. Part II. Accumulation of biomass and nutrients. For. Ecol. Manage. 119:21–38.

Jenkinson, D.S., D.E. Adams, and A. Wild. 1991. Model estimates of CO_2 emissions from soil in response to global warming. Nature (London) 35:304–306.

Jenny, H. 1941. Factors of soil formation. McGraw-Hill Book Co., New York.

Jenny, H. 1980. The soil resource: Origin and behavior. Springer-Verlag, New York.

Johnson, D.W. 1992. The effects of forest management on soil C storage. Water Air Soil Pollut. 64:83–120.

Johnson, D.W. 1993. Effects of forest management on soil carbon storage in eastern Sierra Nevada forests. Rep. Natl. Counc. Pap. Indust. Air Stream Improve., Inc., Univ. Nevada, Reno, NV.

Johnson, D.W., and P. Henderson. 1995. Effects of forest management and elevated carbon dioxide on soil carbon storage. p. 137–145. In R. Lal et al. (ed.) Soil management and greenhouse effect. CRC/Lewis Publ., Boca Raton, FL.

Johnston, M.H., P.S. Homann, J.K. Engstrom, and D.F. Grigal. 1996. Changes in ecosystem carbon storage over 40 years on an old field/forest landscape in east-central Minnesota. For. Ecol. Manage. 83:17–26.

Jug, A., F. Makeschin, K.E. Rehfuess, and C. Hofmann-Schielle. 1999. Short-rotation plantations of balsam poplars, aspen and willows on former arable land in the Federal Republic of Germany: III. Soil ecological effects. For. Ecol. Manage. 121:85–99.

Kätterer, T., A. Fabião, M. Madeira, C. Ribeiro, and E. Steen. 1995. Fine root dynamics, soil moisture and soil carbon content in a Eucalyptus globulus plantation under different irrigation and fertilization regimes. For. Ecol. Manage. 74:1–12.

Khanna, P.K., B. Ludwig, J. Bauhus, I. Serrasesas, and J. Raison. 1998. Changes due to management practices in chemical and biological parameters of soils under eucalypt stands. In Proc. Int. Union Soil Sci., Montpellier, France. (CD-ROM computer file). Int. Union Soil Sci., Montpelier.

Kimble, J.M., H. Eswaran, and T. Cook. 1990. Organic matter in tropical soils. p. 248–255. In Proc. 14th Int. Congr. Soil Sci., Kyoto, Vol. 5, Japan. 12–18 August. Int. Soc. Soil Sci.

Lacelle, B., C. Tarnocai, S. Waltman, J. Kimble, F. Orozco-Chavez, B. Jakobsen. 1997. North American soil organic carbon map. Agric. Agric. Food Canada, USDA, INEGI, Inst. Geogr., Univ. Copenhagen.

Lal, R. 1996. Deforestation and land use effects on soil degradation and rehabilitation in western Nigeria. II. Soil chemical properties. Land Degrad. Develop. 7:87–98.

Lal, R., R.F. Follett, J.M. Kimble, and C.V. Cole. 1999. Managing U.S. cropland to sequester C in soil. J. Soil Water Conserv. 54:374–381.

Lal, R., and J.P. Bruce. 1999. The potential of world cropland to sequester C and mitigatethe greenhouse effect. Environ. Sci. Policy 2:177–185.

Laurance, W.F., P.M. Fearnside, S.G. Laurance, P. Delamonica, T.E. Lovejoy, J.M. Rankin-de Merona, J.Q. Chambers, and C. Gascon. 1999. Relationship between soils and Amazon forest biomass: A landscape-scale study. For. Ecol. Manage. 118:127–138.

Liu, Y., and R.N. Muller. 1993. Above-ground net primary productivity and nitrogen mineralization in a mixed mesophytic forest of eastern Kentucky. For. Ecol. Manage 59:53–62.

Lugo, A.E., and S. Brown 1992. Tropical forests as sinks for atmospheric carbon. For. Ecol. Manage. 54:239–255.

Mäkipää, R. 1995. Effect of N input on C accumulation of boreal forest soils and ground vegetation. For. Ecol. Manage. 79:217–226.

Moraes, J.L., C.C. Cerri, J.M. Melillo, D. Kicklighter, C. Neill, D.L. Skole, and P.A. Steudler. 1995. Soil carbon stocks of the Brazilian Amazon. Soil Sci. Soc. Am. J. 59:244–247.

Morris, D.M., J.P. Kimmins, and D.R. Duckert. 1997. The use of soil organic matter as a criterion of the relative sustainability of forest management alternatives: A modeling approach using FORECAST. For. Ecol. Manage. 94:61–78.

Nambiar, E.K.S., and A. Brown (ed.) 1997. Management of soil, nutrients and water in tropical plantation forests. ACIAR Monogr. 43. ACIAR, Canberra, Australia.

Neary, D.G., C.C. Klopatek, L.F. DeBano, and P.F. Folliott. 1999. Fire effects on below ground sustainability: A review and synthesis. For. Ecol. Manage. 122:51–71.

Nielsen, K.E., U.L. Ladekarl, and P. Nørnberg. 1999. Dynamic soil processes on heathland due to changes in vegetation to oak and Sitka spruce. For. Ecol. Manage. 114:107–116.

Nilsson, L.-O. 1995. Forest biogeochemistry interactions among greenhouse gases and N deposition. Water Air Soil Pollut. 85:1557–1562.

Nye, P.H., and D.J. Greenland. 1960. The soil under shifting cultivation. Tech. Commun. 51. Commonweal. Bur. Soils, Farnham Royal, Bucks, UK.

Oldeman, L.R. 1994. The global extent of soil degradation. p. 99–118. In D.J. Greenland and I. Szabolcs (ed.) Soil resilience and sustainable land use. CAB Int., Wallingford, UK.

Paul, E.A., and F.E. Clark. 1989. Soil microbiology and biochemistry. Acad. Press, San Diego, New York.

Post, W.M., T. Peng, W.R. Emanuel, A.W. King, V.H. Dale, and D.L. DeAngelis. 1990. The global carbon cycle. Am Sci. 78:310–326.

Raich, J.W., and W.H. Schlesinger. 1992. The global carbon dioxide flux in soil respiration an its relationship to vegetation and climate. Tellus 44B:81–99.

Schlesinger, W.H. 1995. An overview of the carbon cycle. p. 9–25. In R. Lal et al. (ed) Soils and global change. CRC Lewis Publ., Boca Raton, FL.

Schroeder, P.E., and J.K. Winjum. 1995. Assessing Brazil's carbon budget: I. Biotic carbon pools. For. Ecol. Manage. 75:77–86.

Schroeder, P.E., and J.K. Winjum. 1999. Assessing Brazil's carbon budget: II. Biotic fluxes and net carbon balance. For. Ecol. Manage. 75:87–99.

Sedjo, R.A. 1992. Temperate forest ecosystems in the global carbon cycle. Ambio 21:274–277.

Silva, J.N.M., J.O.P. de Carvalho, J. do C.A. Lopes, B.F. de Almeida, D.H.M. Costa, L.C. de Oliveira, J.K. Vanclay, and J.P. Skovsgaard. 1995. Growth and yield of tropical rainforest in the Brazilian Amazon 13 years after logging. For. Ecol. Manage. 71:267–274.

Simmons, J.A., I.J. Fernandez, R.D. Briggs, and M.T. Delaney. 1996. Forest floor C pools an fluxes along a regional climate gradient in Maine, USA. For. Ecol. Manage. 84:81–95.

Smethurst, P.J., and E.K.S. Nambiar. 1995. Changes in soil C and N during the establishment of a second crop of *Pinus radiata*. For. Ecol. Manage. 73:145–155.

Turner, J., and M.J. Lambert. 1985. Soil phosphorus forms and related tree growth in a long-term *Pinus radiata* phosphate fertilizer trial. Commun. Soil Sci. Plant Anal. 16:275–288.

Turner, D.P., G.J. Koerper, M.E. Harmon, and J.J. Lee. 1995. A carbon budget for forests of the conterminous United States. Ecol. Appl. 5:421–436.

U.S. Department of Energy 1999. Carbon sequestration: State of the science. A working paper for road mapping future carbon sequestration research and development. U.S. Dep. Energy Rep. Office Sci Office Fossil Energy, Washington, DC.

Waltman, S.W., and N.B. Bliss. 1996. Estimates of soil organic carbon content of the soils of the U.S. USDA-NRCS and EROS Data Center, Sioux Falls, SD.

Wang, D., F.H. Bormann, A.E. Lugo, and R.D. Bowden. 1991. Comparison of nutrient use efficiency and biomass production in five tropical tree taxa. Forest Ecol. Manage. 46:1–21.

Wood, C.W., R.J. Mitchell, B.R. Zufter, and C.L. Lin. 1992. Loblolly pine plant community effects on soil carbon and nitrogen. Soil Sci. 154:410–419.

12 Reconstruction of Soil Inorganic and Organic Carbon Sequestration Across Broad Geoclimatic Regions

L. R. Drees and L. P. Wilding

Texas A&M University
College Station, Texas

L. C. Nordt

Baylor University
Waco, Texas

ABSTRACT

Soil plays a strategic role in the global C balance. It is the biogeochemical interface between the atmosphere, biosphere and the hydrosphere. Soil contains more inorganic C than the atmosphere and more organic C than the biosphere, but estimates of the contributions of soil C to the global C pool are often ignored. The pedogenic processes of carbonate leaching, silicate mineral weathering and organic matter accumulations are natural means of atmospheric C sequestration. Through reconstruction analysis of 29 soil profiles in Ohio and Texas we estimate that soils on a worldwide basis have lost about 105 kg m^{-2} of carbonate C since the end of the last ice age. This equates to a sequestration rate of about 0.36×10^{15} g C yr^{-1}. At least one-half to two-thirds of this total is atmospheric C that is transferred to the hydrosphere for long-term storage.

Carbon sequestration as a strategy in mitigating atmospheric CO_2 concentrations has focused on the role of agriculture, land use practices, and land management in the global C balance (Lal et al, 1995a,b, 1998). This attention to the organic C pool is justified because of the large flux between the atmosphere and plants via photosynthesis and the return of CO_2 to the atmosphere through decomposition (Schlesinger, 1995a). In addition, the organic C pool is strongly affected by soil management practices. It has been estimated that in regions employing mechanized agriculture, 20 to 40% of the native soil organic matter has been depleted in the past 100 yr (Schlesinger, 1995b). The loss of about 68×10^{15} g of C from 1860 to the present has perturbed the biotic C cycle. Undoubtedly, losses may be curtailed or reversed, and additional biotic C will be sequestered with improved land manage-

Table 12–1. Major C reservoirs.

	SIC	SOC	Total C
		10^{15} g	
Atmosphere	760†‡		760
Biosphere		560§	560
Pedosphere	1 700§¶	1 500‡§¶	3 200
Hydrosphere	38 000†§	1 000‡	39 000
Lithosphere	48 000 000#	17 000 000#	65 000 000

† Lal et al. (1995c).
‡ Bolin et al. (1979)
§ Schlesinger (1997).
¶ Eswaran et al. (1995).
# Kemp (1979).

ment, soil quality, alternative cropping systems, and an increased understanding of soil-plant-organic C dynamics. However, Schlesinger (1999) warns that some suggested soil management programs may not be as efficient in sequestering C once ancillary factors are considered. What is commonly neglected in this discussion is the potential role of inorganic C sequestration either by translocation of soluble carbonic acid (H_2CO_3), carbonate (CO_3^{2-}), or bicarbonate (HCO_3^-) to groundwaters and/or formation of pedogenic carbonates from noncalcareous parent materials.

Major C reservoirs are the atmosphere, biosphere, lithosphere, pedosphere, and hydrosphere (Table 12–1). The lithosphere is the largest C reservoir, but participates in C cycling only on a geologic time scale. The pedosphere (soil), including organic and inorganic components, provides the biogeochemical linkage between the atmosphere, biosphere and the hydrosphere. The pedosphere is the earth's weathering medium and sequesters atmospheric CO_2 during the dissolution and leaching of inorganic C (SIC) in the form of soil carbonates, the weathering of silicate minerals in more humid regions, and the synthesis of organic C (SOC). In fact, there is almost three times more SOC than C in the above-ground plant biomass (Table 12–1). The C stored in the pedosphere is transitory in nature with stability ranging from short- to long-term residence times. Because it contains more SIC and SOC than the atmosphere and biosphere combined (Table 12–1), even small changes in this large C pool could have a profound influence on global C cycling and C exchange mechanisms.

Though soil contains an appreciable amount of inorganic and organic C, the role of SIC and soluble bicarbonate in C sequestration is not well understood, is difficult to quantify, and often underestimated or ignored in estimating the available soil C pool and C sequestration potential. Studies have examined C sequestration by organic matter and mineral weathering (Chadwick et al., 1994) and C stored in calcic and petrocalcic soil horizons (Schlesinger, 1982). However, there is a dearth of information on C sequestration from the weathering of calcareous soil profiles. Thus, the aim of this chapter is to examine SIC sequestration pathways, and to make quantitative assessments of the magnitude of SIC and SOC sequestration during pedogenesis of calcareous parent materials over selected geoclimatic regions of the USA employing reconstruction techniques. We will also consider C flux rates over the geological period of pedogenesis using carbonate dissolution and organic matter accumulation as a proxy.

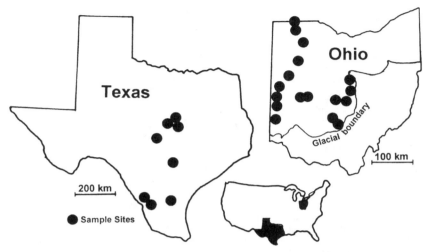

Fig. 12–1. Location of pedon sampling sites in Ohio and Texas.

MATERIALS AND METHODS

Field Methods

Pedons used in this study were obtained from Ohio and Texas (Fig. 12–1, Table 12–2). These pedons were on stable landscape positions, had a moisture regime from aquic to aridic, and a soil temperature regime from mesic to hyper-themic. Geomorphic age was early to late Pleistocene, texture ranged from fine-loamy to clayey skeletal, mineralogy from siliceous to smectitic, calcium carbonate equivalent of parent material from 0 to 80% and solum thickness from 76 to 300 cm.

The pedons in Ohio were obtained from soils developed in late Wisconsin-age glacial till and outwash. They are Aqualfs and Udalfs (Table 12–2) and range in age from 17 500 yr before the present (YBP) in the south to 9000 YBP in the north (Smeck et al., 1968; Rostad et al., 1976). In a north-south transect the mean annual precipitation ranges from 86 to 99 cm and mean annual temperature ranges from 9 to 13°C (long-term averages). In contrast, the Texas study area receives between 82 and 50 cm of mean annual precipitation in a north-south transect, with a mean annual temperature of 19°C. The ages range from recent to >15 000 YBP The soils include Entisols, Vertisols, Aridisols, Mollisols and Alfisols (Table 12–2). Four of the soils of known age developed in Quaternary alluvium in central Texas were evaluated for SIC flux (Nordt et al., 1998, 1999).

Laboratory Methods

Soil organic carbon and SIC were determined for each horizon of each pedon. The SIC was determined using the gasometric procedure of Dreimanis (1962) and

Table 12–2. Soil series and classification of profiles in study.

Pedon no.	Series	Classification
		Ohio profiles
PT-28	Nappanee	Fine, illitic, mesic Aeric Epiaqualfs
HN-97	Nappanee	Fine, illitic, mesic Aeric Epiaqualfs
FT-18	Nappanee	Fine, illitic, mesic Aeric Epiaqualfs
MICH-46	Nappanee	Fine, illitic, mesic Aeric Epiaqualfs
PB-69	Celina Taxadjunct	Fine, mixed, mesic Aquic Hapludalfs
PB-70	Celina	Fine, mixed, active, mesic Aquic Hapludalfs
DK-23	Celina	Fine, mixed, active, mesic Aquic Hapludalfs
DK-24	Morley	Fine, illitic, mesic Oxyaquic Hapludalfs
AG-1	Morley	Fine, illitic, mesic Oxyaquic Hapludalfs
AL-125	Morley	Fine, illitic, mesic Oxyaquic Hapludalfs
FR-57	Celina	Fine, mixed, active, mesic Aquic Hapludalfs
SK-30	Canfield	Fine-loamy mixed, mesic Aquic Fragiudalfs
PG-2	Rittman	Fine-loamy mixed, mesic Aquic Fragiudalfs
LC-21	Cardington Taxadjunct	Fine-loamy, illitic, mesic Aquic Hapludalfs
CH-54	Fox Taxadjunct	Fine-loamy, mixed, mesic Typic Hapludalfs
CH-55	Eldean Taxadjunct	Clayey-loamy-skeletal, mixed, mesic Typic Hapludalfs
KX-2	Fox Taxadjunct	Loamy-skeletal, mixed, mesic Typic Hapludalfs
LC-20	Ockley	Fine-loamy, mixed, active, mesic Typic Hapludalfs
PY-6	Casco	Fine-loamy, mixed, superactive, mesic Typic Hapludalfs
PY-20	Ockley	Fine-loamy, mixed, active, mesic Typic Hapludalfs
		Texas profiles
S77TX-143-1	Windthorst	Fine, mixed, thermic Udic Paleustalfs
S80TX-479-1	Brundage	Fine-loamy, mixed hyperthermic Ustollic Natrargids
S82TX-479-2	Catarina	Fine, smectitic, hyperthermic Typic Torrents
S84TX-299-1	Voca	Fine, mixed, thermic Typic Paleustalfs
S86TX-249-1	Papalote Variant	Fine, mixed, hyperthermic Aquic Paleustalfs
S88TX-187-1	Branyon	Fine, smectitic, thermic Udic Haplusterts
S90TX-09901	Lewisville Variant	Fine, smectitic, thermic Udic Calciusterts
S90TX-099-2	Lewisville Variant	Fine, smectitic, thermic Vertic Calciustolls
S90TX-099-4	Bosque Variant	Coarse-loamy, carbonatic, thermic Typic Ustifluvents
S90TX-099-9	Bosque Variant	Fine-loamy, carbonatic, Pachic Haplustolls
S92-TX-099-1	Bastsil	Fine-loamy, siliceous thermic Udic Paleustalfs
S97-TX-099-1	Lewisville Variant	Fine-loamy, carmonatic, thermic Udic Calciustolls

SOC by the difference between total C and SIC by the dry combustion method (Nelson & Sommers, 1982). Bulk density was determined by the Saran clod method of Brasher et al. (1966).

Quantitative reconstruction (mass balance analysis) techniques (Smeck & Wilding, 1980; Chadwick et al., 1990; Nordt et al., 1999) were used to access the net gain/loss of SIC and SOC for each profile sampled. The advantages of volumetric reconstruction over simple weight reconstruction is that a volume factor is calculated for each horizon indicative of the volume of parent material necessary to form a unit volume of present day soil. The assumptions in quantitative reconstruction are: (i) the soil is derived from uniform parent material, (ii) the presence of an unaltered parent material base horizon, and (iii) an inert and immobile reference index constituent (Smeck & Wilding, 1980; West et al., 1988). Nordt et al. (1999) gives a sample equation for quantitative reconstruction similar to those used in this study.

PATHWAYS FOR SOIL INORGANIC CARBON SEQUESTRATION

The pedosphere is dynamic and rates of weathering and pedogenesis are variable depending on climate and parent material. Pedogenic and inherited carbonates represent one of the more dynamic, reactive and mobile soil minerals and carbonate flux is environmentally, and even seasonally dependent. Before considering actual sequestration scenarios, it is beneficial to review some basic chemical reactions in the global inorganic C cycle.

Dissolution

Global C reactions involve atmospheric CO_2, water and carbonate dissolution or precipitation. The dynamic equilibrium equations leading to carbonate dissolution are complex (Morse, 1983; Morse & Mackenzie, 1990) but can be simplified to a few basic reactions:

$$CO_2 \text{ (g)} + H_2O \leftrightarrow CO_2 \text{ (aq)} + H_2O \quad [1]$$

$$CO_2 \text{ (aq)} + H_2O \leftrightarrow H_2CO_3 \quad [2]$$

$$H_2CO_3 \leftrightarrow HCO_3^- + H^+ \quad [3]$$

If the soil solution contains free CO_3^{2-}, then the proton released in Eq. [3] reacts with the carbonate ion to yield more bicarbonate (Eq. [4])

$$H^+ + CO_3^{2-} \leftrightarrow HCO_3^- \quad [4]$$

$$CaCO_3 + HCO_3^- + H^+ \leftrightarrow 2HCO_3^- + Ca^{2+} \quad [5]$$

The CO_2 in meteoric waters (including rainwater) is usually in equilibrium with atmospheric CO_2 (Eq. [1] and [2]). This equilibrium yields a weak carbonic acid solution (H_2CO_3) with a pH of about 5.7. In a simple CO_2–water reaction there is an almost immediate dissociation of H_2CO_3 to $HCO_3^- + H^+$ (Eq. [3]). In reality the pH of rainwater is lower due to the reaction of other atmospheric constituents. Berner and Berner (1996) cite 64 pH values of worldwide precipitation with an average pH of 5.0 and standard deviation of 0.85. Soil pH is also buffered by inorganic [e.g., carbonate status, electrolyte concentration and species, cation exchange capacity (CEC), etc.] and organic constituents (root and microbial respiration).

The pH of the meteoric or vadose waters and the soil solution is important in determining the dominant carbonate species present. At the pH of most calcareous soil systems, or systems saturated with respect to carbonates (pH about 8.2), HCO_3^- is the dominant inorganic C species (Fig. 12–2). At low pH (pH < 6.4), possibly due to high partial pressure of soil air CO_2, organic acids, or oxidation of pyritic compounds, H_2CO_3 is dominant, while at high pH the carbonate ion CO_3^{2-} is dominant (Fig. 12–2).

It is important to note three fundamental phenomena at this juncture: (i) atmospheric CO_2 is continually sequestered due to its reaction with water (Eq. [1]

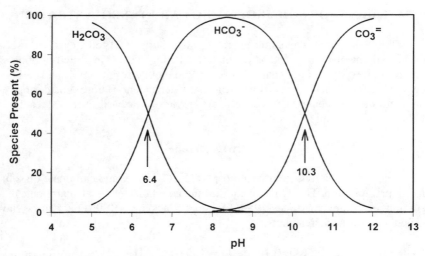

Fig. 12–2. Relative concentration of carbonate species to soil pH and pH where species ratio = 1 (arrow).

and [2]); (ii) unlike strong acids (e.g., HCl) that release CO_2 upon reacting with $CaCO_3$, carbonic acid (H_2CO_3) is a weak acid producing HCO_3^- upon reacting with $CaCO_3$ (Eq. [5]); and (iii) in Eq. [5], of the two moles of HCO_3^- produced during dissolution of calcium carbonate, one comes from atmospheric CO_2 (Eq. [1] and [2]) and the other from $CaCO_3$ (Eq. [5]). The importance of these phenomena relative to C sequestration will be considered later.

Precipitation

Carbonate equilibrium can be simply expressed by:

$$CaCO_3 \leftrightarrow Ca^{2+} + CO_3^{2-} \tag{6}$$

Any reaction that would tend to move the equilibrium to the right will cause dissolution, while a move to the left will result in carbonate precipitation. Traditional pathways for pedogenic carbonate synthesis include supersaturation of $Ca(HCO_3)_2^-$ leachates by dehydration or desiccation, lowering the partial pressure of soil CO_2 (degassing) and in situ alteration of lithogenic carbonates by dissolution-re-precipitation. In addition, micritization and secondary macropore plugging (Rabenhorst & Wilding, 1986; West et al., 1988); and biomineralization of carbonates by organisms (Krumbein, 1979; Jones & Kahle, 1986; Monger et al., 1991; Folk, 1993; Bruand & Duval, 1999) have been shown to be important pathways of pedogenic carbonate formation.

EFFECTS OF CLIMATE

In addition to chemical reactions, water availability is critical in evaluating the fate of pedogenic or lithogenic carbonates and possible C sequestration. The

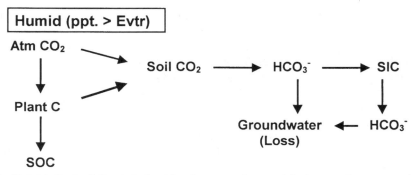

Fig. 12–3. Idealized soil C cycle for humid environments where precipitation exceeds evapotranspiration.

primary source of C is calcareous parent material, calcareous eolian dusts, and atmospheric CO_2 through root respiration. In humid regions, precipitation is greater than evapotranspiration resulting in partial or complete loss of carbonates from the soil profile. Combining Eq. [2] and [5] we arrive at the general reaction:

$$H_2O + CO_2$$
$$\updownarrow$$
$$CaCO_3 + H_2CO_3 \leftrightarrow 2HCO_3^- + Ca^{2+} \qquad [7]$$

In this leaching environment, SIC is lost as HCO_3^-. This process sequesters an equal amount of carbonate CO_2 and atmospheric CO_2 that is transferred to the hydrosphere in groundwater (Fig. 12–3). Discharge of HCO_3^- or $Ca(HCO_3)_2^-$ charged waters from vadose zones into groundwater aquifers, caves, lakes and oceans is a transitory sink for atmospheric CO_2. These groundwaters act as a conduit connecting regions of carbonate dissolution to regions of carbonate redeposition or storage.

In semiarid to subhumid climates precipitation is approximately equivalent to evapotranspiration. In this environment pedogenic carbonate is a transitory phase of sequestered CO_2 (Fig. 12–4). If there is subsequent carbonate dissolution

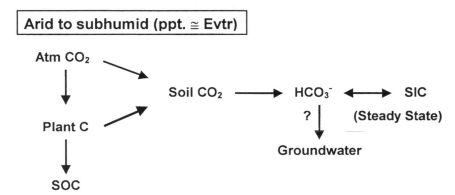

Fig. 12–4. Idealized soil C cycle for semi-arid to subhumid environments where precipitation equals evapotranspiration.

during periods of moisture excess, the resulting HCO_3^- may be transferred to groundwaters. In this environment, only aqueous carbonates and bicarbonates leached to the hydrosphere can be considered as a mechanism for sequestering atmospheric CO_2. However, dissolution of carbonates (pedogenic or lithogenic) in the upper part of the profile and re-precipitation lower in the soil profile or in situ in near proximity to locale of dissolution results in no net change in the SIC pool. Likewise, calcareous eolian dusts do not sequester atmospheric CO_2, but simply act as a transfer agent, moving carbonates to different locales where they may be stored, unless transported to a leaching environment.

In arid and semiarid environments, precipitation is less than evapotranspiration. Hence, soil solutions become saturated with respect to $Ca(HCO_3)_2$ resulting in a net gain in pedogenic carbonate (SIC) (Fig. 12–5). The source of Ca for $Ca(HCO_3)_2$ is critical relative to atmospheric CO_2 sequestration. If the source of Ca is from calcareous dusts, as is common in these environments, the formation of $Ca(HCO_3)_2$ has one mole of HCO_3^- from atmospheric CO_2 and one mole from carbonate dust. Upon precipitation of $CaCO_3$ from this solution one mole of CO_2 is returned to the atmosphere; hence there is no net gain or loss of atmospheric CO_2. However, if the source of Ca is from noncarbonate sources (e.g., $CaSO_4$, $CaCl_2$, etc.) or calcium-silicate mineral weathering, then the precipitation of $CaCO_3$ does result in sequestration of one mole of atmospheric CO_2.

$$2HCO_3^- + Ca^{2+} \leftrightarrow CaCO_3 + H_2O + CO_2\uparrow \qquad [8]$$

This sequestration scenario assumes, of course, that the HCO_3^- originated from either the atmosphere (Eq. [1], [2], [3]) or more likely from plant root respiration through photosynthesis. Dissolution-re-precipitation reactions in calcareous soils consumes atmospheric CO_2 in the dissolution process but liberate an equal mole fraction upon precipitation; thus, there is no net change in the CO_2 pool.

Accumulation rates of SIC in desert soils have been estimated at 0.1 to 0.6 g m^{-2} yr^{-1} (Schlesinger, 1985) that yields an estimated 800 to1000 $\times 10^{15}$ g of SIC

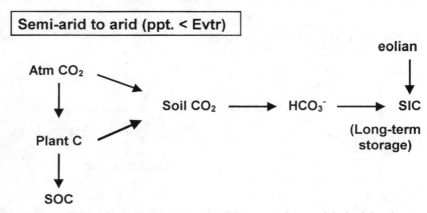

Fig. 12–5. Idealized soil C cycle for arid to semiarid environments where precipitation is less than evapotranspiration.

stored in Aridisols and arid Entisols (Schlesinger, 1982, 1995a). However, in light of the above discussion, it is questionable how much of this SIC truly represents atmospheric CO_2 sequestration or simply the transfer of carbonates from one locale to another via eolian transport. In addition, the CO_2 liberated during precipitation in Eq. [9] may not be released to the atmosphere, but remain in the soil thus increasing the partial pressure of soil air CO_2. Partial pressure of soil air CO_2 in arid regions has been reported to fluctuate seasonally between 0.1 and 1.3% (Parada et al., 1983), considerably higher than the current atmospheric concentration of 0.036%. However, plant and microbial respiration also play a significant role in maintaining high soil CO_2 levels.

MINERAL WEATHERING

To this point we have only considered the role of soil carbonates in C sequestration. Noncarbonate mineral weathering can also sequester atmospheric CO_2. In a leaching environment, silicate mineral weathering involves the process of hydrolysis. A simplistic example is the hydrolysis of fosterite (Mg olivine) (Eq. [9]) or albite (Na feldspar) weathering to kaolinite (Eq. [10]):

$$Mg_2SiO_4 + 4H_2O + 4CO_2 \rightarrow H_4SiO_4 + 4HCO_3^- + 2Mg^{2+} \qquad [9]$$

$$2NaAlSi_3O_8 + 11H_2O + 2CO_2 \rightarrow$$

$$Al_2Si_2O_5(OH)_4 + 4H_4SiO_4 + 2HCO_3^- + 2Na^+ \qquad [10]$$

A more generalized summation of this reaction would be:

primary silicate $+ H_2O + CO_2 \rightarrow$

$$\text{secondary mineral phase} + HCO_3^- + \text{ions in solution} \qquad [11]$$

Unlike carbonate weathering, all the CO_2 in silicate mineral weathering is derived ultimately from atmospheric CO_2. Morse and Mackenzie (1990) estimate that about 30% of the HCO_3^- in river water is derived from the weathering of silicate minerals. If the remaining 70% of the HCO_3^- comes from carbonates (50% of HCO_3^- derived from atmospheric CO_2), then about 64% (Berner & Berner, 1996) to 67% (Schlesinger, 1997) of the HCO_3^- in rivers is derived from atmospheric CO_2.

As with carbonate weathering, HCO_3^- may be leached to the hydrosphere. If there is subsequent precipitation of pedogenic carbonate due to silicate weathering, there is still a net sequestration of atmospheric CO_2. In Eq. [9] and [10] two to four moles of CO_2 are consumed to produce an equal amount of HCO_3^-, but only one mole of CO_2 is released upon carbonate precipitation (Eq. [8]). Chadwick et al. (1994) estimate that the SIC flux due to silicate mineral weathering ranges from 0.06 to 0.11 g m^{-2} yr^{-1} for moist grassland soil profiles. Although the flux may be small, mineral weathering results in the net sequestration of atmospheric CO_2.

CARBON CYCLE

Clearly, either the dissolution of carbonates or the weathering of silicate minerals fosters atmospheric CO_2 sequestration. The rates of sequestration are climate (temperature and moisture) and mineral stability dependent. Any precipitation of pedogenic carbonate can be considered as transitory or long-term storage of CO_2. In the end, the world's oceans are the ultimate sink for weathering products, including Ca^{2+}, HCO_3^- and CO_2. The oceans play a critical role in the global CO_2 balance, although the dynamics are not well understood (Morse & Mackenzie, 1990). Morse and Mackenzie (1990) estimate that only half of the C entering the oceans accumulates as carbonate sediments. A portion of CO_2, due to high pH and buffering capacity, is retained in the oceans as CO_2 with long residence times. The remaining C as HCO_3^- or CO_3^{2-} in oceans is sequestered through photosynthesis, precipitated mineral phases, transformed into lithogenic limestone or dolomite, or in the shells of marine animals through biomineralization. Biomineralization, through enzyme reaction effectively removes carbonic acid from the calcification site (Bolin et al., 1979).

$$Ca^{2+} + 2HCO_3^- \rightarrow Ca(HCO_3^-)_2 \rightarrow CaCO_3 + H_2CO_3 \qquad [12]$$

It is not clear whether biomineralization of calcite releases CO_2 to the atmosphere as does inorganic precipitation of calcite, although Berner and Berner (1996) indicate that the byproducts of biogenic $CaCO_3$ formation are CO_2 and H_2O. If we can assume that about 65% of the C entering the oceans is from atmospheric C (see discussion above) and that only one-half of the HCO_3^- is converted to carbonates, then about 33% of the atmospheric C sequestered through carbonate dissolution and silicate mineral weathering will be sequestered in the oceans in long-term residence.

Although many mechanisms are involved, through marine photosynthesis, biomineralization, carbonate precipitation, and CO_2 retention, a constant HCO_3^- concentration is maintained in the world's oceans. As marine organisms die, their skeletal framework is incorporated into sedimentary rock. Ultimately the lithosphere contains more C than all other sources combined (Table 12–1). This includes the shells of marine organisms and carbonate cementing agents that together form the vast quantity of carbonate sediments found around the world. In addition, abundant organic C is stored in sediments, primarily in shale.

The above discussion may be misleading without completing the C cycle. Although C is continually being sequestered by biological and geochemical means, a renewal vector is present that nearly balances the long-term sequestration vector and completes the geochemical C cycle. Without a return flux, carbonate and mineral weathering would deplete the atmosphere of C in about 10 000 yr (Berner & Lasaga, 1989). Organic matter decomposition, volcanic eruptions and other geological processes release CO_2, roughly balancing the quantity added to the world's oceans (Schlesinger, 1995a). The result is that natural processes have established a well regulated and buffered mechanism to maintain optimum CO_2 levels for sustaining life. The yet unresolved question is whether natural or anthropogenic changes in atmospheric CO_2 concentrations will upset the balance resulting in adverse climatic consequences.

RESULTS AND DISCUSSION

Carbon Sequestered

Organic and inorganic C was determined quantitatively for 20 profiles in Ohio and 12 in Texas (see Table 12–2). Three soil profiles in Texas were noncalcareous and four profiles were only evaluated for SIC flux. The solum thickness of the Ohio soils averaged about 1 m whereas soils in Texas averaged a little over 2 m (Table 12–3). The difference in solum thickness is due mostly to differences in texture and carbonate content of parent material, not weathering intensity or age. In addition, the Ohio soil profiles were separated on the basis of parent material: finer-textured glacial till and coarser-textured glacial outwash. Results for SIC loss and SOC gain per profile (kg m^{-2}) from both studies are given in terms of soil column thickness and to a depth of 1 m (Table 12–3) for comparison among profiles. However, for the Texas profiles, calculations based on solum thickness show that SIC losses are about 40 to 60% greater than calculations to a depth of 1 m. This is a consequence of the greater depth of leaching in these profiles compared to the shallower Ohio profiles.

The net gain of SOC for these soils across a wide climatic range averaged about 10 kg m^{-2} m^{-1} (Table 12–3), with a range from 4 to 23 kg m^{-2} m^{-1}. This gain in SOC represents net sequestration of atmospheric CO_2 through photosynthesis. It is important to note that these values are below-ground net gain in SOC and do not include most roots or above-ground biomass. These values are in close agreement with values reported for Mollisols in Arizona (8.8 kg m^{-2}) (Schlesinger, 1982) and for temperate forest (11.8 kg m^{-2}) and cultivated (12.7 kg m^{-2}) ecosystems (Schlesinger, 1997). The Texas profiles showed a wider range in gain of SOC and a slight positive correlation with mean annual precipitation. No similar correlation was observed for the Ohio profiles because of the more uniform temperature and precipitation. The gain in SOC in Ohio profiles was probably more closely related to parent material composition than to microclimatic effects. The net gain in SOC between Ohio and Texas is rather uniform despite the wide diversity in soils and climate.

The net loss in SIC is more than 10 times greater than the gain in SOC, averaging 105 kg m^{-2} m^{-1} (119 kg m^{-2} if solum considered) (Table 12–3). In the Ohio soil profiles, the loss of carbonates was highly variable and tended to reflect the texture of the parent material. The coarse-textured outwash (>30% coarse fragments) probably had better infiltration rates that allowed for greater dissolution and leaching of carbonates. These soils averaged about 230 kg m^{-2} m^{-1} while the finer textured till samples lost about 58 kg m^{-2} m^{-1} of SIC (Table 12–3). With the exception of one profile (S88TX-187-1), the Texas profiles had SIC losses similar to the finer-textured till soils from Ohio. Profile S88TX-187-1 was a Vertisol in a microlow topographic position, completely leached of carbonates to 120 cm but with ~75% $CaCO_3$ between 165 and 300 cm. This micro-low position represents a leaching environment as contrasted to the adjacent micro-high that is calcareous to the surface. Both the micro-low and micro-high soils have comparable carbonate contents in the parent material, which suggests that there was no significant redistribution of carbonates within the micro-low profile.

Table 12–3. Summary of gain/loss of SIC and SOC per meters squared per column (col) and per meter (m) depth.

Pedon no.	Solum thickness	Gain/loss				Sequestered C SIC/SOC	Ratio SIC/SOC
		SOC	SOC	SIC	SIC		
	cm	kg m^{-2} col^{-1}	kg m^{-2} m^{-1}	kg m^{-2} col^{-1}	kg m^{-2} m^{-1}	kg m^{-2} col^{-1}	kg m^{-2} m^{-1}
				Ohio profiles			
PT-28	97	11	11	−45	−45	4	4
HN-97	86	9	9	−18	−18	2	2
FT-18	102	8	8	−67	−67	8	8
MICH-46	86	7	7	−56	−56	8	8
PB-69	127	9	9	−113	−111	12	12
PB-70	99	8	8	−128	−128	16	16
DK-23	76	8	8	−93	−93	12	12
DK-24	102	10	10	−62	−61	6	6
AG-1	81	9	9	−53	−53	6	6
AL-125	99	7	7	−53	−53	7	7
FR-57	125	16	14	−96	−87	6	6
SK-30	132	12	8	−2	−1	0.2	0.1
PG-2	140	12	10	−14	−8	1	1
LC-21	102	8	8	−29	−29	3	3
CH-54†	99	10	10	−359	−359	35	35
CH-55†	92	5	5	−354	−354	68	68
KX-2†	94	8	8	−99	−99	12	12
LC-20	132	11	8	−131	−107	12	13
PY-6	79	7	7	−209	−209	29	29
PY-20	152	12	9	−329	−254	27	28
All profiles							
Avg.	106	9	9	−116	−110	14	14
Std. dev.	22	2	2	111	105	16	16
Max.	152	16	14	−2	−1	68	68
Min.	76	5	5	−359	−359	0.2	0.1
Coarse textured outwash†							
Avg.	108	9	8	−247	−230	31	31
Std. dev.	28	3	2	116	114	21	20
Max.	152	12	10	−99	−99	68	68
Min.	79	5	5	−359	−359	12	12
Till samples							
Avg.	105	10	9	−59	−58	7	7
Std. dev.	21	2	2	38	37	5	5
Max.	140	16	14	−2	−1	16	16
Min.	76	7	7	−128	−128	0.2	0.1

(continued on next page)

The gain in SOC and loss of SIC both represent sequestration of atmospheric CO_2 although the processes are quite different. The values expressed here are only for 1 m^2 and to a depth of 1 m. However, the depth of leaching of SIC and accumulation of SOC generally occurs within the upper meter for the soils examined.

Table 12–3. Continued.

Pedon no.	Solum thickness	Gain/loss				Sequestered C SIC/SOC	Ratio SIC/SOC
	cm	SOC kg m^{-2} col^{-1}	SOC kg m^{-2} m^{-1}	SIC kg m^{-2} col^{-1}	SIC kg m^{-2} m^{-1}	kg m^{-2} col^{-1}	kg m^{-2} m^{-1}
				Texas profiles			
S77TX-143-1	107	5.4	5.4	0	0	--	--
S80TX-479-1	183	4.2	4.2	−14	−14	3	3
S82TX-479-2	185	12.5	8.1	9	5	1	1
S84TX-299-1	154	7.6	6.8	0	0	--	--
S86TX-249-1	300	9.8	5.7	−109	−51	11	9
S88TX-187-1	250	32.5	23.1	−948	−682	29	30
S90TX-099-1	270	24.5	18.7	−90	−83	3	5
S92TX-099-1	263	6.5	4.3	0	0	--	--
S97TX-099-1	210	27.9	18.2	−19	−15	1	1
Avg.	214	15	11	−130	−93	8	8
Std. dev.	62	11	7	310	223	11	11
Max.	300	33	23	9	5	29	30
Min.	107	4	4	−948	−682	1	1
		Additional Texas sites for SIC flux determination					
S90TX-099-4	75	ND‡	ND	--	--		
S90TX-099-9	250	ND	ND	--	--		
S90TX-099-2	250	ND	ND	--	--		
S90TX-099-1	250	ND	ND	−108	−43		

† Coarse-textured glacial outwash samples.
‡ ND = not determined.

Carbon Flux Rate

Where soil chronology is available, average flux of SIC and SOC can be estimated. However, in these leaching environments there are some basic assumptions and limitations that have to be considered in SIC flux estimates: (i) all the dissolution of carbonate is by the dissociation of carbonic acid to form $HCO_3^- + H^+$ in soil systems (see Eq. [3]); (ii) no carbonate dissolution by organic acids; (iii) no oxidation of sulfides to sulfuric acid that leads to acid sulfate weathering; (iv) dissolution is not a function of fertilization of cultivated soils with anhydrous ammonia or acid producing compounds; and (v) no silicate mineral weathering that would release HCO_3^- (see Eq. [9] and [10]). In addition, there are several assumptions necessary for SOC flux estimates: (i) no additions of sewage sludge, manure or other materials that would tend to elevate SOC levels higher than would be expected under current land management; (ii) minimal erosion of the Ap or A horizon; (iii) rates of accumulation and decomposition have reached near steady state-equilibrium; and (iv) soils of the glaciated region of Ohio contained no SOC until the ice retreated.

In the Ohio soils, and four Texas soils there are good estimates of soil age (Goldthwait, 1959; Forsyth, 1961; Smeck et al., 1968; Rostad et al., 1976; Nordt

Table 12–4. Mass flux for SIC and SOC for calcareous soils.

Pedon no.	Age YBP	SIC	SIC flux	SOC	SOC flux
		kg m^{-2} m^{-1}	g m^{-2} m^{-1} yr^{-1}	kg m^{-2} m^{-1}	g m^{-2} m^{-1} yr^{-1}
			Ohio pedons		
MICH-46	9 000	56	6	7	0.8
FT-18	9 500	67	7	8	0.8
PT-28	10 000	45	5	11	1.1
HN-97	10 000	18	2	9	0.9
AL-125	12 000	53	4	7	0.6
AG-1	13 000	53	4	9	0.7
DK-23	14 000	93	7	8	0.6
DK-24	14 000	61	4	10	0.7
PY-6	14 000	209	15	7	0.5
PY-20	14 000	254	18	9	0.6
SK-30	14 000	1	0.1	8	0.6
PG-2	14 000	8	1	10	0.7
CH-54	14 500	359	25	10	0.7
CH-55	14 500	354	24	5	0.3
LC-20	15 000	107	7	8	0.5
FR-57	15 000	87	6	14	0.9
LC-21	16 000	29	2	8	0.5
PB-70	16 500	128	8	8	0.5
KX-2	16 500	99	6	8	0.5
PB-69	17 500	111	6	9	0.5
Avg.			8		0.7
Std. dev.			7		0.2
Max.			25		1.1
Min.			0.1		0.3
			Texas pedons		
S90TX-099-4	Modern		--		
S90TX-099-9	2 000		--		
S90TX-099-2	5 000		--		
S90TX-099-1	15 000		3		

et al., 1999). In the Ohio soils the SIC flux rate averaged about 8 g m^{-2} m^{-1} yr^{-1} with a range of 0.1 to 25 g m^{-2} m^{-1} yr^{-1} (Table 12–4). The soils examined in Texas were from a subhumid climate and the soils <5000 YBP exhibited mostly transformation of lithogenic to pedogenic carbonate, with no net change in carbonate content (Nordt et al., 1998). The Texas pedon older than 5000 yr exhibited SIC loss with a flux rate of 3 g m^{-2} m^{-1} yr^{-1}. The rate of loss of SIC in the subhumid region is lower than the more humid region, but the depth of profile development in the Texas soil was greater. If the losses of SIC are calculated for the solum, the loss of SIC is comparable in both environments. Although there was probably a greater loss of SIC earlier in the life of these soils, they are still under a leaching regime, and SIC sequestration can be expected to continue. We should reiterate that at least half of these SIC losses represent sequestration of atmospheric CO_2. In arid soils where there is SIC accumulation of carbonates, Schlesinger (1985) calculated a net flux of 0.24 g m^{-2} yr^{-1} over the past 20 000 yr. As pointed out earlier, this flux in arid regions may not be truly representative of the sequestration of atmospheric CO_2.

Table 12–5. Soil orders with potential carbonate leaching, total loss of SIC and SIC sequestration rate.

Order	Total area† $m^2 \times 10^{12}$	Potential area (%) for $CaCO_3$ removal	Leaching area $m^2 \times 10^{12}$	Loss of‡ SIC $g \times 10^{15}$	Flux rate§ loss of C $(g\ C\ yr^{-1})$ $g \times 10^{12}$
Alfisols	12.6	50	6.3	630	50
Andisols	0.91	40	0.4	36	3
Aridisols	15.7	0	0.0	0	0
Entisols	21.1	40	8.4	844	68
Gelisols	11.3	50	5.7	565	45
Histosols	1.5	0	0.0	0	0
Inceptisols	12.8	40	5.1	512	41
Mollisols	9	75	6.8	675	54
Oxisols	9.8	0	0.0	0	0
Spodisols	3.4	0	0.0	0	0
Ultisols	11.1	20	2.2	222	18
Vertisols	3.2	50	1.6	160	13
Other	18.4	50	9.2	920	74
Total	130.81		45.6	4564	365
% of total =			34.9		

† Area measurements from Wilding, 1999.
‡ Value based on average loss of 10 kg m^{-2}.
§ Value based on flux rate of 8 g m^{-2} yr^{-1}.

Accumulation rates of SOC for the Ohio soils averaged 0.7 g m^{-2} yr^{-1} with a range of 0.3 to 1.1 g m^{-2} yr^{-1}. The narrow range is probably because the soils developed on similar parent materials with a uniform climate. These values agree with those of Chadwick et al. (1994) for older temperate grassland soils, but less than the calculated accumulation rate of 1.35 g m^{-2} yr^{-1} for soils developed on glaciated Holocene materials (Schlesinger, 1990, 1997). However, the data reported by Schlesinger (1997) are younger (<10 000 yr), and of different ecosystems than reported here. Both systems probably reached steady-state equilibrium in soil organic matter content well before the current age, probably in <1000 yr. Thus, older soils, such as in Ohio will yield lower flux rates.

Global Perspective

These data give insight to the sequestration of SIC due to leaching of calcareous parent materials. However, the data set is of little use unless it can be extrapolated to other geographic regions in order to gain a better perspective on SIC sequestration on a global scale. Extrapolation is possible if we make some basic assumptions: (i) the SIC sequestration potential is near the calculated mean (~100 kg m^{-2}); (ii) many soils have carbonates susceptible to leaching except those in arid or cryic environments or developed from noncalcareous parent materials; (iii) C is sequestered as HCO_3^- and transported into groundwater with long residence times; and (iv) the HCO_3^- in rainwater is minimal in comparison to the amount derived from mineral weathering. Using the total ice-free land area as a base (130.8 × 10^{12} m^2) (Wilding, 1999), we arrived at a rough estimate of the percentage of each soil order that could have a carbonate leaching potential (Table 12–5). Aridisols were excluded as they represent net C storage. Histosols, Spodisols and Oxisols were excluded because they are generally carbonate-free. The remaining soil orders were adjusted for climatic regimes. Those in aridic or in tropical areas were likewise excluded. The soils remaining were normalized for areal distribution based on their probability of containing carbonates. This yields about 45×10^{12} m^2 (Table 12–5), which is about 35% of the total ice-free land area for potential leaching of carbonates.

Based on these calculations and a sequestration potential of 100 kg m^{-2}, a conservative estimate is that pedogenic processes have sequestered a minimum of 4564 $\times 10^{15}$ g of SIC (Table 12–5), almost 2.5 times as much SIC as current C reserves (Table 12–1). This value does not include weathering of silicate minerals. Clearly, over time, pedogenic processes have been effective C sequesters. Only half of this C is atmospheric C, the other is from carbonates or other sources. Even so, soil weathering can be conservatively estimated to have sequestered over 2280×10^{15} g of atmospheric C, doing its part to maintain CO_2 equilibrium on a global scale.

Soil Inorganic Carbon Global Flux

The flux of SIC on a global scale may be even more speculative. If the estimated ice-free land area for potential leaching is 45×10^{12} m^2 (Table 12–5), and the calculated flux of 8 g C m^{-2} yr^{-1} (Table 12–4) is used as a proxy for SIC loss on a global scale, then a sequestration rate of about 0.36×10^{15} g C yr^{-1} (Table 12–5) is estimated that is just slightly less than the amount of C added to oceans from streams and rivers (0.4×10^{15} g C yr^{-1}; Sarmiento & Sundquist, 1992; Berner & Berner, 1996). This close agreement would seem to indicate that the percentages of potential leaching of carbonates was a reasonable first estimate. However, the percentages may be a little high for the calculations do not account for the estimated 30% of HCO_3^- derived from silicate mineral weathering. But, these calculations probably yield a good first approximation of the relative weathering sequestration potential from each soil order. The actual dissolution potential is temperature and rainfall influenced, and the average flux estimated may not hold for all circumstances. Additional work will be needed to fine-tune these sequestration estimates and partition SIC losses between carbonate dissolution and mineral weathering.

If the total C sequestered (4564×10^{15} g) is extrapolated since glacial retreat, about 10 000 to 15 000 yr, a flux rate of 10 to 6.7 g C m^{-2} yr^{-1} is estimated, which approximates the estimated value of 8 g m^{-2} yr^{-1} based on this study. It is questionable whether this flux will continue. As soils age, the rate of carbonate dissolution will naturally decrease. Many of the calcareous soils in the USA and Europe were first exposed to a leaching environment since the last glacial retreat and are considered young on a geologic time scale. Even so, erosion continues to expose fresh soil material, and new fluvial materials rich in carbonates continue to be deposited in regions subject to leaching environments.

SUMMARY

Understanding C dynamics and exchange reactions on a global scale is not an easy task for there are many interacting variables that influence C release, exchanges between C reservoirs, and C sequestration (short and long-term). A few of the variables are known, but many more are unknown and we are just beginning to appreciate how the many variables operate and interact. The thin veneer of soil over the earth's land surface sustains life through geochemical and biogeochemical linkages between the atmosphere, biosphere and hydrosphere. These linkages are the primary transfer mechanisms that move SIC from the atmosphere to the hydrosphere. In many discussions on global C balance the pedosphere linkage interface is often ignored or at best not fully appreciated. This research has attempted to show that the influx of HCO_3^- from the land surface to the oceans can be accounted for from the dissolution of carbonates from specific soil orders. An overall average dissolution rate of 8 g C m^{-2} yr^{-1} (Table 12–4) does not seem unreasonable. Additional research is needed to better partition weathering rates of carbonates and silicate minerals between soil orders.

The intent herein is to draw attention to the role of soil in a global C perspective. The upper meter of the soil surface contains more SOC than the biosphere and more SIC than the atmosphere. The natural processes of mineral weathering, dissolution of carbonates in calcareous soils, carbonate precipitation, photosynthesis and CO_2 dissolved in water sequesters C at about the same rate as other natural processes cycle C back to the atmosphere to be used as the building block of life. Because the pedosphere is such a large C pool (both SIC and SOC), anthropogenic perturbations to this pool through mismanagement of soil, water and plant resources could adversely influence the global C cycle and feedback mechanisms. Positive feedbacks through erosion control, improved land management and cropping systems, and afforestation can help mitigate C losses and sequester additional C.

REFERENCES

Berner, E.K., and R.A. Berner. 1996. Global environment: Water, air, and geochemical cycles. Prentice Hall, Upper Saddle River, NJ.

Berner, R.A., and A.C. Lasaga. 1989. Modeling the geochemical carbon cycle. Sci. Am. 260(3):74–81.

Bolin, B., E.T. Degens, P. Duvigenaud, and S. Kempe. 1979. The global biogeochemical carbon cycle. p. 1–56. *In* B. Bolin et al. (ed.) The global carbon cycle. John Wiley & Sons, Inc., New York.

Brasher, B., D.P. Franzmeier, V.T. Vallassis, and S.E. Davidson. 1966. Use of Saran resin to coat natural soil clods for bulk density and water retention measurements. Soil Sci. 101:108.

Bruand, A., and O. Duval. 1999. Calcified fungal filaments in the petrocalcic horizons of Eutrochrepts in Beauce, France. Soil Sci. Soc. Am. J. 63:164–169.

Chadwick, O.A., G.H. Brimhall, and D.M. Hendricks. 1990. From a black to a gray box—a mass balance interpretation of pedogenesis. Geomorphology 3:369–390.

Chadwick, O.A., E.F. Kelly, D.M. Merritts, and R.G. Amundson. 1994. Carbon dioxide consumption during soil development. Biogeochemistry 24:115–127.

Dreimanis, A. 1962. Quantitative gasometric determination of calcite and dolomite by using the Chittic apparatus. J. Sediment. Petrol. 32:520–529.

Eswaren, H., E.Van den Berg, P. Reich, and J. Kimble. 1995. Global soil carbon resources. p. 27–43. *In* R. Lal et al. (ed.) Soils and global change. Lewis Publ., Boca Raton, FL.

Folk, R.L. 1993. SEM imaging of bacteria and nannobacteria in carbonate sediments and rocks. J. Sediment. Petrol. 63:990–999.

Forsyth, J.L. 1961. Dating Ohio's glaciers. Ohio Dep. Natur. Resour. Geol. Surv. Inform. Circ. no. 30.

Goldthwait, R.P. 1959. Scenes in Ohio during the last ice age. Ohio J. Sci. 59:193–216.

Jones, B., and C.F. Kahle. 1986. Dendritic calcite crystals formed by calcification of algal filaments in a vadose environment. J. Sediment. Petrol. 56:217–227.

Kempe, S. 1979. Carbon in the rock cycle. p. 343–377. *In* B. Bolin et al. (ed.) The global carbon cycle. John Wiley & Sons, Inc., New York.

Krumbein, W.E. 1979. Calcification by bacteria and algae. p. 47–68. *In* P.A. Trudinger and D.J. Swaine (ed.) Biogeochemical cycling of mineral-forming elements. Elsevier Sci. Publ. Co., New York.

Lal, R., J.M. Kimble, R.F. Follett, and C.V. Cole. 1998. The potential of U.S. cropland to sequester carbon and mitigate the greenhouse effect. Sleeping Bear Press, Chelsea, MI.

Lal, R., J. Kimble, E. Levine, and B.A. Stewart. 1995a. Soils and global change. Lewis Publ., Boca Raton, FL.

Lal, R., J. Kimble, E. Levine, and B.A. Stewart. 1995b. Soil management and greenhouse effect. Lewis Publ., Boca Raton, FL.

Lal, R., J. Kimble, E. Levine, and C. Whitman. 1995c. World soils and greenhouse effect: An overview. p. 1–7. *In* R. Lal et al. (ed.) Soils and global change. Lewis Publ., Boca Raton, FL.

Monger, H.C., L.A. Daugherty, W.C. Lindeman, and C.M. Liddell. 1991. Microbial precipitation of pedogenic calcite. Geology 19:997–1000.

Morse, J.W. 1983. The kinetics of calcium carbonate dissolution and precipitation. p. 227–264. *In* R.J. Reeder (ed.) Carbonates: mineralogy and chemistry. Rev. in Mineral. 11. Mineral. Soc. Am., Blacksburg, VA.

Morse, J.W., and F.T. Mackenzie. 1990. Geochemistry of sedimentary carbonates. Elsevier, New York.

Nelson, D.W., and L.E. Sommers. 1982. Total carbon, organic carbon and organic matter. p. 539-580. *In* A.L. Page et al. (ed.) Methods of soil analysis. Part 2. 2nd ed. Agron. Monogr. 9. ASA and SSSA, Madison, WI.

Nordt, L.C., C.T. Hallmark, L.P. Wilding, and T.W. Boutton. 1998. Quantifying pedogenic carbonate accumulations using stable carbon isotopes. Geoderma 82:115–136.

Nordt, L.C., L.P. Wilding, and L.R. Drees. 1999. Pedogenic carbonate transformations in leaching soil systems: Implications for the global C cycle. p. 43–64. *In* R. Lal et al. (ed.) Global climate change and pedogenic carbonates. Lewis Publ., Boca Raton, FL.

Parada, C.B., A. Long, and S.N. Davis. 1983. Stable-isotope composition of soil carbon dioxide in the Tucson basin, Arizona, U.S.A. Isotope Geosci. 1:219–236.

Rabenhorst, M.C., and L.P. Wilding. 1986. Pedogenesis on the Eswards plateau, Texas: III. New model for the formation of petrocalcic horizons. Soil Sci. Soc. Am. J. 50:693–699.

Rostad, H.P.W., N.E. Smeck, and L.P. Wilding. 1976. Genesis of argillic horizons in soils derived from coarse-textures calcareous gravels. Soil Sci. Soc. Am. J. 40:739–774.

Sarmiento, J.L., and E.T. Sundquist. 1992. Revised budget for the oceanic uptake of anthropogenic carbon dioxide. Nature (London) 356:589–593.

Schlesinger, W.H. 1982. Carbon storage in the caliche of arid soils: A case study from Arizona. Soil Sci. 133:247–255.

Schlesinger, W.H. 1985. The formation of caliche in soils of the Mojave Desert, California. Geochim. Cosmochim. Acta 49:57–66.

Schlesinger, W.H. 1990. Evidence from chronosequence studies for a low carbon-storage potential of soils. Nature (London) 348:232–234.

Schlesinger, W.H. 1995a. An overview of the carbon cycle. p. 9–25. *In* R. Lal et al. (ed.) Soils and global change. Lewis Publ., Boca Raton, FL.

Schlesinger, W.H. 1995b. Soil respiration and changes in soil carbon stocks. p. 159–168. *In* G.M. Woodwell and F.T. Mackenzie (ed.) Biotic feedbacks in the global climate system. Oxford Univ. Press, New York.

Schlesinger, W.H. 1997. Biogeochemistry: An analysis of global change. Acad. Press, New York.

Schlesinger, W.H. 1999. Carbon sequestration in soils. Science (Washington, DC) 284:2095.

Smeck, N.E., L.P. Wilding, and N. Holowaychuk. 1968. Genesis of argillic horizons in Celina and Morley soils of western Ohio. Soil Sci. Soc. Am. Proc. 32:550–556.

Smeck, N.E., and L.P. Wilding. 1980. Quantitative evaluation of pedon formation in calcareous glacial deposits in Ohio. Geoderma 24:1–16.

West, L.T., L.P. Wilding, and C.T. Hallmark. 1988. Calciustolls is central Texas: II. Genesis of calcic and petrocalcic horizons. Soil Sci. Soc. Am. J. 52:1731–1740.

Wilding, L.P. 1999. General characteristics of soil orders & global distributions. p. E175–E183. *In* M.E. Sumner (ed.) Handbook of soil science. CRC Press, Boca Raton, FL.

13 Fate of Eroded Soil Organic Carbon: Emission or Sequestration

Rattan Lal

The Ohio State University
Columbus, Ohio

ABSTRACT

Conversion of natural to agricultural ecosystems accelerates soil erosion that leads to depletion of the soil organic carbon (SOC) content of the surface layer. The increased loss of SOC occurs because it is concentrated in the surface layer, has low density and the labile particulate fraction is relatively unconsolidated. Removal by soil erosion may constitute half of the total loss of SOC pool from agricultural soils, the other half being due to an increase in mineralization caused by change in soil temperature and moisture regimes. The fate of eroded SOC is not clearly understood and depends on the specific stage of the erosional process. Aggregate breakdown by agents of erosion may enhance mineralization of the SOC thus exposed. As much as 20 to 80% of the SOC redistributed over the landscape may be susceptible to mineralization, leading to global emission of an estimated 1.14 Pg C yr^{-1}. Some of the particulate and dissolved organic C (POC and DOC) also may be easily mineralized. Yet, 0.57 to 1.0 Pg of C carried into the aquatic ecosystems may be sequestered annually. The fate of SOC buried in depositional sites depends on the degree of aggregation and anaerobiosis that it causes. While aggregation may suppress mineralization, the labile SOC pool may be mineralized and also be prone to methanogenesis. There is a strong need to strengthen the database and improve scientific knowledge about the fate of eroded SOC.

Natural erosion has shaped the surface of the earth throughout its geologic history. Formation of deep fertile soils of the flood plains and loess plateaus are due to the geologic erosion that has occurred at slow rates for billions of years. The so-called hydric civilizations (ancient civilizations that developed along the flood plains of major rivers e.g., Nile, Indus, Yangtze, etc.) owe their development to fertile soils formed and continuously renewed by the sediments carried from the highlands (Hillel, 1991, 1994). Egypt, as the saying goes, is a gift of the river Nile.

The constructive and slow geologic erosion, however, is accelerated by anthropogenic perturbations. When the rate of erosion exceeds the tolerable limit (about 12 Mg ha^{-1} yr^{-1} for deep soils of mid- and high latitudes and 2 to 5 Mg ha^{-1} yr^{-1} for shallow soils of tropical ecoregions), it becomes a destructive process because the rate of soil renewal lags behind that of soil removal. The top soil layer is

Table 13–1. The loss of soil organic C by accelerated erosion of a plowed and uncropped Alfisol on different slope gradients in western Nigeria (Lal, 1976).

Slope %	1972			1973		
	Season 1†	Season 2‡	Total	Season 1	Season 2	Total
			kg ha^{-1}			
1	25.4	28.0	53.4	269.9	119.6	389.5
5	322.6	254.2	576.8	1005.6	1312.0	2317.6
10	350.0	490.0	840.0	2492.3	1139.4	3631.7
15	892.8	403.7	1296.5	2842.0	939.0	3481.0
Rainfall (mm)	656	168	824	781	416	1197

† Season 1 = April to July.
‡ Season 2 = August to October

thus progressively depleted by the accelerated erosion. The rate of soil erosion is exacerbated by conversion from natural to agricultural ecosystems, and by agricultural practices that cause drastic soil disturbance and lead to removal of crop residue and other protective biomass from the soil surface. Soon after conversion from natural to agricultural ecosystems, the soil loss by erosion is largely derived from the surface layer which is high in SOC content. Most of the organic matter in this layer is poorly consolidated, of low density, and is easily removed by runoff and erosion by water and wind.

While the depletion of the on-site SOC pool of eroded soils is widely recognized, fate of the C transported with the eroded sediments is a matter of much debate. Therefore, the objective of this chapter is to provide a state-of-the-knowledge synthesis about the fate of C entrained with runoff and eroded sediments. The specific discussion includes: (i) the relative on-site loss of SOC by erosion vs. mineralization losses on agricultural soils, (ii) the extent of mineralization of SOC redistributed over the landscape, (iii) the fraction of eroded SOC buried in depositional sites, and (iv) the proportion of eroded SOC carried into the aquatic ecosystems.

DEPLETION OF SOIL ORGANIC CARBON
IN AGRICULTURAL SOILS

Accelerated soil erosion affects environment quality through transport of sediments and pollutants in natural waters (Pimental et al., 1995) and emission of particulate materials and gases into the atmosphere (Lal, 1995, 1999). Being concentrated near the soil surface and of relatively low density (1.2–1.5 Mg m^{-3} for organic matter compared with 2.6 to 2.7 Mg m^{-3} for mineral fraction), SOC is one of the first soil constituents removed by accelerated erosion due to water or wind. Consequently, the transported sediments are usually enriched in SOC for both water erosion (Lal, 1976; Cihacek et al., 1993; Palis et al., 1997) and wind erosion (Zobeck & Fryrear, 1986). Therefore, severe erosion can cause heavy losses of organic matter. The data in Table 13–1 from Nigeria show annual loss of SOC in eroded sediments ranging from 390 kg ha^{-1} for 1% slope to 3781 kg ha^{-1} for 15% slope. As a consequence of the severe losses, the SOC content of eroded soil can be drastically lower than that of the uneroded soil. The data in Table 13–2 show reduction in SOC content by 50 to 56% within 2 yr. Similar data from other experi-

Table 13–2. SOC content of 0- to 5-cm depth of a severely eroded and uneroded Alfisol shown in Table 13–1 (Lal, 1976).

Slope	Under forest cover 1972	Severely eroded 1974	Uneroded (mulched) 1974
%		g kg^{-1}	
1	27.0	13.0	28.5
5	19.1	9.5	19.0
10	21.0	11.5	21.0
15	25.0	11.0	25.0

ments have shown that the SOC pool is lower in eroded than uneroded soils (Lal, 1976; 1996; Voroney et al., 1981; Tiessen et al., 1982; Parton et al., 1987; Nizeyimana & Olson, 1988; Fahnestock et al., 1996). Experiments conducted on a Fragipan soil (fine, silty, mixed, thermic glossy Fragiudalf) in Mississippi by Rhoton and Tyler (1990) showed severe decline in SOC pool by accelerated erosion. The SOC pool to 1-m depth was 60 Mg C ha^{-1} under wooded conditions with none or minimal erosion, 35 Mg C ha^{-1} in a slightly eroded soil, and only 19 Mg C ha^{-1} in moderately and severely eroded soil. While cultivation resulted in a loss of 25 Mg C ha^{-1}, erosion reduced SOC by an additional 16 Mg C ha^{-1}.

SOIL EROSION AND SOIL ORGANIC CARBON DYNAMICS

In its natural state, soil particles are aggregated to form structural units comprising of secondary particles. These secondary particles create a wide range of pores for retention and transmission of water and nutrients, and diffusion of gases between soil and the atmosphere. A well-structured soil comprises aggregates of varying sizes and strength. Macro-aggregates (>250 μm) comprise micro-aggregates bonded by roots and fungal hyphae. Micro-aggregates (<250 μm) comprise clay particles attached to organic matter (OM) and polyvalent cations (P) (Edwards & Bremner, 1967; Tisdall, 1996). These micro-aggregates can be represented as (clay-P-OM) units that in turn are combined to form macro-aggregates schematically represented as (clay-P-OM)$_x$, or [(clay-P-OM)$_x$]$_y$. The C thus encapsulated within stable micro-aggregates is protected, physically and chemically, from microbial activity and is sequestered. Formation of stable micro-aggregates in the subsoil, therefore, takes C out of circulation.

Soil erosion is a three-stage process (Fig. 13–1). In contrast to the process of aggregation, the first stage involves breakdown of aggregates by the kinetic energy of raindrop impact, flowing water or blowing wind. The rate and extent of breakdown of aggregates depend on structural stability of the soil and erosivity of the agents of erosion (e.g., rainfall intensity, wind velocity, etc.) The C contained in the OM physically protected by the aggregate is now exposed to microbial processes. The extent of C mineralized due to the lack of physical protection following the breakdown of aggregates by accelerated erosion is not known for most soils and ecoregions.

Following conversion from natural to agricultural ecosystems, there is little information about the relative loss of SOC by mineralization vs. erosion. In some erosion-prone environments, loss by erosion may be more than that due to miner-

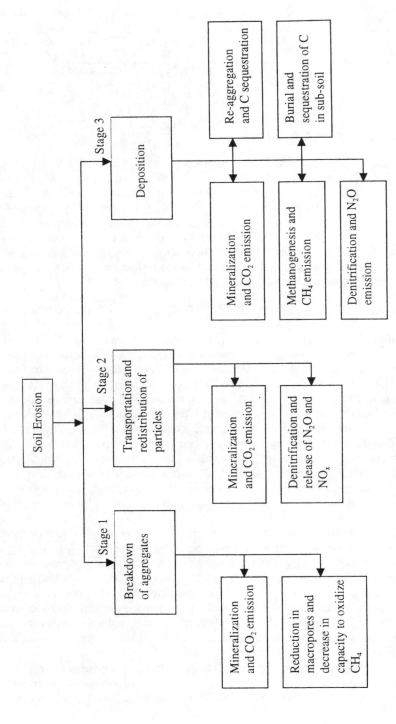

Fig. 13–1. Three stages of soil erosion and the potential for emission of GHG at these stages.

Table 13–3. Impact of erosion and mineralization on SOC content for 0- to 10-cm layer of an Alfisol in western Nigeria.

Slope	Antecedent SOC	SOC content after 2 yr		Δ SOC	
		Plow	No till	Erosion†	Mineralization‡
%		g kg^{-1}		%	
1	27.0	16.0	21.0	18.5	22.2
5	19.0	12.0	16.5	23.6	13.2
10	21.0	19.0	22.0	9.5	(−4.8)
15	25.0	14.0	21.0	19.0	16.0
Mean	23.0 ± 3.7	15.3 ± 3.0	20.1 ± 2.5	17.7 ± 5.9	17.1 ± 4.6

† The loss by erosion − the loss under plow till − loss under no till.
‡ The loss by mineralization − the loss under no till.

alization (Lucas & Vitosh, 1978). In Missouri, Colombia, USA, Slater and Carleton (1938) observed that on a continuously fallowed plot, loss of SOC due to erosion was 1.8 times more than that due to mineralization. Soon after conversion from natural to agricultural ecosystems the loss due to mineralization may be a dominant process (Gregorich & Anderson, 1985). With progressive decline in soil structure, however, the loss of SOC by erosion may ultimately represent 40 to 50% of the total loss (de Jong & Kachanoski, 1988). Schlesinger (1986) synthesized the available literature on SOC dynamics in relation to erosion of soils in midwestern USA and concluded that erosion loss is not responsible for the rapid decline in the labile pool of SOC upon land-use conversion. Since erosional losses of C did not explain the rapid decline of SOC on cultivation, Schlesinger concluded that most of the C must be released to the atmosphere as CO_2 by mineralization. Harden et al. (1999) concluded that in the cropping patterns in the Mississippi basin prior to 1950 cultivation caused 20 to 30% reduction in SOC pool of which erosion may have been responsible for 80% (or 16–24% of the original SOC) of the loss. The data in Table 13–3 from runoff plots in western Nigeria compare the effect of erosion vs. mineralization on SOC content. Over the 2-yr period, the mean loss due to erosion and mineralization was about 17% each. With severe decline in SOC content of the plow till treatment with time, the rate of SOC loss both by erosion and mineralization decreased with time.

The second stage involves transport of detached and entrained particles by splash, overland flow or wind movement (Fig. 13–1). Some of the SOC being transported by wind and water can be mineralized during the transport, especially when it is no longer protected by physical encapsulation. The extent of mineralization depends on climatic conditions and properties of SOC. The rate of mineralization of C in transit may be high for warm and humid conditions.

The third stage involves deposition of eroded material, which occurs when velocity of the fluid (runoff or wind) decreases due to reduction in slope gradient or existence of a barrier (Fig. 13–1). Because of the C enrichment of the eroded sediments, depositional sites may be characterized by relatively high SOC content. The data in Table 13–4 from central Ohio show that the mean SOC content of the 0- to 10-cm layer of the depositional site is 40 to 50% more than that of the erosional phases, and the difference may increase with increase in depth within the deposi-

Table 13–4. SOC content of 0- to 10-cm layer for different erosion of phases of Alfisol in central Ohio (Fahnestuck et al. 1996).

Erosional phase†	Farm A	Farm B	Farm C	Mean
		g kg⁻¹		
Not eroded (wooded)	33.7	30.0	40.4	34.7 ± 5.3
Slightly eroded	10.9	11.6	8.7	10.4 ± 1.5
Moderately eroded	10.3	10.8	8.4	9.8 ± 1.3
Severely eroded	9.4	10.8	8.8	9.7 ± 1.0
Depositional	12.5	16.9	14.1	14.5 ± 2.2
LSD (0.05)	2.8	3.0	3.5	

†All eroded phases are cultivated soils.

tional layer. Similar to C redistributed over the landscape, fate of the C buried in depositional sites also is not understood.

SOIL ORGANIC CARBON REDISTRIBUTED
OVER THE LANDSCAPE

It is widely recognized that eroded soil C ends up in different landscape positions (Fig. 13–2) (Meade et al., 1990). However, there is little experimental data with regards to quantification of the loss of SOC during and following redistribution of the C in eroded material over the landscape. The susceptibility to mineralization depends on the characteristics of the organic material being redistributed. While some eroded C may be protected physically and chemically, other is highly labile. From small agricultural watersheds in north Mississippi, Schreiber and Mc-Gregor (1979) observed that 90% of the total C load in runoff was in the particulate phase, which is a labile pool and thus easily mineralized. In a rainfall simulation study on a Tifton loamy sand soil (fine, loamy, siliceous, thermic Plinthic Palendult) in Georgia, USA, Lowrance and Williams (1988) observed that particulate C accounted for most of the C movement in surface runoff, and that the mag-

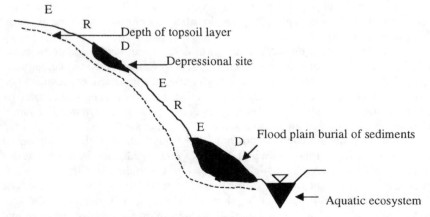

Fig. 13–2. Schematics of a landscape prone to erosional and depositional processes. E, R, and D refer to erosion, redistribution, and deposition of sediments, respectively.

nitude of the loss depended on agricultural management and time of runoff. In Germany, Boyer et al. (1993a,b) reported that about 70% of the original SOC of the Ap horizon is decomposed during translocation or after deposition. For the Mississippi basin, Harden et al. (1999) assumed that 20% of the eroded material is deposited and protected in reservoirs but 80% is decomposed on the landscape. Lal (1995) assumed that 20% of the eroded C is decomposed on the landscape. On a global scale, decomposition by 20% of the eroded C implies emission of 1.14 Pg C yr^{-1} into the atmosphere (Lal, 1995).

FATE OF CARBON TRANSPORTED IN RIVERS

Carbon transport in rivers to the oceans is usually in three forms: (i) POC comprising leaf litter, woody debris, and soil organic matter, (ii) DOC produced from decomposition of C in soil and leaf litter, and (iii) DIC mostly in the form of HCO_3^- and CO_3^- and dissolved CO_2. There are no reliable estimates of the magnitude and interannual variations in transport of POC, DOC and DIC in major rivers of the world. Ludwig et al. (1998) estimated that 0.72 Pg C yr^{-1} is transported by world's rivers, of which 0.62 Pg is organic. Lal (1995) estimated the transport of C into aquatic ecosystem at 0.57 Pg C yr^{-1}, which may well be sequestered.

The fate of C transported in rivers and other aquatic ecosystems also is not well understood. Most of the POC is highly labile and easily mineralized. Estimates of the global flux of POC range from 0.07 to 0.2 Pg C yr^{-1}, and 17% of it may be easily mineralized (Ittekkat & Laane, 1991). In a river in Cameroon, Griesse and Maley (1998) assumed that 50% of POC and DOC is biodegradable. Frankignoulle et al. (1998) estimated the CO_2 flux from European rivers to be at 30 to 60 Tg C yr^{-1}, representing 5 to 10% of the anthropogenic emissions of western Europe.

THE FATE OF CARBON BURIED IN DEPRESSIONAL SITES AND AQUATIC ECOSYSTEMS

The fate of C buried in depositional sites (Fig. 13–2), which also is not widely understood, depends on soil and other ecological factors. In some cases, accumulation of SOC in depositional sites may enhance aggregation (Van Veen & Paul, 1981; Schimel et al., 1985; Elliott, 1986; Woods & Schuman, 1988), and enhance physical protection of the C encapsulated within the micro-aggregates. Carbon sequestration may be further enhanced by increase in aggregation deeper in the soil, the layer not subject to physical disturbance by farm operations. The magnitude of SOC sequestration in depositional sites has been estimated to be 0.57 Pg C yr^{-1} by Lal (1995) and 1.0 Pg C yr^{-1} by Stallard (1998). Goni et al. (1959) also supported the concept of some SOC sequestration in depositional sites.

In contrast to increase in aggregation, the SOC transported to depositional sites may be more biologically active than in erosional sites because of the accumulation of labile and the light fraction (Voroney et al., 1981; Gregorich et al., 1988). The measured rates of SOC loss in depositional sites range from 1 to 1.5 Mg C ha^{-1} yr^{-1} (de Jong, 1981; Gregorich & Anderson, 1985). Prevalence of anaerobic conditions in depositional sites also may lead to methanogenesis and emission of CH_4, and denitrification and emission of N_2O.

The net effect of all these processes on the fate of C transported into depositional sites is hard to assess at the global scale, and requires soil-specific evaluation in contrasting ecoregions.

CONCLUSIONS

Erosional impacts on soil C dynamics are complex, soil-specific and not adequately understood. Fate of the eroded C also differs among three stages of erosion. Review of the state-of-the-knowledge supports the following conclusions:

1. Soil detachment, caused by erosivity of rain and/or wind, breaks down soil aggregates, exposes SOC to microbial processes, and increases its mineralization with an attendant increase in emission of CO_2 to the atmosphere.
2. Redistribution of eroded SOC over the landscape may enhance its mineralization. Estimates of mineralization range from 20 to 80% of the SOC redistributed over the landscape. Global CO_2 emission from mineralization of redistributed SOC may be as much as 1.14 Pg C yr^{-1}.
3. Some of the SOC buried in depositional sites may be sequestrated because of the increase in aggregation caused by enhancement of SOC content. Estimates of C sequestration in depositional sites range from 0.57 to 1 Pg C yr^{-1}. In contrast, prevalence of the labile fraction under anaerobic conditions may lead to methanogenesis, with emission of CH_4 and of denitrification with emission of N_2O.
4. The fate of SOC transported in world's rivers is not known. The POC fraction is highly labile with a global flux of 0.07 to 0.2 Pg C yr^{-1}. Some estimates indicate that 17 to 50% of the POC and DOC transported in world rivers may be mineralized.

The knowledge about processes involved in dynamics of SOC is weak at plot, soilscape, landscape, watershed and global scales. There is a strong need to conduct basic and applied research to strengthen the database and fill in the knowledge gaps.

REFERENCES

Beyer, L., C. Köbbemann, J. Finnern, D. Elsner, and U. Schleuß. 1993a. Colluvisols under cultivation in Schleswig Holstein, J. Genesis, definition and geo-ecological significance. J. Plant Nutr. Soil Sci. 156:197–202.

Beyer, L., R. Fründ, U. Schleuß, and C. Wachendorf. 1993b. Celluvisols under cultivation in Schleswig-Holstein 2. Carbon distribution and soil organic matter composition. J. Plant Nutr. Soil Sci. 156:213–217.

Cihacek, L.J., M.D. Sweeney, and E.J. Diebert. 1993. Characterization of wind erosion sediments in the red river valley of North Dakota. J. Environ. Qual. 22:305–310.

de Jong, E. 1981. Soil aeration as affected by slope position and vegetative cover. Soil Sci. 131:34–43.

de Jong, E., and R.G. Kachanoski. 1988. The importance of erosion in the carbon balance of prairie soils. Can. J. Soil. Sci. 68:111–119

Edwards, A.P., and J.H. Bremner. 1967. Microaggregates in soils. J. Soil Sci. 18:64–73.

Elliott, E.T. 1986. Aggregate structure and carbon, nitrogen and phosphorus in native and cultivated soils. Soil Sci. Soc. Am. J. 50:627–633.

Fahnestock, P., R. Lal, and G.F. Hall. 1996. Land use and erosional effects on two Ohio Alfisols. I. Soil properties. J. Sust. Agric. 7:85–100.

Goni, M.A., K.C. Ruttenberg, and T.I. Englinton. 1997. Sources and contributions of terrigenous organic carbon to surface sediments in the Gulf of Mexico. Nature (London) 389:275–278.

Gregorich, E.G., and D.W. Anderson. 1985. The effects of cultivation and erosion on soils of four toposequences in the Canadian prairies. Geoderma 36:343–354.

Griesse, P., and J. Maley. 1998. The dynamic of organic carbon in south Cameroon: Fluxes in a tropical river system and a lake system as a varying sink on a glacial-interglacial time scale. Glob. Planet. Change 16–17:5374.

Harden, J.W., J.M. Sharpe, W.J. Parton, D.S. Ojima,, T.L. Fries, T.G. Huntington, and S.M. Dabney. 1999. Dynamic replacement and loss of soil carbon on eroding cropland. Glob. Biogeochem.Cycles 13:885–901.

Hatcher, P.G., and E.C. Spikes. 1988. Selective degradation of plant biomolecules. p. 59–74. In F.H. Frimmel and R.F. Christman (ed.) Humic substances and their role in the environment. J. Wiley & Sons, Inc., New York.

Hillel, D. 1991. Out of the Earth: Civilization and the life of the soil. The Free Press, New York.

Hillel, D. 1994. Rivers of Eden. Oxford Univ. Press, New York.

Ittekkot, V., and R.W.P.M. Laane. 1991. Fate of riverine particulate organic matter. p. 233–243. In E.T. Degens et al. (ed.) Biogeochemistry of major rivers. J. Wiley & Sons, Inc., New York.

Lal, R. 1976. Soil erosion on Alfinols in western Nigeria. IV. Nutrient losses in runoff and eroded sediments. Geoderma 16:403–417.

Lal, R. 1995. Global soil erosion by water and carbon dynamics. p. 131–142. In R. Lal et al. (ed.) Soils and global change. CRC/Lewis Publ., Boca Raton, FL.

Lal, R. 1996. Deforestation and land use effects on soil degradation and rehabilitation in western Nigeria. II. Soil chemical properties. Land Degrad. & Dev. 7:87–96.

Lal, R. 1998. Soil erosion impact on agronomic productivity and environment quality. Crit. Rev. Plant Sci. 17:319–464.

Lowrance, R., and R.G. Williams. 1988. Carbon movement in runoff and erosion under simulated rainfall conditions. Soil Sci. Soc. Am. J. 52:1445–1448.

Lucas, R.E., and M.L. Vitosh. 1978. Soil organic matter dynamics. Univ. Michigan Agric. Exp. Res. Rep. Stn. 358.

Ludwig, W., P. Amiotte-Suchet, G. Munhoven, and J-L. Probst. 1998. Atmospheric CO_2 consumption by continental erosion: Present day controls and implications for the last glacial maximum. Glob. Planet. Change 16–17:107–120.

Meade, R.N., T.R. Yuzyk, and T.J. Day (ed.) 1990. Movement and storage of sediment in rivers of the U.S. and Canada. p. 255–280. In Surface water hydrology: The geology of North America. Geol. Soc. Am., Boulder, CO.

Nizeyimana, E., and K.R. Olson. 1988. Chemical, mineralized and physical properties differences between moderately and severely eroded Illinois soils. Soil Sci. Soc. Am. J. 52:1740–1748.

Palis, R.G., H. Ghadiri, C.W. Rose, and P.G. Saffigna. 1997. Soil erosion and nutrient loss: III. Changes in the enrichment ratio of total nitrogen and organic carbon under rainfall detachment and entrainment. Aust. J. Soil Res. 35:891–905.

Parton, W.J., J.W.B. Stewart, and C.V. Cole. 1987. Analysis of factors controlling soil organic matter levels in Great Plains grasslands. Soil Sci. Soc. Am. J. 51:1173–1179.

Pimental, D., C. Harvey, P. Resodudarmo, K. Sinclair, D. Jurtz, M. McNair, S. Crist, L. Shpritz, L. Fitton, R. Saffouri, and R. Blair. 1995. Environmental and economic costs of soil erosion and conservation benefits. Science (Washington, DC) 169:1117–1123.

Rhoton, F.E., and D.D. Tyler. 1990. Erosion-induced changes in the properties of a Fragipan soil. Soil Sci. Soc. Am. J. 54:223–228.

Schlesinger, W.H. 1986. Changes in soil carbon storage and associated properties with disturbance and recovery. p. 194–220. In J.R. Trabalka and D.E. Reichle (ed.) The changing carbon cycle: A global analysis. Springer, New York:

Schimel, D.S., D.C. Coleman, and K.A. Horton. 1985. Soil organic matter dynamics in paired rangeland and cropland toposequences in North Dakota. Geoderma 36:210–214.

Schreiber, J.D., and K.C. McGregor. 1979. The transport and oxygen demand of organic carbon released to runoff from crop residues. Prog. Water Technol. 11:253–261.

Stallard, R.F. 1998. Terrestrial sedimentation and the carbon cycle: Coupling weathering and erosion to carbon burial. Glob. Biochem. Cycles 12:231–257.

Tiessen, H., J.W.B. Stewart, and J.R. Betany. 1982. Cultivation effects on the amount and concentration of carbon, nitrogen and phosphorus in grassland Soils. Agron. J. 74:831–834.

Tisdall, J.M. 1996. Formation of soil aggregates and accumulation of soil organic matter. p. 57–96. *In* M.R. Carter and B.A. Stewart (ed.) Structure and organic matter range in agricultural soils. CRC/Lewis Publ., Boca Raton, FL.

Van Veen, J.A., and E.A. Paul. 1981. Organic carbon dynamics in grassland soils. I. Background information and computer simulation. Can. J. Soil. Sci. 61:185–201.

Veroney, R.P., J.A. Van Veen, and E.A. Paul. 1981. Organic C dynamics in grassland soils. 2 Model validation and simulation of the long-term effects of cultivation and rainfall erosion. Can. J. Soil Sci. 61:211–224.

Woods, L.E., and G.E. Schuman. 1988. Cultivation and slope position effects on soil organic matter. Soil Sci. Soc. Am. J. 52:1371–1376.

Zobeck, T.M., and D.W. Fryrear. 1986. Chemical and physical characteristics of wind blown sediment. Trans. ASAE 29:1037–1041.

14 Carbonaceous Materials in Soil-Derived Dusts

R. Scott Van Pelt

Wind Erosion and Water Conservation Laboratory
Big Spring, Texas

Ted M. Zobeck

Wind Erosion and Water Conservation Laboratory
Lubbock, Texas

ABSTRACT

Wind erosion causes 6.1 billion t of soil loss on non-federal cultivated, pasture, and range land annually in the USA. Wind erosion can produce soil loss from valuable agronomic systems and fragile natural ecosystems that may have great economic impact. In addition to soil loss from valuable agronomic systems and fragile natural ecosystems, wind erosion is a significant source of fine dust that represents the most active portion of soil for plant growth. When this fine dust blows off fields, it often carries C of different forms. This carbonaceous portion of the soil material that is lost due to wind erosion is the most chemically active portion of the soil. A large body of literature deals with carbonaceous aerosols of anthropogenic and other non-soil related sources such as industry, combustion, and volatile organic materials released from vegetation and geological sources. In this chapter we will explore the processes of entrainment and transport as well as the forms, sources, transformations, and fate of soil-derived carbonaceous material in atmospheric particulates. The discussion includes structured, amorphous, and soluble organic C and inorganic C including elemental C, charcoal, and carbonate minerals.

Wind erosion causes 6.1 billion t of soil loss on non-federal cultivated, pasture, and range land annually in the USA (USDA-NRCS, 1994). In the Great Plains alone, about 2 million ha (5 million acres) are moderately to severely damaged by wind erosion each year. Rangeland produced almost twice the total amount of wind-blown material as cropland (3.9 vs. 2.1 billion mt) in 1992.

In addition to soil loss from valuable agronomic systems and fragile natural ecosystems, wind erosion creates several other problems of great economic impact. Moving soil particles can produce abrasive forces that inflict serious injury to crop plants, especially in the seedling stage (Armbrust, 1968; Fryrear & Downes, 1975). In addition, the deposition of wind-blown soil on crops decreases their yield and value and hinders their processing (Farmer, 1993). Although soil particles moved by wind erosion range in size up to about 1 mm, particles traveling great distances are usually quite small (<100 μm) and kept aloft for long distances by turbulent eddies in the wind (Bagnold, 1941, p. 5). Wind erosion is a significant source of fine dust and PM_{10} (airborne particles with diameters of less than 10 μ; Gertler et al., 1995; Gillette et al., 1978). This fine material represents the most active portion of the soil for plant growth (Stetler et al., 1994; Zobeck & Fryrear, 1986). Dust also obscures visibility, pollutes the air, fouls machinery, and imperils animal and human health.

The carbonaceous portion of soils material that is lost due to wind erosion is the most chemically active portion of the soil (Bohn et al., 1985, p. 135) and perhaps the most important for soil fertility and physical characteristics such as structure and tilth. Soil-derived carbonaceous material exists in many organic and inorganic forms in soils including but not limited to living organisms, biotic remains or metabolites in various stages of decomposition including relatively stable forms such as humus, sorbed anthropogenic chemicals such as pesticides, herbicides, and incidental contaminants, charcoal, and carbonate minerals. The literature reflects the role that wind erosion plays in the reduction of the organic matter in the surface of the solum and in certain cases to the role that dust deposition plays in the limited enrichment of organic matter to certain select soils. A large body of literature deals with carbonaceous aerosols of anthropogenic and other non-soil related sources such as industry, combustion, volatile organic materials released from vegetation, etc., and natural geological emissions. In this chapter we will explore the processes of entrainment and transport as well as the forms, sources, transformations, and fate of soil-derived carbonaceous material in atmospheric particulates. The discussion will include structured, soluble, and amorphous (noncrystalline) organic C and inorganic forms of C including elemental C, charcoal, and carbonates.

DYNAMICS OF WIND EROSION

As the name implies, wind erosion results from the interaction of the wind and the soil surface. The *Encyclopedia of Climatology* (Oliver & Fairbridge, 1987, p. 929) defines wind as "a stream of air flowing relative to the earth's surface, usually more or less parallel to the ground." The earth's surface exerts a drag on the wind resulting in a vertical profile of wind speeds that is accurately described as logarithmic decay function (Geiger, 1961, p. 117). There is a sharp decrease of mean horizontal wind speed as the surface is approached, resulting in turbulent eddies (gusts) that may impact the soil surface with forces four to six times as great as the mean wind speed just a few centimeters above (Oke, 1978, p. 44–46).

The mean wind speed at the soil's surface is a function of the roughness of the interface of the earth and atmosphere (Geiger, 1961, p. 118) and thus any departure from a plane surface at the interface lessens the wind's shearing energy contacting the soil and lessens the probability of erosion (Bagnold, 1941, p. 52–55). Standing plant material is an important contributor to surface roughness (Fryrear, 1963). Tillage of the soil tends to increase the roughness of the soil surface and thus reduce its susceptibility to wind erosion (Zobeck & Popham, 1998). In row crop systems, an oriented soil roughness is affected by the creation of ridges when beds are listed, and this roughness is particularly effective in reducing erosion as long as the wind is not blowing parallel to the direction of the beds (Armbrust et al., 1964). A more or less random roughness is created when soil aggregates or clods are created and preserved with specialized tillage implements. This random roughness also provides protection against erosion (Zobeck, 1991). Soil amendments such as incorporated stubble or crop processing wastes may also reduce erosion (Fryrear & Koshi, 1971).

Soils vary in their susceptibility to erosion. Additionally, individual soils may vary in their susceptibility to wind erosion from one day to the next due to their surface water content (Chepil, 1956). Perhaps the most important soil property in determining soil erodibility is dry aggregate stability (Zobeck, 1991; Fryrear et al., 1994). Clay content, organic matter content, and calcium carbonate content all tend to increase dry aggregate stability (Usman, 1995). However, the stability of the aggregates often tends to decrease when the calcium carbonate content of the soil approaches and exceeds 40%. The USDA-NRCS soil survey specialists know that when calcium carbonate levels approach 40%, depending on the amount and type of clay, the soil particles will disperse, increasing erodibility (Fryrear et al., 1994). In the absence of dry aggregate stability or in the case of a soil whose aggregates have been dispersed by intense rainfall or mechanical means, a large number of individual erodible soil particles may be scattered on the surface. If the surface is unprotected by crops or surface roughness, these particles may start moving when the wind velocity at the surface is sufficient.

The Physics of Blown Sand and Desert Dunes (Bagnold, 1941) is still widely regarded as the definitive treatise on the physics of sand movement. (Unless otherwise noted, the principles outlined in the following three paragraphs come from this work.) The minimum wind velocity initiating particle movement is known as the threshold velocity. Particle susceptibility to wind-induced movement is affected by size, density, and shape. The quartz sand-size particle most susceptible to wind-induced movement is approximately 100 μm and this size particle will begin to move with an incident wind velocity of approximately 0.15 m s^{-1}. Particles larger and smaller than this size require greater wind velocities to initiate movement from rest (Fig. 14–1).

Wind-blown materials move in three modes; creep, saltation, and suspension. In general, the largest soil particles (1000–2000 μm) moving will move in the creep mode, as they are too massive to leave the soil surface. Particles between 100 and 1000 μm in size tend to move in saltation mode, alternately leaving the surface, flowing with the wind, and returning via gravity to strike the surface. The third mode of particulate transport, suspension, is composed primarily of soil particles less than 100 μm in diameter. These materials may travel great distances before returning to earth.

As previously mentioned, the fine materials carried in suspension are less susceptible to direct entrainment by the wind than are the fine sand particles that initiate the erosion process. Saltating particles return to the soil surface with a force that is a function of their mass and the speed of lateral and vertical movement. These saltating grains are credited with the disruption of soil aggregates and surface crusts that results in the release of finer particles. The linkage between the movement of sand-size grains to the emissions of fine dust is one aspect of the wind ero-

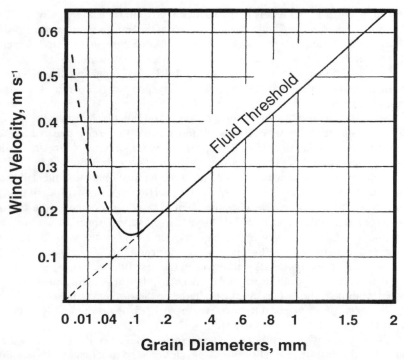

Fig. 14–1. Relationship between quartz sand particle diameter and threshold wind velocity (after Bagnold, 1941).

sion process that is imperfectly understood (Porch, 1974; Gillette, 1977; Shao et al., 1993). Sandblasting has been proposed as the primary causative process (Gillette et al., 1974; Gillette, 1978; Gomes et al., 1990; Alfaro et al., 1997). Depending upon the classification system used, the lower limit of particle diameter for material classed as sand ranges from 20 to 60 μm. We casually tend to think of sand-size material as coarser than dust, and thus immune to long-term suspension. While in most cases sand-size particles are redeposited in the source field or nearby, in extreme events fine sands may enter true suspension mode and be transported great distances. Bagnold (1941, p. 6) offers an interesting erosion-related definition of sand size, "We can thus define the lower limit of size of sand grains, without reference to their shape or material, as that at which the terminal velocity of fall becomes less than the upward eddy currents within the average surface wind. Particles of smaller size tend to be carried up into the air and to be scattered as dust."

SUSPENDED PARTICLES

Magnitude of Suspended Sediments

Soil material in the suspension mode may travel long distances and constitutes most of the airborne dust that settles in our homes and on our automobiles.

Estimates of the annual flux of airborne dust deposition over land areas range from less than 10 to about 200 t $km^{-2} yr^{-1}$ (Pye, 1987, p. 91). As expected, these estimates vary with respect to distance and direction from the major source regions. Pye (1987, p. 80) also reports dust deposition rates for the oceans to range from 0.37 to 4.5 t $km^{-2} yr^{-1}$ and a total annual flux to the oceans to range from 532 to 851 million t yr^{-1}. Dust deposition rates were probably greater in past geological history as the current rates are insufficient for the formation of loess common to many areas of the earth (Tsoar & Pye, 1987).

Individual storms may contribute a large amount of the total annual dust flux for selected geographical regions (Fryrear, 1981). Woodruff and Hagen (1972) collected historical data from the years of the Dust Bowl and from the 1950s and compared these data to the 1960s, a decade during which wind erosion was relatively slight. From accounts of visibility and dust load/visibility indices developed during the 1950s, they conclude that as much as 2133 t (9800 tons) of dust may have been suspended per cubic kilometer (cubic mile) of atmosphere during an individual "black blizzard" of the 1930s.

Sources and Transport Distances of Suspended Particles

The source areas of suspended dust are primarily deserts (Pewe, 1981; Rahn et al., 1981; Shutz et al., 1981; Chester et al., 1984; Zhang & An, 1998), agricultural operations in arid and semiarid environments (McCauley et al., 1981; Tsoar & Pye, 1987; Clausnitzer & Singer, 1996; Wagoner, 1997), dry lake beds (Cahill et al., 1996; Reheis, 1997), braided streams, deltas, and glacial outwash (McGowan & Sturman, 1997), and unpaved roads (Tsoar & Erell, 1995; Chow & Watson, 1997). These areas can be located on any mid-latitude continent, and the literature is composed of studies of dust source regions on all continents. Interestingly, a very small source region, the Owens dry lake bed in California is the single greatest source of PM_{10} in North America (Cahill et al., 1996), accounting for approximately 10% of the total PM_{10} load (M. Reheis, 1999, personal communication).

Dust carried into the atmosphere on one continent is often transported with global circulation patterns to other continents. Dust from the Saharan Desert has been documented to have fallen in Europe (Goudie, 1978), South America (Talbot et al., 1990), the Carribean Sea (Delaney et al., 1966), the North Atlantic Ocean (Glaccum & Prospero, 1980; Schutz et al., 1981; Prospero, 1996a,b), and to the interior of North America, a distance of over 9000 km from the source region (Perry et al., 1997; Gatz & Prospero, 1996). It has been estimated that up to one-half of the southeastern U.S. PM_{10} originates from Africa (Prospero, 1999). Similarly, Asian dust from the deserts and semiarid regions of northern China have been documented in Alaska (Rahn et al., 1981) and Hawaii (Shaw, 1980; Braaten & Cahill, 1986). The direction of transport from a given erosion event is dependent upon the seasonal variation in circulation patterns.

Factors Affecting Transport Distances and Deposition

The factors affecting the distances over which dust may be transported from its source region are meteorological conditions, the nature of the surface over which it is being transported, and particle size, shape, and density. In order for dust to be carried great distances, it must either become entrained in rapidly moving air aloft or carried in strong vertical updrafts to great altitude (Pye, 1987). Although microscale whirlwinds such as dust devils may lift soil materials hundreds of meters into the air, most of the suspended materials are deposited locally (Gillette & Sinclair, 1990). Most suspended dust is entrained by much larger-scale events. McCauley et al. (1981) describe the large-scale erosion in the southern High Plains and transport of dust across the USA to the mid-Atlantic Ocean that resulted from a large-scale cyclonic disturbance in February 1977. Downdraft winds at the leading edge of thunderstorms and squall lines, often called *Haboobs*, are also significant in entraining dust to heights in excess of 15 km, allowing for long distance transport (Takemi, 1999).

The distance that dust entrained by the leading edge of thunderstorms is actually transported may be cut short, however, if the dust is incorporated into raindrops and falls to the ground near its point of origin (Pan & Prather, 1998). Dust particles are also important as condensation nuclei (Rogers, 1979). The effectiveness of rainfall at scavenging dust from the atmosphere is evidenced by the "blood rains" or red rains over Europe (Avila et al., 1997). Fryrear (1987) cites an 1847 account of a storm that deposited 0.025 m (1 in.) of dust in 0.225 m (9 in.) of rain at Lucarno and 2.74 m (9 ft) of red snow in the Alps above the city.

In the absence of rainfall for wet deposition and turbulent eddies to keep it aloft, airborne dust settles by gravity or by impaction on surfaces projecting into the slipstream. Goldenberg and Brook (1997) have successfully modeled dry deposition rates of particles with different size, shape, and density over smooth planar surfaces, sea surfaces, and surfaces with micro-roughness. Krauss and Smith (1969) conclude that for scales exceeding 1.6 km (1 mi), variations in wind velocity and gustiness, vegetation, and rainfall promote differences in dust deposition. Plant surfaces can be efficient collectors of airborne particles (Chamberlain, 1975) and Bauman et al. (1998) noted greater rates of aeolian sediment accumulation under juniper (*Juniperus* spp.) communities growing on a lava flow than adjacent lower-growing plant communities and bare lava. It is probable that the world's great loess deposits were trapped by vegetated surfaces (Tsoar & Pye, 1987).

The ultimate fate of suspended particles is to return to the earth's surface (Seinfeld, 1986). Approximately 70% of the earth's surface is covered by water, providing a surface for deposition that offers a relatively undisturbed resting palace for aeolian sediments. Numerous studies have used marine sediment, lacustrine sediment, and glacial ice cores to study the historical sources and magnitudes of dust deposition (Windom, 1969; Johnson, 1979; Savoie & Prospero, 1980; Carder et al., 1986; Loye-Pilot et al., 1986, Gao et al., 1992; Wagoner, 1997; Biscaye & Grousset, 1998). Recently, Prospero (1996a) has modeled dust deposition rates to several marine environments using Group of Experts on Scientific Aspects of Marine Pollution (GESAMP) and reports estimates of 170 Tg yr^{-1} for the North Atlantic, 25 Tg yr^{-1} for the Mediterranean, and 5 Tg yr^{-1} for the Carribean.

Pye (1987, p. 91) states that continental dust deposition rates should be two to three orders of magnitude greater than that for the oceans due to the proximity to the source, but also notes that this rate may not hold for marine environments close to source regions. Numerous studies have documented the value of aeolian materials as sources of soil parent material (Yaalon & Ganor, 1973; Gile & Grossman, 1979; Bennett, 1980; Pewe et al., 1981; Rabenhorst et al., 1984; Simonson, 1995; Herwitz et al., 1996; Modaihsh, 1996; Feng, 1998; Mikkelson & Langohr, 1998), as well as sources of soil fertility for plant growth (Wilke et al., 1984; Talbot et al., 1990; Moberg et al., 1991; Tiessen et al., 1991; Offer et al., 1995; Littman, 1997). Unless the dust is incorporated by deep rainfall following deposition, it is probable that it will become entrained in the wind again and be deposited elsewhere (Gile & Grossman, 1979; Reheis, 1997; Loosmore & Hunt, 1998). Goosens and Offer (1994) conclude that resuspension is the reason that a free water surface consistently collects more dust than any of the other dust collection devices they tested.

While the nature of the surface has great impact on the deposition of aeolian sediments from air masses close to the ground, the particles at altitudes greater than 100 to 200 m must also settle. Larger grains, given the same density, have higher terminal velocities and are the first to be lost from suspension (Pye, 1987, p. 61). Studies tracking individual dust events as well as long-term trends note the decrease in mean particle diameter and mineral density with time and distance from the source region (Glaccum & Prospero, 1980; McGowan & Sturman, 1997; Prospero, 1996b). Figure 14–2 depicts an idealized curve for the terminal velocity of different diameter quartz grains. Tsoar and Pye (1987) present a diagram relating the likely maximum distances of particle transport under constant meteorological conditions based solely upon particle diameter.

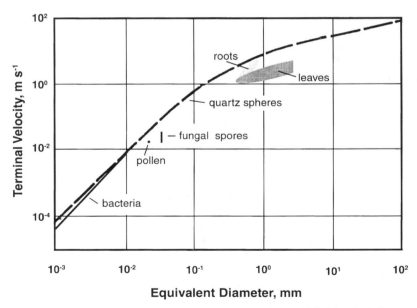

Fig. 14–2. Relationship between particle diameter and terminal velocity of fall. Of particular importance is the lower terminal velocities for organic material as compared to quartz spheres of the same diameter [adapted from Malcolm & Raupach (1991) with information on bacteria from Seinfeld (1986) and information on pollen and fungal spores from Sehmel (1980)].

Many authorities have presented data on the size distribution of dust deposits (Swineford & Frye, 1945; Zeuner, 1949; Patterson & Gillette, 1977; Pewe, 1981; Pewe et al., 1981; Wilshire et al., 1981; D'Almeida & Schutz, 1983). Few authors have presented mathematical formulations of dust particle size distributions. Patterson and Gillette (1977) described aeolian dust using a log-normal distribution. A recent analysis by Zobeck et al. (1999) suggests a Weibull distribution is the most appropriate distribution to mathematically describe airborne dust.

Particle shape and density also affect the terminal velocity of the particles and their rate of fall from the atmosphere as well as their tendency to adhere to surfaces (Goldenberg & Brook, 1997). Tsoar and Pye (1987) cite Green and Lane (1964) and present the following equation for calculating the terminal velocity of particles:

$$U_f = KD^2 \qquad [1]$$

where U_f is settling velocity (cm s^{-1}); D is the grain diameter (cm); $K = \rho_g g/18\mu$, where ρ_g is the grain density, g is the acceleration due to gravity, and μ is the dynamic viscosity of air. The strong dependence of U_f upon the grain density and particle diameter is obvious from the equation. Organic materials are rarely spherical. For particle shapes approximating cylinders (roots, stems, fibers, etc.), D should be taken as the diameter of the cylinder and for particle shapes approximating plates (leaves, suberin flakes, humic and fulvic acids on clay particles, etc.) D should be taken as 0.5 of the square root of the plate area (Malcolm & Raupach, 1991). Density of organic materials range from 0.3 g cm^{-3} for dry root material to 1.6 g cm^{-3} for pure sucrose and cellulose. Figure 14–2 shows the calculated terminal velocities of bacteria as well as common organic plant parts and metabolic products as compared to quartz spheres with a density of 2.6 g cm^{-3}. Sehmel (1980) reports terminal velocities for 20-μm pollen grains to be 1.5 cm s^{-1} and for 32-μm *Lycopodium* spores to be between 1.4 and 2.1 cm s^{-1}. As might be expected, soil organic matter is more easily lifted to greater heights than mineral particles of the same diameter and organic C enrichment of aeolian dust samples compared to the source soils is the rule rather than the exception. The ultimate demonstration of this phenomenon is the documented cases of lichen fragments of from 5- to 15-mm diam. traveling far from the source area and covering the ground up to 15 cm deep following a dust storm (Donkin, 1981).

FORMS OF CARBON FOUND IN WIND-BLOWN SEDIMENTS

Carbon exists in many forms in soils, from elemental C in charcoal and soot to some of the most complex organic molecules in living organisms. Carbon, especially when in amorphous organic forms, is often closely associated with and may be chemically bound to soil fines, most notably clays (Bohn et al., 1985, p. 149). Inorganic forms of C include elemental C and carbonate minerals. Organic forms of C include structural materials such as plant and animal parts in various stages of decomposition, pollen, fungal spores, living organisms such as bacteria, algae, fungi, single cell animals, and viruses, soluble organics such as sugars, proteins, and metabolic products, amorphous organics such as humic acids, fats, waxes, oils, lignin, and polyuronides (Brady, 1974, p. 137–163), and anthropogenic organic chemical compounds and daughter products of decomposition (Bohn et al., 1985, p. 196). Since these organic materials are formed by processes on or near the surface, they are highly susceptible to removal and transport by wind erosion. Table 14–1 contains a synthesis of reported carbonaceous materials in suspended atmospheric particulates. Considering their abundance in surface soils, carbonaceous materials are probably present in all aerosols. Chow and Watson (1997) found that every sample collected from a long-term study of numerous PM_{10} sampling sites in the Las Vegas valley contained some form of C. They, like many researchers, did not identify the exact nature of the carbonaceous materials found. It is probable that in addition to soil-derived carbonaceous materials, elemental black C and organics from anthropogenic sources were present in varying fractions. These anthropogenic forms of C are, in general, outside the scope of this chapter but since the eventual fate of all atmospheric aerosols is deposition onto the Earth's surface, we may expect them to be present in soils and should briefly review their forms and interactions with soil-derived carbonaceous materials.

INORGANIC CARBON

Elemental Carbon and Charcoal

Elemental C and charcoal are predominantly combustion products and are, for the most part, chemically inert. Much of the literature concerning elemental C is concerned with the effect it has on light attenuation and resultant effects on the global energy budget (Liousse et al., 1993, 1996), areas of interest that are outside the scope of this chapter. While elemental C and charcoal are very resistant to being changed by reaction, they offer excellent sorption sites that may catalyze reactions of many other compounds, including organic compounds (Charlson & Ogren, 1981). Charlson and Ogren (1981) also point out that since condensation nuclei may go through several sequences of hydration and drying, there may be many chances for reactions with surface groups and recrystalization of the formed compounds before the core C atom and its host molecules get rained out of the atmosphere and into the soil. Elemental C is often found as a heterogenous mixture of C, sulfates, and other compounds, allowing for a hydrophylic behavior that makes them effective as condensation nuclei (Liousse et al., 1996). Carbon and organic particles are very important in catalyzing the oxidation of sulfur dioxide in the atmosphere (Liberti et al., 1978). Pinnick et al. (1993) found submicron-sized particles of C attached to clay minerals, hydrocarbons, and $(NH_4)_2SO_4$ in aerosol samples collected at several remote sites in southern New Mexico. The $^{13}C/^{14}C$ have been used to determine if particulate C aerosol originates from the burning of biogenic or fossil materials (Currie, 1981). Fossil C containing compounds also occur naturally in the weathering products of oil shales dust, coal seams, and oil seeps. In many cases, elemental C and organic C are found in the same aerosol sample. Using thermal oxidation of 350 °C and 550 °C, Mueller et al. (1981) were able to separate the elemental C fraction from the organic fraction of carbonaceous aerosols. They found that elemental C percentages were higher in urban areas than in rural areas.

Table 14–1. Ranges of concentrations (actual and percentage) of carbonaceous suspended solids reported for various regions worldwide.

Location	Total solids	Carbonates	Organics	Reference
China Sea	0.15–0.31 µg m^{-3}	2%	13–22%	Aston et al., 1973
Indian Ocean	0.01–4.4 µg m^{-3}	<2–2%	4–47%	Aston et al., 1973
South Atlantic	0.42–1.8 µg m^{-3}	<2%	3–30%	Aston et al., 1973
North Atlantic	0.55–22.3 µg m^{-3}	<2–2%	2–16%	Aston et al., 1973
North Atlantic	10.0–>120 µg m^{-3}	0.9–9.5%	--	Glaccum % Prospero, 1980
North Atlantic	5×10^{-12} cm^3 cm^{-3}	1–8%	--	Delaney et al., 1966
North Mediterranean	0.01–1.42 g m^{-2} d^{-1}	<40%	Present	Pye, 1992
East Mediterranean	4.6–37 µg m^{-3}	<5–45%	22–42%	Chester et al., 1977
Italy	--	1–28%	--	Liberti et al., 1978
Spain	30–19 435 mg m^{-2}	--	--	Avila et al., 1997
France	--	--	5.48%	Fryrear, 1987
Turkey	212–551.6 kg m^{-2} 10^4 yr^{-1}	9–48%	--	Irmak & Aydemir, 1998
Israel	41–550 t km^{-3}	8.7–75%	<2 – 22%	Yaalon & Ginzbourg, 1966
Israel	0–>550 mg m^{-2} d^{-1}	--	--	Littman, 1997 / Offer et al., 1995
Saudi Arabia	95–183.4 µg m^{-2} mo^{-1}	22.1–47.2%	--	Modaihsh, 1997
Saudi Arabia	0.023–0.061 g m^{-2} d^{-1}	Present	Present	Behairy et al., 1985
Egypt	--	Present	5.36%	Moharram & Sowelim, 1980
Niger	--	--	1.52–2.22%	Sterk et al., 1996
Nigeria	99 g m^{-2} season^{-1}	--	4.0%	Wilke et al., 1984
Nigeria	700–1000 kg ha^{-1} season^{-1}	--	2.5 –3.7%	Moberg et al., 1991
Northwestern Africa	220–13 421 µg m^{-3}	Present	--	Lepple & Brine, 1976
Tadzhikistan (USSR.)	--	Present	--	Gomes & Gillette, 1993
China	--	--	Oxalates, Acetates	Sheng et al., 1981
Amazon Basin (South America)	--	--	--	Talbot et al., 1990
California, Nevada, USA	5.7–114.0 g m^{-2} yr^{-1}	7.9–30.7%	--	Reheis & Kihl, 1995
Southern Nevada, USA	--	9.0–14.5%	0.9–30.1%	Reheis et al., 1995
Southern California, USA	7.8–2170 mg m^{-2} d^{-1}	7.7–35.7%	0.3–3.0 kg ha^{-1} yr^{-1}	Reheis, 1997
Nevada, USA	21.1–140.3 t ha^{-1} yr^{-1}	--	7.1–15.1%	Blank et al., 1999
Arizona, USA	54 g m^{-2} yr^{-1}	1.12–3.87%	0.6–12.0%	Pewe et al., 1981
Southern New Mexico, USA	9.3–125.8 g m^{-2} yr^{-1}	<0.4–8.0%	--	Gile & Grossman, 1979
Southern New Mexico, USA	--	10%	--	Hoidale & Smith, 1968
Western Texas, USA	--	13.3–45.6%	3.0–14.5%	Laprade, 1957
Western Texas, USA	--	0.51–15.31%	4.92–16.57%	Gill et al., 1999
Western Texas, USA	--	--	10.2–29.2 g kg^{-1}	Zobeck et al., 1989
Western Texas, USA	285.7 kg m^{-2} yr^{-1}	--	4.5 kg m^{-2} yr^{-1}	Zobeck & Fryrear, 1986
Southwestern Texas USA	10.8–13.7 g m^{-2}	Trace	--	Rabenhorst et al., 1984

Carbonate Minerals

Carbonate minerals are another important form of inorganic C commonly found in atmospheric aerosols. Carbonate materials of all aerosol size classes are found in suspended sediments worldwide. Desert areas in North Africa and the eastern Mediterranean are an important and well-documented source area for carbonate-rich aeolian materials. Mineralogical analysis of solids collected from red rain events in Spain indicated a variation in percentage calcite ranging from 0.5 to 21% and variation in percentage dolomite ranging from 0.5 to 7% (Avila et al., 1997). The authors were able to trace the dust back to its northwestern African source regions using the mineralogical analysis. They also found that the source regions had lower concentrations of carbonate minerals in the soils, indicating an enrichment of carbonate materials in the suspended sediments. Pye (1991) noted high percentages of calcite, dolomite, and gypsum in dust collected over Crete and concluded that this dust was not of local origin, but from Tunisia and adjacent areas of North Africa where dry lake beds and areas of exposed dolomitic materials are common. Calcite noted in the dust over Jeddah on the west coast of Saudi Arabia is assumed to be of local origin, but Behairy et al. (1985) do not note the concentrations in either the suspended materials or source region. Modaihsh (1997) reports an annual average calcium carbonate percentage of 31.8 for dust falling on Riyadh, Saudi Arabia. Irmak and Aydemir (1998) modeled the deposition of calcium carbonate from dust originating in Syria and deposited on volcanic soils in Turkey that had accumulated between 16 and 27% calcium carbonate in the solum. They concluded that during the last 10 000 yr, a total flux of 212 kg m^{-2} of calcium carbonate had been deposited in the profiles furthest from the source region and that at least 467.5 kg m^{-2} had been deposited in the profiles nearest the source region, with intermediate amounts at locations in-between the extremes. From mineralogical analyses of dust collected over the Negev desert in Israel, calcium carbonate was found to compose between 9 and 75% of the deposited fines (Yaalon & Ginzbourg, 1966; Littman, 1996).

In the USA, several studies of suspended solids have reported calcium carbonate percentages, but the dust on this continent contains significantly less carbonate than the dust from North Africa and the eastern Mediterranean. Gile and Grossman (1979) report concentrations ranging from a trace to a maximum of 7% calcium carbonate amounting to a 10-yr average of less than 1 g m^{-2} yr^{-1}. In a nearby area of southern New Mexico, Hoidale and Smith (1966) report an average 10% calcium carbonate component in deposited dust samples. Laprade (1957) found that suspended dust collected near Lubbock, Texas, consisted, on average, of 24% calcium carbonate. Modern samples collected near Lubbock range from 0.5 to 15% calcium carbonate (Gill et al., 1999). Brown (1956) attributes this aeolian input with the formation of an extensive petrocalcic horizon on Lubbock area soils that were otherwise derived of igneous outwash from the southern Rocky mountains. Interestingly, Rabenhorst et al. (1984) report low concentrations of calcium carbonate in dust sample collected in a nearby area of west Texas dominated by calcareous soils. Petrocalcic horizon formation in Nevada has also been attributed to historical and present deposition of carbonate rich aeolian sediments (Lattman, 1973; Reheis et al., 1995). In a study of dust emanating from the Owens dry lake bed in California, calcium carbonate concentrations varied from nondetectable to over 35% (Reheis, 1997). Gypsum (calcium sulfate) was also noted in the samples, an interesting anomaly since gypsum is scarce to absent in the lake bed crustal materials. Aeolian gypsum has been noted in several areas devoid of surface gypsum sources (Moharram & Sowelim, 1980; Reheis et al., 1995; Avila et al., 1997). One explanation offered is the influence of atmospheric sulfuric acid from anthropogenic sources. Litaor (1987) attributes aeolian carbonates in alpine soils of Colorado with a buffering effect on acid rain.

Acid rain is a worldwide problem resulting from the dissolution of anthropogenic nitrates and sulfates in cloud droplets. Calcium carbonate's ability to neutralize acid aerosols in the atmosphere has generated interest in the role it may play in ameliorating this global problem. Gillette and Sinclair (1990) estimated the carbonate fluxes into the atmosphere caused by dust devils as part of an acid rain modeling effort and concluded that the contribution of dust devils to the annual carbonate flux could be quite high locally. Loye-Pilot et al. (1986) showed that Saharan dust significantly increased the pH of rainwater over the northern Mediterranean. Hedin and Likens (1996) conclude acid rain remains a problem because although emissions of nitrates and sulfates have decreased over the past several years, the emissions of carbonates have also decreased in like fashion. As carbonates react with sulfuric and nitric acid in aqueous aerosols, sulfates and nitrate salts and carbon dioxide are the resulting materials. Ultimately, many of the nitrate and sulfate salts (primarily with calcium as the base cation) are deposited in the soil where they may be leached below the surface. These soluble calcium salts may be microbially transformed back into calcium carbonate. Gardner (1972) proposed that as much as 25% of the caliche on Mormon Mesa in Nevada may have been microbially precipitated from soluble salts of calcium. Monger et al. (1991) successfully showed that live microbes could precipitate calcium carbonate from calcium chloride, air, and a carbohydrate energy source.

Calcium carbonate in dust originating from soil wind erosion does not always enter pathways that lead back to the soil. Glaccum and Prospero (1980) report calcium carbonate concentrations in aerosols over the North Atlantic that range from 0.0 to 30.7%. They also noted gypsum concentrations in the range of from 0.1 to 13.1% and attributed this gypsum to interaction of calcium carbonate with S-containing aerosols. Comparisons of the composition of deep sea sediments with the composition and trajectories of aeolian dusts produced a strong correlation between the locations of sediments containing from 0.25 to greater than 2% dolomite and trade wind belts between 3° and 35° N lat containing dust with an average 0.8% dolomite (Johnson, 1979). Chester et al. (1977) found carbonate concentrations to range from less than 5 up to 45% in dust samples collected over the Mediterranean Sea and proposed that sea bottom sediments in certain areas of the Mediterranean may contain a significant amount of aeolian carbonates. Aston et al. (1973) investigated the composition of atmospheric dust over several oceans and connected seas and found less than 2% carbonate in all cases but used an infrared technique developed for marine sediments that may have been less sensitive to crystalline calcite.

ORGANIC CARBON

Wind erosion results in the net loss of organic matter from the surface of affected soils. Lyles and Tatarko (1986) report that of 10 Kansas soils studied, 8 had lost organic matter during the period 1948 to 1984 at an average rate of about 0.01 percentage points yr^{-1}. They attributed wind erosion to part of this decline as they also noted similar losses in the silt and clay fractions of these soils. Suspended dusts are significantly richer in organic matter than are the soils of provenance (origin) and samples taken at progressively greater heights immediately downwind of eroding fields are increasingly richer in organic C. Laprade (1957) attributes this enrichment phenomenon with the low specific gravity and large surface area common to many organic materials. He found an average organic content of 7% at 3.05 m (10 ft) above the ground and 9% at 19.83 m (65 ft) above the ground from soils that contained an average of 2.6%. Modern dust samples from this same area average 11% (Gill et al., 1999). Sterk et al. (1996) report organic C concentrations in 2-m height dust samples as high as 5.36% from soil that contained less than 0.2%, an enrichment ratio of 31.53. They also noted enrichment in saltating materials, but the ratios were much smaller at 1.33 to 4.74. It is probable that the increased organic concentrations with height are caused by the close association of organic C with the soil fines. Delany and Zenchelsky (1976) investigated aerosols derived from organic poor soils on the southern high plains of Texas and found that organic C contributed from 0.2 to 4.0% of the total mass of aerosols with diameters from 1 to 40 μm, but composed from 5 to 20% of the total mass of the aerosols with diameters from 0.4 to 1.0 μm. Zobeck and Fryrear (1986) estimated that in an average year, almost 4.5 t of organic matter blow across a 1-km length between the heights of 0.5 and 2 m in West Texas. Since wind erosion results in the net loss of soil organic C in soils of provenance (origin), it may also be expected that soils affected by the deposition of these materials will show a net gain of organic C, a process shown to be locally significant by Moberg et al. (1991) in northern Nigeria.

Organic material is also found in the atmosphere from sources other than eroding soils. Liousse et al. (1996) present estimates of 44 Tg yr^{-1} from biomass burning, 28.5 Tg yr^{-1} from fossil fuel combustion, and 7.8 Tg yr^{-1} from above-ground natural sources including the release of volatile organics by living plants. These volatile organics are primarily terpenoids that, like anthropogenic volatile organics, are rapidly converted to particulates in the presence of O_3 (ozone) or NO_x (nitrous oxides) (Duce, 1978). Anthropogenic organic C in the particulate form also consists of, among other materials, asphalt road dust, tire debris, and organometallic brake lining dust (Rogge et al., 1993). Even though these nonsoil derived organic materials are outside the scope of this chapter, they will eventually be deposited and incorporated into the soil and thus cannot be entirely ignored.

Structured Organic Materials

Little information exists on the exact nature of organic matter in suspended dusts. On occasion, the context of the chapter gives insight into the nature of the organic material present. Zobeck et al. (1989) compare the total amounts and composition of dust for several land management systems. From the higher organic matter contents for dust emanating from fields with vegetative cover vs. the clean tilled field, it may be surmised that a significant portion of that organic material was composed of plant parts and surface litter. Leathers (1981) lists anthers, cotton (*Gossypium hirsutum* L.) lint, leaf, stem, and flower fragments, seeds and small fruits, and trichomes as plant parts recognizable from dust samples collected in central Arizona. He also notes the pollen of many plants and estimates as much as 4 t of pollen may be released each year in Phoenix by olive (*Olea europaea* L.) trees alone. The total pollen from all plant species released to the atmosphere each year in the Phoenix area is probably at least an order of magnitude higher. Chamberlain (1975) cites studies in which certain plant taxa preferentially release pollen at times of day or in wind conditions that will insure maximum dispersal distances. Pollen is also a component of dust samples collected in the Mediterranean area (Pye, 1992) as well as dust collected in the eastern Carribean Sea (Delaney et al., 1966). Delaney et al. (1966) also make note of the presence of fragments of vascular plants in the samples they collected at Barbados and, considering the constant direction of the trade winds at that latitude, conclude that they must have come from North Africa. Large pollen and fungal spores are as large as 200 to 300 μm (Duce et al., 1983).

Delaney et al. (1966) also found several lower life forms in the dust samples collected at Barbados including fungal hyphae, freshwater diatoms from mountain streams, bacteria, and algae. Soil algae have also been noted in analyzed dust samples from Arizona (Leathers, 1981) and the Mediterranean (Pye, 1992). Fungal spores are also a component of the dust collected in Arizona and Leathers (1981) identified 16 species that are human pathogens. Valley fever, caused by the fungus *Coccidioides immitis* is responsible for more than 27 human deaths in Arizona each year, as well as economic losses to income and medical costs of approximately $320 million. Bacteria occur in soils in numbers of 10^9 to 10^{12} kg^{-1} of soil. It is not surprising that these organisms are a component of suspended dusts. Bacteria are extremely small, have low specific gravities, terminal velocities in still air of 0.001 to 1 cm s^{-1} and are capable of being transported very long distances by even moderate winds (Seinfeld, 1986). Bacteria, fungi, and viruses pose significant threats to the human health and economic agriculture. Viruses and mycoplasmas may be found as small as 0.1 μm in diameter (Duce et al., 1983). Recently, researchers have used the fatty acids present in dust samples to provide biological fingerprints of the dust that would elucidate the area of provenance (Kennedy & Busacca, 1998).

Soluble Organic Materials

Soluble forms of organic C have also been noted in dust aerosols worldwide. Manna Lichen occasionally falls in recognizable pieces in areas of western Asia along with crystals of calcium oxalate formed from the interaction of oxalic acid produced by the lichen and the calcium in the host rock at the source area (Donkin, 1981). Acid-soluble oxalates have also been documented in dust from a saline playa surface in Nevada (Blank et al., 1999). The authors attributed the presence of the oxalates to the high content (15–20 % dry weight) of oxalates in saline-adapted plants that might have been washed into the playa and decomposed. Talbot et al.

(1990) found both oxalate and acetate ions in aerosols over the Amazon basin that compared favorably with concentrations found in Saharan dust that frequents the region. Offer et al. (1995) report mean monthly dissolvable organic concentrations as high as 2200 mg kg^{-1} in dust samples collected at selected locations of the Negev desert in Israel, but do not elaborate on the composition of these materials. Lepple and Brine (1976) investigated dust samples in North Africa and found that approximately 5% of the total organic C present was dissolvable in 50 °C seawater. They concluded that these soluble organics could be readily assimilated by certain surface organisms. Unpublished data (1999) provided by J. Bischoff indicate that between 1 and 55% of the soluble material in Owens dry lake bed dust is organic. His data also indicate that as the pH increases in the aqueous solutions, so does the soluble organic component, indicating a possible dispersion of structural and amorphous organics by the alkaline lake bed sediments. Laboratory analyses of three of these samples indicated that the dissolved organics were predominantly composed of hydrophobic forms at pH 10.9, while more hydrophilic than hydrophobic forms were present at pH 5.7 and 5.8 (M. Reheis, 1999, personal communication).

Amorphous Organic Materials

The final decomposition product of soil organic matter is commonly termed humus. Humus is a dark brown or black amorphous material that has a large surface area with many charged exchange sites that give it a cation exchange capacity of from 1500 to 3000 meq kg^{-1} (Brady, 1974, p.148). The exchange sites are primarily deprotonated carboxyl and phenol groups that may be affected to some extent by the pH of the environment (Bohn et al., 1985, p. 145). Humus is composed of colloids with weak cohesion and low plasticity, allowing it to easily attach to soil mineral colloids such as clays and silts. Attempts to unravel the exact composition of humus have gone unrewarded, but it is generally agreed that humic acid and polysaccharides make up over 90% of soil humus (Bohn et al., 1985, p. 143). The large surface areas, high charge densities, and variability of chemical forms make humus the most chemically active portion of the soil.

Most analyses for organic matter in suspended dusts has been performed by calculating the percentage lost on ignition of the sample, making determination of the exact nature of the organic constituent a matter of deduction. Since soil humus is found in close association with clays and silts, it may be assumed that if humus is present in the soil of provenance, a certain amount will be transported with the mineral fines. Gile and Grossman (1979) found that dust samples from southern New Mexico contained between 0.6 and 9.0% organic C. They also state that visual examination of the dust samples revealed that most of the organic C in their samples was in the humified form, closely associated with clay, and that a certain portion was water soluble. Selected solvent extraction and chromatographic analysis of marine sediments and aeolian dusts has revealed that the lipids found in these materials originated from the waxes of higher plants that are resistant to degradation (Simoneit, 1977; Simoneit & Eglinton, 1977). More recently, Simoneit (1997) has been able to use the proportion of ^{13}C to ^{14}C in the organic fraction of aerosols to determine the photosynthetic pathway of the plants producing the lipids, allowing for a determination of source area climate. Wilke et al. (1984) also performed ignition analysis of the Harmattan dust samples, but report that the sample contained enriched concentrations of trace elements N and P, elements commonly associated with humus (Bohn et al., 1985, p. 143). Windom (1969) studied the atmospheric dust records captured in permanent snowfields and found that greater than 50% of the organic compounds present were in the amorphous form. He also found the trend to be more predominant in the northern hemisphere. This would seem to substantiate the importance of humus/clay interactions in wind-blown sediments since it may be deduced that the hemisphere with the greatest erodible land mass would also have an atmosphere with more mineral dust and that this mineral dust would seem to contain a higher percentage of organic C in the amorphous form.

As important a role as humus serves in soils, it may also serve a very important role in the atmosphere. Studies have shown that hydrophobic organic chemicals are preferentially sorbed on soil humus over soil mineral colloids (Chiou, 1989; Hassett & Banwart, 1989) and that soil humus and biota are instrumental in catalyzing hydrolytic transformations of organic chemicals (Wolfe et al., 1989). Seinfeld (1986, p. 620) describes the dry deposition and precipitation scavenging of submicron particles in air. Assuming that the mean diameter of soil organic C in the atmosphere is supermicron and possesses a surface that is attractive to hydrophobic organic species in the atmosphere, it may well be that soil derived humus in the atmosphere is an effective scavenger of smaller organic molecules that, due to their size, are resistant to dry deposition. Once sorbed onto the humus surface, processes described by Wolfe et al. (1989) would tend to transform the organic contaminants, especially in the presence of liquid water. Additionally, Miller et al. (1989) have shown humus in the presence of sunlight to be an effective catalyst for creating highly reactive free radicals of O that are instrumental in the oxidation of organic pollutants. They also point out that insufficient sunlight penetrates the upper millimeter of soil for this phenomenon to occur other than at the surface. Suspended dust particles are constantly insolated by high intensity sunlight during daylight hours, providing ample opportunity for photolysis and photooxidation reactions to occur.

Very recently, researchers began investigating the role that dust emanating from polluted soils serves in the process of contaminant dispersal. Larney et al. (1999) investigated the dust mediated transport of endosulfan (hexachlorohexahydromethano-2,4,3-benzodioxathiepin 3-oxide) into waterways that had been contaminated with the pesticide. Although their results were inconclusive, they point out that inefficient dust traps were the main cause of inconclusive results and that toward the end of the study degradation products of endosulfan were collected during a period of several weeks after active pesticide application had ceased. They also cite studies in which wind erosion did contribute to the spread of agricultural chemicals away from the field of application, including a study in which one of the contributing authors showed that farm road surfaces produced 50 times as much dust as the adjacent fields when blown with a portable tunnel. Farm roads often receive as much spray material as the fields themselves due to overspray and drift by aerial applicators.

CONCLUSIONS

Wind erosion almost universally results in the loss of organic matter from the surface horizon of eroded soils. Soils in deposition areas may receive a net increase of organic matter, but the small amounts received and the relative unimportance of these soils to the world's food and fiber supply would make this gain of minor importance. Additionally, much of the suspended soil materials are transported over rivers, lakes, and oceans where deposition results in the loss of the incorporated carbonaceous materials from the terrestrial ecosystem. It is probable that the more unstable carbonaceous materials (water soluble materials and plant detritus) enter the food chain in these limnotic environments and are ultimately oxidized, resulting in the release of carbon dioxide. Carbonaceous soil materials do appear to serve an important role in catalyzing as well as neutralizing the reactions of anthropogenic pollutants. In the case of eolian carbonates neutralizing acidic aerosols of sulfates and nitrates, carbon dioxide is released to the atmosphere.

Reliable global estimates of soil organic matter and carbonates are not available at this time. There are several reasons for the lack of reliable estimates including the year-to-year variability in the total dust flux, the difficulty with estimating the total suspended solids in the global atmospheric system on any given day, and the variability of reported percentages of organic matter and carbonates in the suspended load. As satellite technology and the global network of ground-based reporting stations improve, we may be able to make reliable estimates in the near future. Duce (1978) estimates atmospheric particulate organic C load from soil and crustal weathering sources to be approximately 10.8 MT yr^{-1}. This estimate was based on a total annual dust flux from these sources of 250 MT yr^{-1}. and a weighted average of 4.3% particulate organic C in eolian dust. If we apply their technique to estimate carbonate fluxes, we may assume the same total dust flux and from Table 14–1 an average carbonate content of 14.7%, resulting in an estimated annual flux of 36.7 MT. It is important to consider that since estimates of total annual global dust flux vary by more than one order of magnitude and the reported percentages of organic matter and carbonate vary by the same amount, the above estimates of annual carbonaceous material flux may be off by as much as an order of magnitude or more in either direction.

In spite of the amount of literature available on aeolian soil materials, several questions remain largely unresolved. In order for global budgets of soil-incorporated C to be feasible, more information concerning the amount of soil C lost to erosion by soil type and region must be determined. Equally important is the need for more information about the forms of soil carbonaceous materials lost to erosion and how management practices affect these losses. More information is also needed concerning the fate of carbonaceous materials once they have been entrained in the atmosphere, especially regarding the regions of deposition and the amounts deposited over different landforms and bodies of water. Finally, more research is needed as to the role that aeolian materials serve in the transport and transformations of anthropogenic organic compounds.

REFERENCES

Alfaro, S.C., A. Gaudichet, L. Gomes, and M. Maille. 1997. Modeling the size distribution of a soil aerosol produced by sandblasting. J. Geophys. Res. 102 (D10):11239–11249.

Armbrust, D.V. 1968. Windblown soil abrasive injury to cotton plants. Agron. J. 60:622–625.

Armbrust, D.V., W.S. Chepil, and F.H. Siddoway. 1964. Effects of ridges on erosion of soil by wind. Soil Sci. Soc. Am. Proc. 28:557–560.

Aston, S.R., R. Chester, L.R. Johnson, and R.C. Padgham. 1973. Eolian dust from the lower atmosphere of the eastern Atlantic and Indian Oceans, China Sea, and Sea of Japan. Mar. Geol. 14:15–28.

Avila, A., I. Qeuralt-Mitjans, and M. Alarcon. 1997. Mineralogical composition of African dust delivered by red rains over northeastern Spain. J. Geophys. Res. 102 (D18):21 977–21 996.

Bagnold, R.A. 1941. The physics of blown sand and desert dunes. Methuen, London.

Bauman, S.L., P.R. Kyle, and J.B.J. Harrison, 1998. Characterization of the desert loess on the Carrizozo lava flow, south-central New Mexico. p. 128–130. In A. Busacca (ed.) Dust aerosols, loess soils and global change. College Agric. Home Econ. Misc. Publ. MISC0190. Washington State Univ., Pullman, WA.

Behairy, A.K.A., M.Kh. El-Sayed, and N.V.N. Durgaprasada Rao. 1985. Eolian dust in the coastal area north of Jedda, Saudi Arabia. J. Arid Environ. 8:89–98.

Bennet, J.G. 1980. Aeolian deposition and soil parent materials in northern Nigeria. Geoderma 24:241–255.

Biscaye, P.E., and F.E. Grousset. 1998. Ice-core and deep-sea records of atmospheric dust. p. 101–103. In A. Busacca (ed.) Dust aerosols, loess soils and global change. College Agric. Home Econ. Misc. Publ. MISC0190. Washington State Univ., Pullman, WA.

Blank, R.R., J.A. Young, and F.L. Allen. 1999. Aeolian dust in a saline playa environment, Nevada U.S.A. J. Arid Environ. 41:365–381.

Bohn, H.L., B.L. McNeal, and G.A. O'Connor. 1985. Soil chemistry. 2nd ed. John Wiley & Sons, Inc., New York.

Braaten, D.A., and T.A. Cahill. 1986. Size and composition of dust transported to Hawaii. Atmos. Environ. 20(6):1105–1109.

Brady, N.C. 1974. The nature and property of soils. 8th ed. McMillan, New York.

Brown, C.N. 1956. The origin of caliche on the northeastern Llano Estacado, Texas. J. Geol. 64:1–15.

Cahill, T.A., T.E. Gill, J.S. Reid, E.A. Gearhart, and D.A. Gillette. 1996. Saltating particles, playa crusts, and dust aerosols at Owens (dry) lake, California. Earth Surf. Process. Landforms 21:621–639.

Carder, K.L., R.G. Steward, and P.R. Betzer. 1986. Dynamics and composition of particles from an aeolian input event to the Sargasso Sea. J. Geophys. Res. 91(D1):1055–1066.

Chamberlain, A.C. 1975. The movement of particles in plant communities. p. 795–802. In J.L. Monteith (ed.) Vegetation and the atmosphere. Vol. 1. Acad. Press, London.

Charlson, R.J., and J.A. Ogren. 1981. The atmospheric cycle of elemental carbon. p. 3–16. In G.T. Wolff and R.L Klimisch (ed.) Particulate carbon atmospheric life cycle. Plenum, New York.

Chepil, W.S. 1956. Influence of moisture on erodability of soil by wind. Soil Sci. Soc. Am. Proc. 20:288–292.

Chester, R., G.G. Baxter, A.K.A. Behairy, K. Connor, D. Cross, H. Elderfield, and R.C. Padgham. 1977. Soil-sized eolian dusts from the lower troposphere of the eastern Mediterranean Sea. Mar. Geol. 24:201–217.

Chester, R., E.J. Sharples, G.S. Sanders, and A.C. Saydam. 1984. Saharan dust incursion over the Tyrrhenian Sea. Atmos. Environ. 18(5):929–935.

Chiou, C.T. 1989. Theoretical considerations of the partition uptake of nonionic organic compounds by soil organic matter. p. 1–30. In B.L. Sawhney and K. Brown (ed.) Reaction and movement of organic chemicals in soils. SSSA Spec. Publ. 22. SSSA and ASA, Madison, WI.

Chow, J.C., and J.G. Watson. 1997. Fugitive dust and other source contributions to PM10 in Nevada's Las Vegas Valley. Vol. 2. DRI Document no. 4039.2F1. Desert Res. Inst., Reno, NV.

Clausnitzer, H., and M.J. Singer. 1996. Respirable-dust production from agricultural operations in the Sacramento Valley, California. J. Environ. Qual. 25:877–884.

Currie, L.A. 1981. Contemporary particulate carbon. p. 245–260. In G.T. Wolff and R.L Klimisch (ed.) Particulate carbon atmospheric life cycle. Plenum, New York.

D'Almeida, G.A., and L. Schutz. 1983. Number, mass, and volume distributions of mineral aerosols and soils of the Sahara. J. Am. Meteorol. Soc. 22:233–243.

Delany, A.C., A.C. Delany, D.W. Parkin, J.J. Griffin, E.D. Goldberg, and B.E.F. Reimann. 1966. Airborne dust collected at Barbados. Geochim. Cosmochim. Acta 31:885–909.

Delany, A.C., and S.T. Zenchelsky. 1976. The organic component of wind-erosion-generated soil-derived aerosol. Soil Sci. 121:146–155.

Donkin, R.A. 1981. The "Manna Lichen": *Lecanora esculenta*. Anthropos 76:562–576.

Duce, R.A. 1978. Speculations on the budget of particulate and vapor phase non-methane organic carbon in the global troposphere. Pure Appl. Geophys. 116:244–273.

Duce, R.A., V.A. Mohnen, P.R. Zimmerman, D Grosjean, W. Cautreels, R. Chatfield, R. Jaenicke, J.A. Ogren, E.D. Pellizzari, and G.T. Wallace. 1983. Organic material in the global troposphere. Rev. Geophys. Space Phys. 21(4):921–952.

Farmer, A.M. 1993. The effects of dust on plants: a review. Environ. Pollut. 79:63–75.

Feng, Z. 1998. Eolian environments in Nebraska during the marine isotope stage 2. p. 107–110. *In* A. Busacca (ed.) Dust aerosols, loess soils and global change. College Agric. Home Econ. Misc. Publ. MISC0190. Washington State Univ., Pullman, WA.

Fryrear, D.W. 1963. Annual crops as wind barriers. Trans. ASAE 6:340–342, 352.

Fryrear, D.W. 1981. Dust storms in the southern great plains. Trans. ASAE 24:991–994.

Fryrear, D.W. 1987. Aerosol measurements from 31 dust storms. *In* T. Ariman and T.N. Veziroglu (ed.) Particulate and multiphase processes. Vol. 2. Hemisphere Publ., Washington, DC.

Fryrear, D.W. 1999. Wind erosion. p. 191–211. *In* Soil science handbook. CRC Press, Boca Raton, FL.

Fryrear, D.W., and J.D. Downes. 1975. Estimating seedling survival from wind erosion parameters. Trans. ASAE 18:888–891.

Fryrear, D.W., and P.T. Koshi. 1971. Conservation of sandy soils with surface mulch. Trans ASAE 14:492–495, 499.

Fryrear, D.W., C.A. Krammes, D.L. Williamson, and T.M. Zobeck. 1994. Computing the wind erodible fraction of soils. J. Soil Water Conserv. 49:183–188.

Gao, Y., R. Arimoto, R.A. Duce, D.S. Lee, and M.Y. Zhou. 1992. Input of atmospheric trace elements and mineral matter to the Yellow Sea during the spring of a low-dust year. J. Geophys. Res. 97(D4):3767–3777.

Gardner, L.R. 1972. Origin of the Mormon Mesa caliche, Clark County, Nevada. Geol. Soc. Am. Bull. 83:143–156.

Gatz, D.F., and J.M. Prospero. 1996. A large silicon-aluminum aerosol plume in central Illinois: North African Dust? Atmos. Environ. 30:3789–3799.

Geiger, R. 1961. The climate near the ground. 4th ed. Harvard Univ. Press, Cambridge, MA.

Gertler, A.W., D.A. Lowenthal, and W.G. Coulombe. 1995. PM_{10} source apportionment study in Bullhead City, Arizona. J. Air Waste Mgt. Assoc. 45:75–82.

Gile, L.H., and R.B. Grossman. 1979. The desert project soil monograph: Soils and landscapes of desert region astride the Rio Grande Valley near Las Cruces, New Mexico. USDA-SCS, and U.S. Gov. Print. Office, Washington, DC.

Gill, T.E., M.C. Reheis, and T.M. Zobeck. 1999. Present-day aeolian deposition in the Southern High Plains of Texas: first results of a dustfall study. Geol. Soc. Am. Abstr. Progr. 31(1)(abstract):A8.

Gillette, D.A. 1977. Fine particulate emissions due to wind erosion. Trans. ASAE 20:890–897.

Gillette, D.A. 1978. A wind-tunnel simulation of the erosion of soil: Effect of soil moisture, sandblasting, wind speed and soil consolidation on dust production. Atmos. Environ. 12:1735–1743.

Gillette, D.A., I.H. Blifford, Jr., and D.W. Fryrear. 1974. The influence of wind velocity on the size distributions of aerosols generated by the wind erosion of soils. J. Geophys. Res. 79:4068–4075.

Gillette, D.A., R.N. Clayton, T.K. Mayeda, M.L. Jackson, and K. Sridhar. 1978. Tropospheric aerosols from major dust storms of the southwestern United States. J. Appl. Meteorol. 17:832–845.

Gillette, D.A., and P.C. Sinclair. 1990. Estimation of suspension of alkaline material by dust devils in the United States. Atmos. Environ. 24A:1135–1142.

Glaccum, R.A., and J.M. Prospero. 1980. Saharan aerosols over the tropical North Atlantic—mineralogy. Mar. Geol. 37:295–321.

Goldenberg, M., and J.R. Brook. 1997. Dry deposition of nonspherical particles on natural and surrogate surfaces. Aerosol Sci. Technol. 27:22–38.

Gomes, L., G. Bergametti, G. Coude-Gaussen, and P.Rognon. 1990. Submicron desert dusts: A sandblasting process. J. Geophys. Res. 95(D9):13 927–13 935.

Gomes, L., and D.A. Gillette. 1993. A comparison of characteristics of aerosol from dust storms in central Asia with soil-derived dust from other regions. Atmos. Environ. 27A:2539–2544.

Goosens, D., and Z.Y. Offer. 1994. An evaluation of the efficiency of some eolian dust collectors. Soil Technol. 7:25–35.

Goudie, A.S. 1978. Dust Storms and their geomorphological implications. J. Arid Environ. 1:291–310.

Green, H.L., and W.R. Lane. 1964. Particulate clouds: dusts, smokes, and mists. Spon, London.

Hagen, L.J., E.L. Skidmore, and J.B. Layton. 1988. Wind erosion abrasion: effects of aggregate moisture. Trans. ASAE 31:725–728.

Hassett, J.J., and W.L. Banwart. 1989. The sorption of nonpolar organics by soils and sediments. p. 31–45. *In* B.L. Sawhney and K. Brown (ed.) Reaction and movement of organic chemicals in soils. SSSA Spec. Publ. 22. SSSA and ASA, Madison, WI.

Hedin, L.O., and G.E. Likens. 1996. Atmospheric dust and acid rain. Sci. Am. 275(6):88–92.

Herwitz, S.R., D.R. Muhs, J.M. Prospero, S. Mahan, and Bruce Vaughn, 1996. Origin of Bermuda's clay-rich Quaternary paleosols and their paleoclimatic significance. J. Geophys. Res. 110(D18):23 389–23 400.

Hoidale, G.B., and S.M. Smith. 1968. Analysis of the giant particle component of the atmosphere over an interior desert basin. Tellus 20(2):251–268.

Irmak, S., and S. Aydemir. 1998. Calculation of the amount of sediment deposited by wind transport onto volcanic soils of arid regions in Turkey. p. 215–219. *In* A. Busacca (ed.) Dust aerosols, loess soils and global change. College Agric. Home Econ. Misc. Publ. MISC0190. Washington State Univ., Pullman, WA.

Johnson, L.R. 1979. Mineralogical dispersal patterns of North Atlantic deep-sea sediments with particular reference to eolian dusts. Mar. Geol. 29:335–345.

Kahle, C.F. 1977. Origin of subaerial Holocene calcareous crusts: Role of algae, fungi, and sparmicritisation. Sedimentology 24:413–435.

Kennedy, A.C. 1998. Biological fingerprinting of dust aerosols. p. 215–219. *In* A. Busacca (ed.) Dust aerosols, loess soils and global change. College Agric. Home Econ. Misc. Publ. MISC0190. Washington State Univ., Pullman, WA.

Kiefert, L., G.H. McTainsh, and W.G. Nickling. 1996. Sedimentological characteristics of saharan and Australian dusts. p. 183–190. *In* S. Guerzoni and R Chester (ed.) The impact of desert dust across the Mediterranean. Kluwer Acad. Publ., The Netherlands.

Krauss, R.K., and R.M. Smith. 1969. Variability in atmospheric dust sampling. Trans. Kansas Acad. Sci. 72(2):167–176.

Laprade, K.E. 1957. Dust-storm sediments of the Lubbock area, Texas. Bull. Am. Assoc. Petrol. Geol. 41(4):709–726.

Larney, F.J., J.F. Leys, J.F. Muller, and G.H. McTainsh. 1999. Dust and endosulfan deposition in cotton-growing area of Northern New South Wales, Australia. J. Environ. Qual. 28:692–701.

Lattman, L.H. 1973. Calcium carbonate cementation of alluvial fans in southern Nevada. Geol. Soc. Am. Bull. 84:3013–3028.

Leathers, C.R. 1981. Plant components of desert dust in Arizona and their significance for man. p. 191–206. *In* T.L. Pewe (ed.) Desert dust: Origin, characteristics, and effect on man. Geol. Soc. Am. Spec. Publ. 186. Geol. Soc. Am., Boulder, CO.

Lepple, F.K., and C.J. Brine. 1976. Organic constituents in eolian dust and surface sediments from Northwest Africa. J. Geophys. Res. 81:1141–1147.

Liberti, A., D. Brocco, and M. Possanzini. 1978. Adsorption and oxidation of sulfur dioxide on particles. Atmos. Environ. 12:255–261.

Liousse, C., H. Cachier, and S.G. Jennings. 1993. Optical and thermal measurements of black carbon aerosol content in different environments: variation of the specific attenuation cross-section sigma (σ). Atmos. Environ. 27A:1203–1211.

Liousse, C., J.E. Penner, C. Chuang, J.J. Walton, H. Eddleman, and H. Cachier. 1996. A global three-dimensional model study of carbonaceous aerosols. J. Geophys. Res. 101(D14):19 411–19 432.

Litaor, M.I. 1987. The influence of eolian dust on the genesis of alpine soils in the Front Range, Colorado. Soil Sci. Soc. Am. J. 51:142–147.

Littmann, T. 1997. Atmospheric input of dust and nitrogen into the Nizzana sand dune ecosystem, northwestern Negev, Israel. J. Arid Environ. 36:433–457.

Loosmore, G., and J.R. Hunt. 1998. Predicting dust resuspension by considering saltator availability. p. 63–66. *In* A. Busacca (ed.) Dust aerosols, loess soils and global change. College Agric Home Econ. Misc. Publ. MISC0190. Washington State Univ., Pullman, WA.

Loye-Pilot, M.D., J.M. Martin, and J. Morelli. 1986. Influence of Saharan dust on the rain acidity and atmospheric input to the Mediterranean. Nature (London) 321:427–428.

Lyles, L., and J. Tatarko. 1986. Wind erosion effects on soil texture and organic matter. J. Soil Water Conserv. 41:191–193.

Malcolm, L.P., and M.R. Raupach. 1991. Measurements in an air settling tube of the terminal velocity distribution of soil material. J. Geophys. Res. 96(D8):15 275–15 286.

McCauley, J.F., C.S. Breed, M.J. Grolier, and D.J. MacKinnon. 1981. The U.S. dust storm of February 1977. p. 123–147. *In* T.L. Pewe (ed.) Desert dust: Origin, characteristics, and effect on man. Geol. Soc. Am. Spec. Publ. 186. Geol. Soc. Am., Boulder, CO.

McGowan, H.A., and A.P. Sturman. 1997. Characteristics of aeolian grain transport over a fluvio-glacial lacustrine braid delta, Lake Tekapo, New Zealand. Earth Surf. Process. Landforms 22:773–784.

Mikkelsen, J.H., and R. Langohr. 1998. Impact of dust on soil evolution and fertility, case studies from Northern Ghana and the Altai Mountains, SW Siberia. *In* A. Busacca (ed.) Dust aerosols, loess soils and global change. College Agric. Home Econ. Misc. Publ. MISC0190. Washington State Univ., Pullman, WA.

Miller, G.C., V.R. Herbert, and W.W. Miller. 1989. Effect of sunlight on organic contaminants at the atmosphere-soil interface. p. 99–110. *In* B.L. Sawhney and K. Brown (ed.) Reaction and movement of organic chemicals in soils. SSSA Spec. Publ. 22. SSSA and ASA, Madison, WI.

Moberg, J.P., I.E. Esu, and W.B. Malgwi. 1991. Characteristics and constituent composition of Harmattan dust falling in Northern Nigeria. Geoderma 48:73–81.

Modaihsh, A.S. 1997. Characteristics and composition of the falling dust sediments on Riyadh city, Saudi Arabia. J. Arid Environ. 36:211–223.

Moharram, M.A., and M.A. Sowelim. 1980. Infrared study of minerals and compounds in atmospheric dustfall in Cairo. Atmos. Environ. 14:853–856.

Monger, H.C., L.A. Daugherty, and W.C. Lindemann. 1991. Microbial precipitation of pedogenic calcite. Geology 19:997–1000.

Mueller, P.K., K.K. Fung, S.L. Heisler, D. Grosjean, and G.M. Hidy. 1981. Atmospheric particulate carbon observations in urban and rural areas of the United States. p. 343–368. *In* G.T. Wolff and R.L Klimisch (ed.) Particulate carbon atmospheric life cycle. Plenum, New York.

Offer, Z.Y., S. Sarig, and Y. Steinberger. 1995. Dynamics of nitrogen and carbon content of aeolian dry deposition in an arid region. Arid Soil Res. Rehabilat. 10:193–199.

Oke, T.R. 1978. Boundary layer climates. Methuen, London.

Oliver, J.E., and R. W. Fairbridge (ed.) 1987. Encyclopedia of climatology. Encyclopedia of earth sciences. Vol. 11. Van Nostrand Reinhold Co., New York.

Pan, D.M., and M.J. Prather. 1998. Aerosol scavenging by cumulus precipitation in global transport modeling. p. 79–80. *In* A. Busacca (ed.) Dust aerosols, loess soils and global change. College Agric. Home Econ. Misc. Publ. MISC0190. Washington State Univ., Pullman, WA.

Patterson, E.M., and D.A. Gillette. 1977. Commonalities in measured size distributions for aerosols having a soil-derived component. J. Geophys. Res. 82:2014–2082.

Perry, K.D., T.A. Cahill, R.A. Eldred, D.D. Dutcher, and T.E. Gill. 1997. Long-range transport of North African dust to the eastern United States. J. Geophys. Res. 102:11 225–11 238.

Pewe, T.L. 1981. Desert dust: an overview. *In* T.L. Pewe (ed.) Desert dust: Origin, Characteristics, and effect on man. Geol. Soc. Am. Spec. Publ. 186. Geol. Soc. Am., Boulder, CO.

Pewe, T.L., E.A. Pewe, R.H. Pewe, A. Journaux, and R.M. Slatt. 1981. Desert dust: Characteristics and rates of deposition in central Arizona. *In* T.L. Pewe (ed.) Desert dust: Origin, characteristics, and effect on man. Geol. Soc. Am. Spec. Publ. 186. Geol. Soc. Am. Spec., Boulder, CO.

Pinnick, R.G., G. Fernandez, E. Martinez-Andazola, B.D Hinds, A.D.A. Hansen, and K. Fuller. 1993. Aerosol in the arid southwestern United States: measurements of mass loading, volatility, size distribution, absorption characteristics, black carbon content, and vertical structure to 7 km above sea level. J. Geophys. Res. 98:2651–2666.

Porch, W.M. 1974. Fast-response light scattering measurements of aerosol suspension in a desert area. Atmos. Environ. 8:897–904.

Prospero, J.M. 1981. Arid regions as sources of mineral aerosols in the marine environment. p. 71–86. *In* T.L. Pewe (ed.) Desert dust: Origin, characteristics, and effect on man. Geol. Soc. Am. Spec. Publ. 186. Geol. Soc. Am., Boulder, CO.

Prospero, J.M. 1996a. Saharan dust transport over the North Atlantic Ocean and Mediterranean: and overview. p. 133–151. *In* S. Guerzoni and R. Chester (ed.) The impact of desert dust across the Mediterranean. Kluwer Acad. Press, Netherlands.

Prospero, J.M. 1996b. The atmospheric transport of particles to the Ocean. *In* V.S. Ittekkott et al. (ed.) Particle flux in the ocean. SCOPE Rep. John Wiley & Sons, LTD, New York.

Prospero, J.M. 1999. Long-term measurements of the transport of African mineral dust to the southeastern United States: Implications for regional air quality. J. Geophys. Res. 104:15 917–15 927.

Pye, K. 1987. Aeolian dust and dust deposits. Acad. Press, London.

Pye, K. 1991. Aeolian dust transport and deposition over Crete and adjacent parts of the Mediterranean Sea. Earth Surf. Process. Landforms 17:271–288.

Rabenhorst, M.C., L.P. Wilding, and C.L. Girdner. 1984. Airborne dusts in the Edwards Plateau region of Texas. Soil Sci. Soc. Am. J. 48:621–627.

Rahn, K.A., R.D. Borys, and G.E. Shaw. 1981. Asian dust over Alaska: Anatomy of an Arctic haze episode. p. 37–70. *In* T.L. Pewe (ed.) Desert dust: Origin, characteristics, and effect on man. Geol. Soc. Am. Spec. Publ. 186. Geol. Soc. Am., Boulder, CO.

Reheis, M.C. 1997. Dust deposition downwind of Owens (dry) Lake, 1991–1994: Preliminary findings. J. Geophys Res. 102:25 999–26 008.

Reheis, M.C., J.C. Goodmacher, J.W. Harden, L.D. McFadden, T.K. Rockwell, R.R. Shroba, J.M. Sowers, and E.M. Taylor. 1995. Quaternary soils and dust deposition in southern Nevada and California. Geol. Soc. Am. Bull. 107:1003–1022.

Reheis, M.C., and R. Kihl. 1995. Dust deposition in southern Nevada and California, 1984–1989: Relations to climate, source area, and source lithology. J. Geophys. Res. 100:8893–8918.

Rogers, R.R. 1979. A short course in cloud physics. 2nd ed. Pergamon Press, Oxford.

Rogge, W.F., L.M. Hildemann, M.A. Mazurek, G.R. Cass, and B.R.T. Simoneit. 1993. Sources of fine organic aerosol. 3. Road dust, tire debris, and organometallic brake lining dust: Roads as sources and sinks. Environ. Sci. Technol. 27:1892–1904.

Savoie, D.L., and J.M. Prospero. 1980. Water-soluble potassium, calcium, and magnesium in the aerosols over the tropical North Atlantic. J. Geophys. Res. 85:385–392.

Schutz, L., R. Jaenicke, and H. Pietrek. 1981. Saharan dust transport over the North Atlantic. In T.L. Pewe (ed.) Desert dust: Origin, characteristics, and effect on man. Geol. Soc. Am. Spec. Publ. 186. Geol. Soc. Am., Boulder, CO.

Sehmel, G.A. 1980. Particle and gas dry deposition: A review. Atmos. Environ. 14:983–1011.

Seinfeld, J.H. 1986. Atmospheric chemistry and physics of air pollution. John Wiley & Sons, Inc., New York.

Shao, Y., M.R. Raupach, and P.A. Findlater. 1993. Effects of saltation bombardment on the entrainment of dust by wind. J. Geophys. Res. 98:12 719–12 726.

Shaw, G.E. 1980. Transport of Asian desert aerosol to the Hawaiian Islands. J. Appl. Meteorol. 19:1254–1259.

Sheng, L.T., G.X. Fei, A.Z. Sheng, and F.Y. Xiang. 1981. The dustfall in Beijing, China on April 18, 1980. In T.L. Pewe (ed.) Desert dust: Origin, characteristics, and effect on man. Geol. Soc. Am. Spec. Publ. 186. Geol. Soc. Am., Boulder, CO.

Simoneit, B.R.T. 1977. Organic matter in eolian dusts over the Atlantic Ocean. Marine Chem. 5:443–464.

Simoneit, B.R.T. 1997. Compound-specific carbon isotope analyses of individual long-chain alkanes and alkanoic acids in Harmattan aerosols. Atmos. Environ. 31:2225–2233.

Simoneit, B.R.T., and G. Eglinton. 1977. Organic matter of eolian dusts and itsinput to marine sediments. p. 415–430. In R. Campos and J. Goni (ed.) Advances in organic geochemistry 1975. ENADIMSA, Madrid.

Simonson, R.W. 1995. Airborne dust and its significance to soils. Geoderma 65:1–43.

Sterk, G., L Hermann, and A. Bationo. 1996. Wind-blown nutrient and soil productivity changes in southwest Niger. Land Degradat. Develop. 7:325–335.

Stetler, L.D., K.E. Saxton, and D.W. Fryrear. 1994. Wind erosion and PM_{10} measurements from agricultural fields in Texas and Washington. p. 1–14. In Air Waste Manage. Assoc. 87th Annu. Meet., Pap. 94-FA145.02, Cincinnati, OH. 19–24 June. Air Waste Manage. Assoc., Pittsburgh, PA.

Swineford, A., and J.C. Frye. 1945. A mechanical analysis of wind-blown dust compared with the analysis of loess. Am. J. Sci. 161:616–632.

Takemi, T. 1999. Structure and evolution of a severe squall line over the arid region in northwest China. Monthly Weath. Rev. 127:1301–1309.

Talbot, R.W., M.O. Andreae, H. Berresheim, P. Artaxo, M. Garstang, R.C. Harriss, K.M. Beecher, and S.M. Li. 1990. Aerosol chemistry during the wet season in central Amazonia: The influence of long-range transport. J. Geophys. Res. 95:16 955–16 969.

Tiessen, H., H.-K. Hauffe, and A.R. Mermut. 1991. Deposition of Harmattan dust and its influence on base saturation of soils in northern Ghana. Geoderma 49:285–299.

Tsoar, H., and E. Erell. 1995. The effect of a desert city on aeolian dust deposition. J. Arid Land Stud. 55:115–118.

Tsoar, H., and K. Pye. 1987. Dust transport and the question of desert loess formation. Sedimentology 34:139–153.

U.S. Department of Agriculture Natural Resources Conservation Service. 1994. Summary report 1992 national resources inventory. Iowa State Univ. Stat. Lab Rep. EI&D 94-920. USDA, NRCS, Washington, DC.

Usman, H. 1995. Wind erosion in northeastern Nigeria. 1. Erodibility factors. Arid Soil Res. Rehabil. 9:457–466.

Wagoner, L. 1997. Modern and geologic rates of dust flux into Fourth of July Lake, Columbia Plateau, southeastern Washington. p. 42. In Wind erosion. Proc. Int. Symp./Workshop Manhattan, Kansas. 3–5 June. USDA-ARS, NPA Wind Eros. Res. Unit, Manhattan, KS.

Wilke, B.M., B.J.D. Bayreuth, W.L.O. Lancaster, and K Jimoh. 1984. Mineralogy and chemistry of Harmattan dust in northern Nigeria. Catena 11:91–96.

Wilshire, H.H., J.K. Nakata, and B. Hallet. 1981. Field observations of the December 1977 wind storm San Joaquin Valley, California. p. 233–252. *In* T.L. Pewe (ed.) Desert dust: Origin, characteristics, and effect on man. Geol. Soc. Am. Spec. Publ. 186. Geol. Soc. Am., Boulder, CO.

Windom, H.L. 1969. Atmospheric dust records in permanent snowfields: Implications to marine sedimentation. Geol. Soc. Am. Bull. 80:761–782.

Windom, H.L. 1970. Contribution of atmospherically transported trace metals to South Pacific sediments. Geochim. Cosmochim. Acta 34:509–514.

Wolfe, N.L., M.E. Metwally, and A.E. Mofta. 1989. Hydrolytic transformations of organic chemicals in the environment. p. 229–242. *In* B.L. Sawhney and K. Brown (ed.) Reaction and movement of organic chemicals in soils. SSSA Spec. Publ. 22. SSSA and ASA, Madison, WI.

Woodruff, N.P., and L.J. Hagen. 1972. Dust in the Great Plains. p. 241–258. Great Plains Agric. Counc. Publ. 60. Kansas Agric. Exp. Stn., Manhattan, KS.

Yaalon, D.H., and E. Ganor. 1973. The influence of dust on soils during the Quaternary. Soil Sci. 116:146–155.

Yaalon, D.H., and D. Ginzbourg. 1966. Sedimentary characteristics and climatic analysis of easterly dust storms in the Negev (Israel). Sedimentology 6:315–332.

Zeuner, F.E. 1949. Frost soils on Mount Kenya, and the relation of frost soils to aeolian deposits. J. Soil Sci. 1:20–30.

Zhang, X.Y., and Z.S. An. 1998. Sources, emission, regional-and global-scale transport of Asian dust. *In* A. Busacca (ed.) Dust aerosols, loess soils and global change. College Agric. Home Econ. Misc. Publ. MISC0190. Washington State Univ., Pullman, WA.

Zobeck, T.M. 1991. Soil properties affecting wind erosion. J. Soil Water Conserv. 46:112–118.

Zobeck, T.M., and D.W. Fryrear. 1986. Chemical and physical characteristics of windblown sediment. II. Chemical characteristics and total soil and nutrient discharge. Trans. ASAE 29:1037–1041.

Zobeck, T.M., D.W. Fryrear, and R.D. Pettit. 1989. Management effects on wind-eroded sediment and plant nutrients. J. Soil Water Conserv. 44:160–163.

Zobeck, T.M., T.E. Gill, and T.W. Popham. 1999. A two-parameter Weibull function to describe airborne dust particle size distributions. Earth Surf. Process. Landforms 24:1–13.

Zobeck, T.M., and T.W. Popham. 1998. Wind erosion roughness index response of observation spacing and measurement distance. Soil Till. Res. 45:311–324.

Zobeck, T.M., J.E. Stout, T.E. Gill, and D.W. Fryrear. 1996. Midwest plan service. p. 49–56. *In* Airborne dust and sediment measurements in agricultural fields. Int. Conf. on Air Pollution from Agricultural Operations, Kansas City, MO. 7–9 February. Iowa State Univ., Ames, IA.

15 Modeling and Regional Assessment of Soil Carbon: A Case Study of the Conservation Reserve Program

Keith Paustian, Kendrick Killian, and Jan Cipra

Colorado State University
Fort Collins, Colorado

George Bluhm

USDA-Natural Resources Conservation Service
Davis, California

Jeffrey L. Smith

USDA-Agricultural Research Service
Pullman, Washington

Edward T. Elliott

University of Nebraska
Lincoln, Nebraska

ABSTRACT

Regional assessments of soil C sequestration are needed to quantify potentials at state and national levels and to identify how C sequestration rates may vary regionally as a function of climate, soil, land use history, and other factors. This information can help guide policy and management decisions to maximize the environmental and societal benefits from C sequestration. We employ an approach combining an ecosystem simulation model and a set of geographically distributed databases that provide the driving variables for the model. In an application of this approach, we assessed C sequestration in Conservation Reserve Program (CRP) grasslands during the first 10-yr contract period, for a 16-state region in the central and northwestern USA, containing about 70% of the CRP acreage nationwide. Climate, soil and land use data were used to delineate 59 climate/land use zones each including multiple soil textural classes. The model results suggest that CRP lands are sequestering C and that regional differences in rates of C storage are controlled mainly by differences in primary production rates and secondarily by abiotic and soil factors influencing decomposition. Simulated rates of soil organic matter (SOM) C accumulation under CRP ranged from less than 10 to more than 40 g C m^{-2} yr^{-1}, with the highest rates in the most humid regions. Total below-ground C accumulation rates (including roots) ranged from 25 to 135 g C m^{-2} yr^{-1}. For the study region, predicted C increases for the first 10-yr period of the CRP were on the order of 25 Tg (10^{12} g) for SOM C alone or up to 69 Tg if plant and litter C stocks are included. These levels of C sequestration could contribute significantly to the overall sequestration potential in the agricultural sector. However, C sequestered on CRP grasslands is vulnerable to loss with reversion to annual cropping and the effects of future management of previous and newly contracted CRP lands on C sequestration needs further evaluation.

The role of soils as a sink for C and thus the potential for C sequestration to mitigate the build-up of atmospheric CO_2 has attracted increasing interest from scientists as well as from governments, agricultural producers and industry. Global assessments of the potential contribution of C sequestration in agricultural soils, through changes in land management, are on the order of 600 to 900 Tg yr^{-1} (Cole et al., 1996). A limited number of assessments of C sequestration potential have been made for individual countries or regions (Kern & Johnson, 1993; Smith et al., 1997; Lee et al., 1993; Dumanski et al., 1998) but few analyses have been made considering the influence of specific management practices over a range of climate, soil and land use conditions at a regional scale (Donigian et al., 1998).

A robust approach to regional analysis needs to integrate modeling, spatial data and relevant field and process-level information into a common framework (Elliott & Cole, 1989). We have adopted this approach in analyzing soil C sequestration at the regional scale. Briefly, a simulation model provides the integrating "engine" for the many complex interactions controlling soil C dynamics. Reliable models are necessarily based on extensive process-level research and it is crucial that they be validated for the range of conditions that exist within the region. For soil organic matter dynamics, regional networks of long-term field experiments provide the most suitable source of validation and testing data (Paustian et al., 1995; Paul et al., 1997). Driving variables, such as climate, soil properties and management practices, control the dynamics of soil C over space and time. This information is inherently spatial in nature and is often stored and accessed with geographic information system (GIS) software. Many of the key types of data, e.g., climate, soil maps, digital elevation models, crop distributions, etc., are now readily available through various state and federal agencies. For simulating ecosystem processes across a region, these data constitute the inputs for the model and as such they also define the spatial units (polygons) to which the model is applied. By performing simulations for each polygon, the outcome for any of the processes simulated by the model (e.g., changes in soil C storage) can be integrated for the entire region as well as displayed spatially in the form of a map.

The Conservation Reserve Program (CRP) is a U.S. Government-funded program that was established in 1985 to remove highly erodible agricultural lands from production for periods of 10 yr or more by converting them to perennial grass or tree vegetation. Interest in assessing the overall benefits of the CRP program has grown beyond the initial focus on erosion and water quality to include other amenities such as wildlife habitat and the effect of CRP on greenhouse gases—specifically the sequestration of C in soil and biomass. Land planted to grasses under the CRP occurs in 47 states and encompasses a wide range of climate and soil conditions. Consequently, rates of soil C change would be expected to vary considerably across the USA. While no systematic monitoring program was established at the onset of the program to detect changes in soil properties, there have been several studies comparing CRP with adjacent or nearby cropped land (Gebhart et al., 1994; Huggins et al., 1998; Burke et al., 1995; Robles & Burke, 1997, 1998; Reeder et al., 1998; Follett et al., 2001). However, with the exception of the studies by Follett et al. (2001) and Huggins et al. (1998), most have been restricted to a few sites, limiting their value for making regional extrapolations.

To complement the information from field studies, we conducted a model-based regional assessment of potential C sequestration under the CRP program for the initial 10-yr contract period (1985–1995). The objective of the study was to quantify the changes in soil and ecosystem C due to conversion of cropland to grassland CRP, including estimates of regional differences associated with climate, soil and previous land use history. We focused on a 16-state region where most of the CRP lands were located.

METHODOLOGY FOR REGIONAL ANALYSES

Overview of methodology

Our 16-state study region comprised most of the central USA plus the Pacific Northwest, where 87% of the grassland CRP and about 70% of the total CRP acreage is located. We assembled regional databases of climate and soil properties and the distribution of CRP acreage and these data were then used as inputs to a series of simulation model runs for 59 multicounty "climate zones" within the region. Results from the model runs were then mapped using a GIS and changes in C stocks for CRP lands were calculated. We used the Century model, Agroecosystem Version 4.0 (Metherell et al., 1993; Parton et al., 1994). The model calculations apply only to the top 20 cm of soil.

Geographic Information System Databases

The function of the GIS was to process spatial information used for model input and to map model outputs. Geographic areas were delineated in which model driving variables (e.g., climate, land use, soil conditions) could be aggregated and used as input to the model simulations. The choice of spatial resolution for the regional analysis was constrained by several factors, including the resolution of the original databases, the need to minimize aggregation error, and practical limits on the number of simulations that could be run and interpreted.

Climate Data and Climate Zone Derivation

Long-term (1961–1991) normal data from approximately 2500 weather stations in the 16-state region were obtained from the USDA-NRCS Western National Technical Center, Portland, Oregon (P. Pasteris, 1996, personal communication). Only mean monthly minimum, average, and maximum temperature and monthly precipitation were used for model input. Climate station data were converted on a state-by-state basis from latitude and longitude to a Lambert Azimuthal projection and joined spatially using ARC/INFO.

Annual precipitation and average maximum monthly temperature data were chosen as reference values for the derivation of "climate zones" (Fig. 15–1). The climate zones comprised the basic geographical unit for simulation in the regional analyses, for which temperature and precipitation data of all weather stations within the zone were averaged to represent that subregion.

To delineate the climate zones, weather station data were first processed to yield temperature and precipitation vector point coverages (Fig. 15–1). Mountainous areas with minimal agricultural area, mainly in the Rocky Mountains and in the Pacific Northwest, were excluded from the study region and thus data from weather stations located in these areas were deleted. Point coverages were then con-

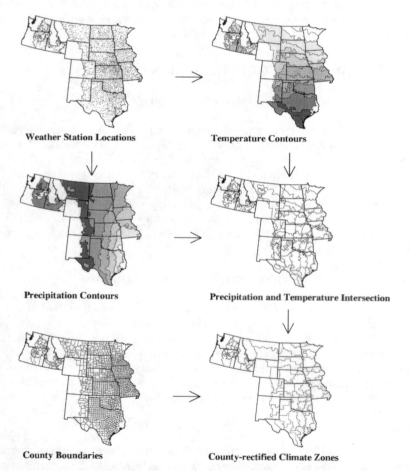

Fig. 15–1. The derivation of climate-management zones used long-term records from weather stations (upper left) to derive temperature (upper right) and precipitation (middle left) contour maps. The intersection of contours (middle right) formed polygons that were aligned with county boundaries (lower left) to form the climate zones (lower right).

verted to temperature and precipitation contour maps using an inverse distance weighting (IDW) contouring algorithm of ARC/INFO's GRID module. The IDW algorithm is a raster-based function that determines cell values by using a linearly weighted combination of a set of sample points, with the weight being a function of inverse distance.

Raster maps resulting from the IDW analysis were converted back to vector format, and edited to delete polygons containing a single weather station. Resultant maps of temperature (spanning 11 intervals) and precipitation contours (spanning 6 intervals) were overlaid to produce intersecting temperature and precipitation isolines that defined discrete climate zones (Fig. 15–1). Climate zones boundaries were then aligned with county boundaries to ensure compatibility with the CRP land use database that is georeferenced by county only. A total of 59 climate zones were delineated and used in the regional modeling analysis.

The choice of the intervals used to delineate climate zones, i.e., 2 °C differences in mean maximum monthly temperature and 20-cm differences in annual precipitation, was based on a need to minimize aggregation error while keeping a practical limit on the number of simulations required (Paustian et al., 1997). Implicit in the aggregation is the assumption that the effects of temperature and precipitation on soil C changes within a climate zone can be accurately represented by the average monthly climate variables for the zone. The error associated with this aggregation is dependent on the characteristics of the model's response to climate driving variables. In general, the degree of aggregation error increases with the degree of nonlinearity of the model response (Rastetter et al., 1992). For purely linear response functions, the aggregation error is zero.

For the region as a whole, the strongest influence of precipitation on soil C dynamics is through its effect on primary production and the amount of plant residue added to the soil. In Century, the functional response of soil C to C input rate is linear (Paustian et al., 1995) and, for temperate grasslands, the model's production response to precipitation is also quite linear (Parton et al., 1987). Therefore, there is relatively little error associated with using averaged precipitation values. This was demonstrated by Burke et al. (1991) who found little difference in simulated soil C levels using county-level precipitation data vs. using averaged precipitation values for multicounty aggregations. In contrast, temperature effects on production and decomposition rates are both nonlinear (Metherell et al., 1993) and therefore aggregation error increases as the range of temperatures being averaged increases. Based on these considerations, temperature intervals were chosen to be relatively small (yielding 11 intervals across the region) in comparison to precipitation intervals (6 across the region).

Conservation Reserve Program Contract Data

A database containing information on each individual CRP contract in the first 12 sign-ups (from 1986–1992) was obtained from USDA/Farm Service Agency (FSA; A. Barbarika, 1996, personal communication). The contract information used in our analysis included the county FIPS code (Federal Information Processing Standards—a unique identifier for each county in the USA), acreage and type of conservation practice (CP), plus soils information from all contracts in sign-ups 10 to

12 and from a special resampling of contracts in sign-ups 1 to 9 (discussed below). Only contracts having CP1 (introduced grass) and CP2 (native grass) were included in the study. More than 90% of the CRP in the study area belonged to these two categories and over 80% of the total CRP nationwide was in CP1 or CP2. The CRP area represented in our regional analysis was 10.2 million ha.

Data on CRP acreage were linked to the 16-state county-level map by FIPS code. Areas having the highest concentration of CRP were: (i) the southwestern Great Plains (in eastern Colorado, western Kansas and the Panhandles of Oklahoma and Texas), (ii) the northern Great Plains (Montana, North Dakota and northwestern Minnesota), (iii) parts of the Palouse region of eastern Washington and Oregon, and (iv) along the Iowa-Missouri border.

Soils Data

Soils information, soil texture in particular, is an important input to the Century model. In CRP sign-ups 10 to 12 (held during 1991–1992), soil series designations were recorded for each CRP contract. However, soil classification information was not included in the original information for sign-ups 1 to 9 (during 1986–1989). To obtain additional soils information, FSA re-surveyed a random sample of 5% of the contracts in sign-ups 1 to 9 and identified the soil series for those contracts. Thus for our study region, approximately 10% of all contracts (sign-ups 1–12) included soils information at the series level.

To obtain the soil textures for the soils given in the contract database, we linked the soil series identifications for CRP contracts to the MUIR (Mapping Unit Interpretive Record) database (NRCS, 1994). About 93% of the map units given in the contract database were matched by the MUIR. The remaining 7% could not be identified, due either to data entry errors on CRP contracts or because the map unit identifiers used had not yet been incorporated into MUIR (J. Vrana, personal communication).

Soil types that represented less than 200 ha in the 10% contract database were excluded, as were muck (organic) soils, which represented about 2000 ha. Thus 13 soil textural types were included in the analysis. To reduce the number of simulation runs, we grouped soils into four textural classes: (i) sands, loamy sands and sandy loams, (ii) loams, (iii) silt loams, (iv) clay loams and clays (Fig. 15–2).

For each climate zone, we estimated soil physical parameters (i.e., percentage sand, percentage silt, percentage clay, volumetric water content at field capacity, wilting point and bulk density) for each of four soil classes, based on an area-weighted calculation of all soil textural types occurring in the climate zone. The mean attributes for each soil textural type were derived from an analysis of 2500 pedons for the central USA (Donigian et al., 1994). Thus for each climate zone, we ran separate simulations for up to four soil classes. Model results were then aggregated by climate zone based on the percentage area of CRP within each soil class.

Model Initialization

In the regional analysis, a set of preparatory simulations were run in order to initialize the model prior to simulating soil C changes under CRP. As described

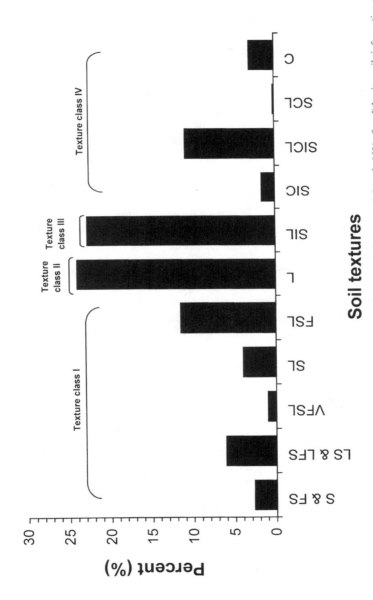

Fig. 15–2. Distribution of CRP acreage by soil texture type for the study region, based on CRP contracts (approximately 10% of total) having soils information. Muck soils and other textures for which there less than 200 ha reported on within the total study area were excluded. Soil textures were grouped into four classes for model simulations.

above, input data included the climate zone averages for weather variables and soil characteristics estimated from the CRP contract and Natural Resources Conservation Service (NRCS) soils databases. Since present organic matter levels are highly influenced by past management practices, we needed to include land use history in the analysis. To do this, climate zones were grouped into 16 land management categories. Land use histories and management practices for each category were based on MLRA descriptions (SCS, 1981) and advice from Dr. Gary Peterson, Department of Soil and Crop Sciences, Colorado State University, and Dr. Paul Rasmussen, USDA ARS, Pendleton, Oregon.

The standard initialization procedures for Century (Paustian et al., 1992) were used for estimating the amount and composition of SOM prior to grassland conversion under CRP. First, the model was run to equilibrium (7000 yr) to estimate precultivation soil C levels under native grassland conditions. Native vegetation type, expressed as relative composition of warm-season (C_4) and cool-season (C_3) grasses, and fire frequency were specified for each of the 16 land use zones. We assumed moderate grazing during each month of the growing season for the precultivation grassland simulations. Next, we simulated soil C changes from the onset of cultivation to the initiation of CRP using the information compiled on land use histories. The purpose of the equilibrium and prior land use simulations was to provide an unbiased estimate of the initial organic matter composition at the time CRP is established.

To summarize the methodology for the regional analysis: we derived 193 sets (i.e., climate zones × soil types) of input parameters to represent climate, soil and land use conditions across the region. For each set, the model was run to establish precultivation equilibrium conditions (7000-yr simulation), followed by a 65- to 115-yr period of cropping prior to establishment of CRP. We then simulated soil C changes under CRP for a 10-yr period.

RESULTS AND DISCUSSION

Primary Production

The regional simulation results are strongly influenced by the pattern of primary production of CRP grasslands (Fig. 15–3), which increases along the west to east precipitation gradient across the Great Plains. Simulated above-ground productivity ranged from around 1 Mg ha^{-1} yr^{-1} (dry matter) in the semiarid western Great Plains to as much as 6 Mg ha^{-1} yr^{-1} along the eastern border of the study region. Simulated production levels on CRP grasslands in the Pacific Northwest were mostly below 1.5 Mg ha^{-1} with greater productivity in the higher rainfall areas along the Washington-Idaho border.

We did a first-order evaluation of the model estimates of aboveground production, in terms of relative magnitudes and regional differences in production. Grassland production estimates for a number of sites in the central USA obtained from the literature were compared with model estimates. To standardize for the different measurement methodologies used, we chose only those studies in which production was estimated by harvesting peak standing biomass. For multiple measured values within a larger climate zone, mean values were calculated and plotted. For

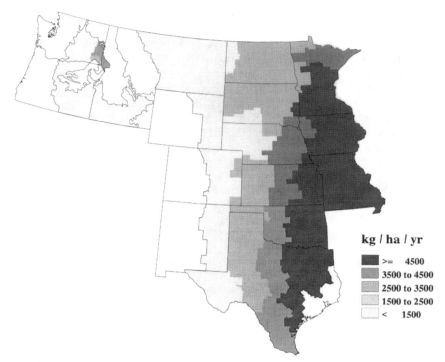

Fig. 15–3. Map of annual simulated above-ground primary production (kg dry matter ha^{-1} yr^{-1}) for CRP grasslands in the study area.

the comparison, we then took the average simulated peak standing biomass of 10-yr old CRP grassland for the climate zone corresponding to the measurement location.

The simulated biomass showed a reasonable correspondence with overall magnitudes and range of observed production over the region (Fig. 15–4). It should be noted that the observed values represent production estimates by different investigators, in different years and at specific sites within a climate zone. Thus it is not surprising that there is considerable scatter around the one-to-one prediction line, given that the model values are based on average climate and soil conditions for the climate zone. Overall, the simulated values are somewhat lower than the literature estimates, which for the most part derive from studies of native grassland. Somewhat lower production from CRP grassland compared to native grassland is not unreasonable, considering the loss of native fertility that has generally occurred due to past cropping.

Carbon Sequestration

There are different interpretations as to what constitutes C sequestration in the context of CO_2 mitigation policies. We calculated three alternative expressions of C sequestration: (i) change in soil organic matter C (i.e., excluding litter), (ii) change in soil organic matter plus soil litter and roots (i.e., referred to as total below-ground C), and (iii) change in below-ground C plus the minimum standing stocks of above-ground litter and biomass over the year (i.e., referred to as ecosystem C level). Rates were calculated as difference in C stocks (as defined above) between the end and the beginning of the CRP contract period, divided by 10 (the duration of the contract period). While differing in magnitude, the regional patterns for each of these different quantities were relatively similar.

The model predicted that increases in SOM per se generally made up less than half of the total increase in C stocks. Much of the increases in below-ground C stocks were as root material and soil litter that had not yet been fully decomposed. Considering the short periods of time under CRP management, these results are rea-

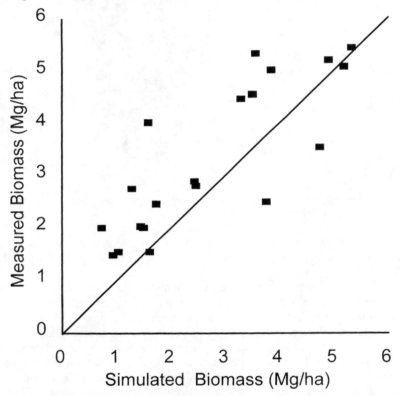

Fig. 15–4. Simulated vs. observed above-ground primary production of grassland, where production was estimated as peak standing crop. Simulated values were matched with field estimates by taking the average production for the climate zone within which the field estimates were located (literature values from: Abrams et al., 1986; Adams & Anderson, 1978; Anderson et al., 1970; Beebe & Hoffman, 1968; Boutton et al., 1980; Branson, 1956; Dodd et al., 1982; Dyksterhuis & Schmutz, 1947; Hadley, 1970; Hadley & Buccos, 1967; Hadley & Kieckhefer, 1963; Hazell, 1967; Heitschmidt et al., 1982; Heitschmidt & Dowhower, 1986; Hulbert, 1969; Hulett & Tomanek, 1968; Kelting,1954; Koelling & Kucera, 1965; Kucera & Dahlman, 1967; Lauenroth & Sala, 1992; Lauenroth & Whitman, 1977; Lura et al., 1988; McMurphy & Anderson, 1963; Moir, 1969; Oldham et al., 1982; Owensby & Anderson, 1967; Potvin & Harrison, 1984; Reardon & Merrill, 1976; Redmann, 1975; Rice & Parenti, 1978; Risser et al., 1981; Sims et al., 1978; Smeins & Olsen, 1970; Smeins et al., 1976; Tomanek & Albertson, 1957; Towne & Owensby, 1984; and Weaver & Tomanek, 1951).

sonable. In comparing CRP and wheat (*Triticum aestivum* L.) fields in southeastern Wyoming, Robles and Burke (1986) found greater amounts of particulate organic matter and higher rates of C mineralization under CRP, although total soil C was not significantly different.

For SOM C accumulation alone, the average rates for the 10-yr period ranged from less than 10 g C m^{-2} yr^{-1} in eastern Wyoming to more than 40 g C m^{-2} yr^{-1} in parts of Iowa and Minnesota. Simulated rates of below-ground C accumulation under CRP (Fig. 15–5) ranged from less than 30 g C m^{-2} yr^{-1} along the western edge of the Great Plains and in central Washington to more than 110 g C m^{-2} yr^{-1} in the western Corn Belt (Iowa, southern Minnesota, and Missouri). Eastern Oklahoma and Texas are predicted to have among the highest production rates (Fig. 15–3), but they have only intermediate C storage rates due to concomitantly greater decomposition rates (Fig. 15–5). If total ecosystem C stores are considered then annual rates of accumulation range from 30 to 160 g C m^{-2} yr^{-1} across the region.

There are some published measurements of soil C accumulation under grassland conversions that can be used to evaluate the model results (Table 15–1). The range of SOM increases predicted for the region, between 5 to 50 g C m^{-2} yr^{-1}, are in line with values reported for CRP lands and other conversions of cultivated lands to grasslands in temperate regions. The relative difference between warm-dry and cool-wet environments predicted by the model is in line with observed differences

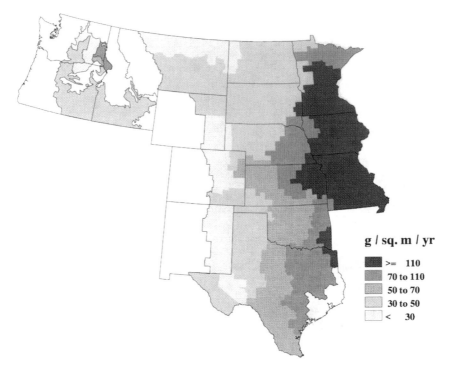

Fig. 15–5. Map of below-ground C (SOM + soil litter + roots) accumulation rate (g C m^{-2} yr^{-1}) under CRP.

Table 15–1. Some estimates of soil C accumulation in surface soils following conversion of cultivated lands to grassland.

Region	Land use conversion	Time period	Soil depth	Method	Mean soil C gain	Reference
		yr	cm		g m^{-2} yr^{-1}	
Eastern Colorado	Abandoned to grassland	50	10	Paired plots	3	Burke et al., 1995
Eastern Wyoming	Grassland CRP	6	5	Paired plots	0	Robles & Burke, 1998
Eastern Wyoming	Grass(80%)–legume(20%) CRP	6	5	Paired plots	16	Robles & Burke, 1997
Eastern Wyoming	Grass(20%)–legume(80%) CRP	6	5	Paired plots	32	Robles & Burke, 1997
Alberta	Abandoned to grassland	24	15	Paired plots	7–13	Dormaar & Smoliak, 1985
Central USA	Grassland CRP (5 sites)	5	20	Paired plots	2–185	Gebhart et al., 1994
Central USA	Grassland CRP (14 sites)	5–10	20	Paired plots	–46–473	Follett et al., 2001
Illinois	Abandoned to grassland	17	10	Chronosequence	100	Jastrow, 1996
Southern England	Planted grassland (managed)	15	15	Time series	75	Tyson et al., 1990
Southern England	Abandoned to grass	83	23	Time series	55	Jenkinson (1971, p. 113–137)
New Zealand	Planted grassland (managed)	18	20	Chronosequence	100	Haynes et al., 1991

between regions. Estimates by Gebhart et al. (1995) of soil C accumulation rates in the range of 5 to 15 g m^{-2} yr^{-1} for two paired CRP vs. cultivated sites in Texas are very similar to those obtained with the model. In contrast, Gebhart et al. (1995) estimated considerably higher rates, about 40 and 180 g C m^{-2} yr^{-1} (0–20 cm), for two paired sites in Kansas. For corresponding locations, our model predicted accumulation rates of around 15 to 20 g C m^{-2} yr^{-1} for SOM C but up to 100 g m^{-2} yr^{-1} increases in total below-ground C. In field experiments in Colorado (Peterson et al., 1998), where winter wheat-fallow plots have been converted to perennial grasses, field estimates of soil C increase for three locations averaged 25 g C m^{-2} yr^{-1} over 10 yr. These values are midway between model estimates for SOM C accumulation (10–15 g C m^{-2} yr^{-1}) and total below-ground C accumulation rates (25–50 g C m^{-2}) for the same region. Mixed legume and grass CRP in eastern Wyoming had apparent C accumulation rates of 16 g C m^{-2} yr^{-1} with a low proportion of legumes (20%) and higher rates, 32 g C m^{-2} yr^{-1}, where legumes (80%) dominated (Robles & Burke, 1997). On wheat-fallow cropland converted to perennial grasses in eastern Wyoming, Reeder et al. (1998) reported apparent increases (0–10 cm) of 63 (for unfertilized) and 94 g C m^{-2} yr^{-1} (for N fertilized) on a sandy loam soil (Phiferson series—coarse-loamy, mixed, mesic, Aridic Haplustoll). However, C increases on a clay loam soil (Ulm—fine montmorillonitic, mesic, Ustollic Haplargid) were only significant for the top 2.5 cm layer, with rates of 14 and 32 g C m^{-2} yr^{-1} for unfertilized and fertilized treatments, respectively.

In one of the two regional studies, Huggins et al. (1998) compared CRP and cropland at a number of locations in Iowa, Minnesota, North Dakota, and Washington. State averages for C stocks under 6 to 8 yr old CRP (0–20 cm) were 2 to 5 t ha^{-1} higher (corresponding to accumulation rates of 25–85 g C m^{-2} yr^{-1}) than in corresponding cropland, however the differences were generally not statistically significant. However, for the 0- to 7.5-cm depth, there were more statistically significant differences, with generally higher values under CRP for total SOC as well as for SOC fractions such as microbial biomass and particulate organic matter.

Follett et al. (2001) used a paired plot design to assess C changes under CRP for 14 sites distributed across a 13-state region comprising the historic grasslands of the central USA. Most of the sites had been under CRP for 7 to 10 yr. For the 0- to 20-cm depth, they found apparent rates of change ranging from −46 to 473 g C m^{-2} yr^{-1} for cropland converted to CRP. The data were grouped into a matrix of three thermal and three moisture regimes to analyze for climate-related trends and to integrate the results to the regional level. They found the highest rates of change for the coolest, wettest region as was predicted by our model. However, within thermal regimes there were no clear trends with varying moisture, except for the coldest region, where rates of C accumulation increased with increasing moisture. The area weighted average for C accumulation calculated by Follett et al. (2001) was about 90 g m^{-2} yr^{-1} for in the 13-state region, which is more than three times the average rate predicted by our model (25 g m^{-2} yr^{-1}). Part of the discrepancy may be due to the modeled region (10.2 Mha) including a greater area of semiarid CRP (with lower rates of C increase) than in the 5.6-Mha region studied by Follett et al. (2001). Second, the designation of C fractions making up SOM may not exactly correspond between the model and measurements, depending on sample preparation and analysis, i.e., some of the measured soil C may include material that are

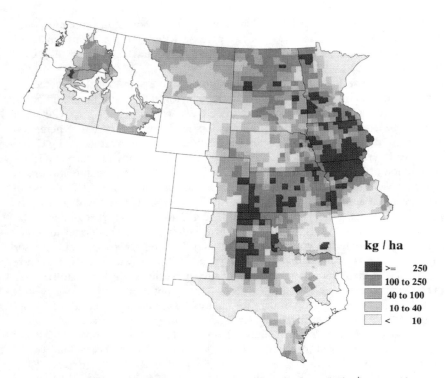

Fig. 15–6. Map of below-ground C accumulation per unit total land area (kg ha⁻¹), over a 10-yr contract period, which is attributable to the CRP.

designated as part of the litter and fine root fractions in our model. Our mean simulated rate of accumulation for total below-ground C (57 g m^{-2} yr^{-1}) is closer to the estimate of Follett et al. (2001). Finally, the mean C accumulation for the region estimated by Follett et al. (2001) is heavily weighted by the values for the cold, wet climate regime (473 g C m^{-2} yr^{-1} average for 2 sites), which accounts for 70% of the total C accumulation in the region on only 13% of the area. If this zone is excluded, the average rate of C gain for the remaining 87% of the region is 33 g C m^{-2} yr^{-1}.

Our estimates of potential rates of soil C storage under CRP and the density of CRP acreage and its regional distribution can be combined to produce a map of soil C change per unit area (Fig. 15–6). The map shows areas of high C storage in the west-central Great Plains (i.e., low accumulation rates but high CRP densities) and in the Corn Belt region (i.e., high accumulation rates with moderate CRP density). Moderate rates of C storage occur in the northern and central Great Plains. The low C accumulation rates per county that were predicted for most of Texas outside the Panhandle, and for eastern Oklahoma, southern Missouri, and northeastern Minnesota were due to the low amounts of CRP acreage.

Summed for the CRP acreage included in our study (10.2 million ha), C accumulations over the initial 10-yr contract period were 24.6 Tg (Tg = 10^{12} g or million metric tonnes) for soil organic matter C, 57.6 Tg in total below-ground C (soil

Table 15–2. Estimates of total increases in total C storage attributable to CRP and comparisons with C equivalents for other proposed mitigation options.

Category	Tg yr^{-1} C equivalents	Source
Conservation Reserve Program		
10.1 million ha CRP grassland	6.9	This study
3.6 million ha (other) CRP grassland	2.5	Extrapolation from this study
1 million ha forested CRP (total C stocks including wood)	4.4	Dunn et al., 1993
Total	13.8	
U.S. climate action plan		
Commercial demand sector	10.6	U.S. Dep. of State (1994)
Residential demand sector (e.g., appliance improvement, home improvements)	16.3	U.S. Dep. of State (1994)
Industrial demand sector (e.g., accelerated efficiency, pollution prevention)	19.0	U.S. Dep. of State (1994)
Transportation demand sector	8.1	U.S. Dep. of State (1994)
Energy supply sector (e.g., enhanced natural gas utilization, improved energy efficiency)	10.8	U.S. Dep. of State (1994)

+ root C), or up to 69.2 Tg if total C accumulated in the ecosystem is counted (Table 15–2). Extrapolating our analysis to the entire acreage in CRP and including biomass accumulated on forested CRP yielded a total of 138 Tg.

Carbon Dioxide Mitigation Potential of Conservation Reserve Program

Both the model analysis and available data suggest that the CRP has led to an increase in the stocks of C contained in soils and vegetation. Given the short time period since the inception of the CRP a considerable proportion of the increases in C should be in the form of plant biomass (especially roots) and labile forms of organic matter (i.e., litter and particulate organic matter). Where lands are maintained in perennial grasses, C stocks can be expected to increase further for a number of years, although at a gradually diminishing rate. Over time, the majority of the accumulated C will be found in more recalcitrant SOM fractions. During this period, the lands converted from annual crops to grassland will function as a net sink for CO_2 and the total net change in ecosystem C stocks is the most appropriate measure of C sequestration in the context of CO_2 mitigation. However, if CRP land is plowed and reverted to annual cropping then the above-ground stocks of biomass, litter and root materials that have built up during the CRP period will be quickly lost—SOM will decrease more slowly, but would be expected to revert to previous levels within a few years to decades. Losses of soil organic matter C would probably be less rapid (and less overall) if reduced tillage, in particular no-tillage, was employed following conversion of grassland back to annual cropping (Fenster & Peterson 1979; Lyon et al., 1997). How much of the C gained during 10 yr of CRP could be preserved with direct conversion to no-tillage cropping is uncertain. However, there is likely to still be a net decline in ecosystem C stocks and thus CRP land reverted to cropland would become a net source of CO_2.

How important is the CRP for greenhouse gas mitigation? To put these values into perspective, consider that emissions of C from fossil fuel combustion in the USA are currently on the order of 1400 Tg yr^{-1}—more than two orders of magnitude greater than the C storage rates predicted for CRP in the study area. From this perspective, the value of the CRP—or indeed of any single mitigation option for increasing C sinks—appears insignificant. However, for the purposes of stabilizing or reducing (e.g., by 7% from 1990 levels) CO_2 levels as called for in the U.N. Framework Convention on Climate Change and the Kyoto Protocol, the CRP and other agricultural C sequestration strategies (Bruce et al., 1999; Lal et al., 1998) can make a significant contribution. The amounts we estimate for 1995, represent an offset of about 35% of the 40 Tg yr^{-1} C emitted from the entire U.S. agricultural sector (Lal et al., 1998) and are of a similar magnitude as that for mitigation strategies proposed for other sectors of the economy (Table 15–2). Since 1995, some former CRP land has returned to cropland, new areas have gone into CRP and some areas have remained in CRP with renewed contracts. These management changes are important in determining the long-term role of the CRP for C sequestration but to our knowledge the net effects of this complex pattern of land use change have yet to be analyzed.

ACKNOWLEDGMENTS

Support for the research reported here was provided by USDA/National Resource Conservation Service and by the EPA. We thank Phil Pasteris, Alex Barbarika, and Jon Vrana of USDA for assistance with databases. We also thank Rudy Bowman, Indy Burke, Ron Follett, Dick Gebhart, Dave Huggins, Doug Karlen, Gary Peterson, Jean Reeder, Mary Rosek, and Jerry Schuman for providing additional information on field studies and to Gary Peterson and Paul Rasmussen for information on current and historical management systems.

REFERENCES

Abrams, M.D., A.K. Knapp, and L.C. Hulbert. 1986. A ten-year record of aboveground biomass in a Kansas tallgrass prairie: Effects of fire and topographic position. Am. J. Bot. 73:1509–1515.

Adams, D.E., and R.C. Anderson. 1978. The response of a central Oklahoma grassland to burning. S.W. Natur. 23:623–632.

Anderson, K.L., E.F. Smith, and C.E. Owensby. 1970. Burning bluestem range. J. Range Manage. 23:81–91.

Beebe, J.D., and G.R. Hoffman. 1968. Effects of grazing on vegetation and soils in southeastern South Dakota. Am. Mid. Natur. 80:96–110.

Boutton, T.W. A.T. Harrison, and B.N. Smith. 1980. Distribution of biomass of species differing in photosynthetic pathway along an altitudinal transect in southeastern Wyoming. Oecologia 45:287–298.

Branson, F.A. 1956. Range forage production changes on a water spreader in southeastern Montana. J. Range Manage. 9:187–191.

Bruce, J.P., M. Frome, E. Haites, H. Janzen, R. Lal and K. Paustian. 1999. Carbon sequestration in soils. J. Soil Water Conserv. 54:382–389.

Burke, I.C., T.G.F. Kittel, W.K. Lauenroth, P. Snook, C.M. Yonker and W.J. Parton. 1991. Regional analysis of the Central Great Plains: sensitivity to climate variability. Bioscience 41:685-692.

Burke, I.C., W.K. Lauenroth and D.P. Coffin. 1995. Recovery of soil organic matter and N mineralization in semiarid grasslands: Implications for the Conservation Reserve Program. Ecol. Appl. 5:793–801.

Cole, V., C. Cerri, K. Minami, A. Mosier, N. Rosenberg, D. Sauerbeck, J. Dumanski, J. Duxbury, J. Freney, R. Gupta, O. Heinemeyer, T. Kolchugina, J. Lee, K. Paustian, D. Powlson, N. Sampson, H. Tiessen, M. van Noordwijk and Q. Zhao. 1996. Agricultural options for mitigation of greenhouse gas emissions. p. 745–771. In Climate change 1995. Impacts, adaptations and mitigation of climate change: Scientific-technical analyses. IPCC Work. Group II. Cambridge Univ. Press, Cambridge.

Dodd, J.L., W.K. Lauenroth, and R.K. Heitschmidt. 1982. Effects of controlled SO_2 exposure on net primary production and plant biomass dynamics. J. Range Manage. 35:572–579.

Donigian, A.S., Jr., T.O. Barnwell, R.B. Jackson IV, A.S. Patwardhan, K.B. Weinrick, A.L. Rowell, R.V. Chinnaswamy and C.V. Cole. 1994. Assessment of alternative management practices and policies affecting soil carbon in agroecosystems of the central United States. EPA/600/R-94/067. Office Res. Develop., Washington, DC.

Donigian, A.S., A.S. Patwardhan, R.V. Chinnaswamy, and T.O. Barnwell. 1998. Modeling soil carbon and agricultural practices in the central US: An update of preliminary study results. p. 499–518. In R. Lal et al. (ed.) Soil processes and the carbon cycle. CRC Press, Boca Raton, FL.

Dormaar, J.F., and S. Smoliak. 1985. Recovery of vegetative cover and soil organic matter during revegetation of abandoned farmland in a semi-arid climate. J. Range Manage. 38:487–491.

Dumanski, J., R.L. Desjardins, C. Tarnocai, C Monreal, E.G. Gregorich, C.A. Campbell, and V. Kirkwood. 1998. Possibilities for future carbon sequestration in Canadian agriculture in relation to land use changes. Climat. Change 40:81–103.

Dunn, C.P., F. Stearns, G.R. Guntenspergen, and D.M. Sharpe. 1993. Ecological benefits of the Conservation Reserve Program. Conserv. Biol. 7:132–139.

Dyksterhuis, E.J., and E.M. Schmutz. 1947. Natural mulches or "litter" of grasslands: With kinds and amounts on a southern prairie. Ecology 28:163–179.

Elliott, E.T., and C.V. Cole. 1989. A perspective on agroecosystem science. Ecology 70:1597–602.

Fenster, C.R., and G.A. Peterson. 1979. Effects of no-tillage fallow as compared to conventional tillage in a wheat-fallow system. Nebraska Exp. Stn. Publ. RB 289.

Follett, R., S.E. Samson-Liebig, J.M. Kimble, E.G. Pruessner and S.W. Waltman. 2001. Carbon sequestration under the CRP in the historic grassland soils of the USA. p. 27–49. In R. Lal and K. McSweeney (ed.) Soil management for enhancing carbon sequestration. SSSA Spec. Publ. 57. SSSA Madison, WI.

Gebhart, D.L., H.B. Johnson, H.S. Mayeux, and H.W. Polley. 1994. The CRP increases soil organic carbon. J. Soil Water Conserv. 49:488–492.

Hadley, E.B. 1970. Net productivity and burning responses of native eastern North Dakota prairie communities. Am. Mid. Natur. 84:121–135.

Hadley, E.B., and R.P. Buccos. 1967. Plant community composition and net primary production within a native eastern North Dakota prairie. Am. Mid. Natur. 77:116–127.

Hadley, E.B., and B.J. Kieckhefer. 1963. Productivity of two prairie grasses in relation to fire frequency. Ecology 44:389–395.

Haynes, R.J., R.S. Swift, and R.C. Stephen. 1991. Influence of mixed cropping rotations (pasture-arable) on organic matter content, water stable aggregation and clod porosity in a group of soils. Soil Till. Res. 19:77–81.

Hazell, D.B. 1967. Effect of grazing intensity on plant composition, vigor, and production. J. Range Manage. 20:249–253.

Heitschmidt, R.K., D.L. Price, R.A. Gordon, and J.R. Frasure. 1982. Short duration grazing at the Texas Experimental Ranch: Effects on aboveground net primary production and seasonal growth dynamics. J. Range Manage. 35:367–372.

Heitschmidt, R.K., and S.L. Dowhower. 1986. Long-term vegetation response to grazing treatments. Texas Agric. Exp. Stn. Publ. PR-4423.

Huggins, D.R., D.L. Allan, J.C. Gradner, D.L. Karlen, D.F. Bezdicek, M.J. Rosek, M.J. Alms, M. Flock, B.S. Miller, and M.L. Staben. 1998. Enhancing carbon sequestration in CRP-managed land. p. 323–350. In R. Lal et al. (ed.) Management of carbon sequestration in soil. CRC Press, Boca Raton, FL.

Hulbert, L.C. 1969. Fire and litter effects in undisturbed bluestem prairie in Kansas. Ecology 50:874–877.

Hulett, G.K., and G.W. Tomanek. 1968. Forage production on a clay upland range site in western Kansas. J. Range Manage. 22:270–276.

Jastrow, J.D. 1996. Soil aggregate formation and the accrual of particulate and mineral-associated organic matter. Soil Biol. Biochem. 28:665–676.

Jenkinson, D.S. 1971. The accumulation of organic matter in soil left uncultivated. Rothamsted Exp. Stn. Annu. Rep. 1970. Part 2. Rothamsted Exp. Stn., Harpenden, Herts, UK.

Kelting, R.W. 1954. Effects of moderate grazing on the composition and plant production of a native tall-grass prairie in central Oklahoma. Ecology 35:200–207.

Kern, J.S., and M.G. Johnson. 1993. Conservation tillage impacts on national soil and atmospheric carbon levels. Soil Sci. Soc. Am. J. 57:200–210.

Koelling, M.R., and C.L. Kucera. 1965. Productivity and turnover relationships in native tallgrass prairie. Iowa State J. Sci. 39:387–392.

Kucera, C.L., and R.C. Dahlman. 1967. Total net productivity. Ecology 48:536–541.

Lal, R., R.F. Follett, J.M. Kimble, and C.V. Cole. 1998. The potential for U.S. cropland to sequester carbon and mitigate the greenhouse effect. Ann Arbor Press, Chelsea, MI.

Lauenroth, W.I., and O.E. Sala. 1992. Long-term forage production of North American shortgrass steppe. Ecol. Appl. 2:397–403.

Lauenroth, W.K., and W.C. Whitman. 1977. Dynamics of dry matter production in a mixed-grass prairie in western North Dakota. Oecologia 27:339–351.

Lee, J.J., D.L. Phillips, and R. Liu. 1993. The effect of trends in tillage practices on erosion and carbon content of soils in the U.S. corn belt. Water Air Soil Pollut. 70:389–401.

Lura, C.L., W.T. Barker, and P.E. Nyren. 1988. Range plant communities of the Central Grasslands Research Station in south central North Dakota. Prairie Nat. 20:177–192.

Lyon, D.J., C.A. Monz, R.E. Brown, and A.K. Metherell. 1997. Soil organic matter changes over two decades of winter wheat-fallow cropping in Western Nebraska. p. 343–351. *In* E.A. Paul et al. (ed). Soil organic matter in temperate agroecosystems: Long term experiments in North America. CRC Press, Boca Raton, FL.

McMurphy, W.E., and K.L. Anderson. 1963. Burning bluestem range—forage yields. Trans. Kansas Acad. Sci. 66:49–51.

Metherell, A.K., L.A. Harding, C.V. Cole, and W.J. Parton. 1993. CENTURY Soil organic matter model environment. Agroecosyst. Vers. 4.0. Tech. Document. GPSR Tech. Rep. 4. USDA ARS, Fort Collins, CO.

Moir, W.H. 1969. Steppe communities in the foothills of the Colorado Front Range and their relative productivities. Am. Mid. Natur. 81:331–340.

Natural Resources Conservation Service. 1994. National map unit interpretation record (MUIR) database. Available at http://www.statlab.iastate.edu/soils/muir/metadata.html.

Oldham, T.W., C.J. Scifres, and D.L. Drawe. 1982. Gulf Coast tick response to prescribed burning on the coastal prairie. Texas Agric. Exp. Stn. Publ. PR-3993.

Owensby, C.E., and K.L. Anderson. 1967. Yield responses to time of burning in the Kansas Flint Hills. J. Range Manage. 20:12–16.

Parton, W.J., D.S. Schimel, C.V. Cole, and D.S. Ojima.1987. Analysis of factors controlling soil organic matter levels in Great Plains grasslands. Soil Sci. Soc. Am. J. 51:1173–1179.

Parton, W.J., D.S. Ojima, C.V. Cole, and D.S. Schimel. 1994. A general model for soil organic matter dynamics: Sensitivity to litter chemistry, texture and management. p. 147–167. *In* Quantitative modeling of soil forming processes. SSSA Spec. Publ. 39. SSSA, Madison, WI.

Paul, E.A., K. Paustian, E.T. Elliott, and C.V. Cole (ed.) 1997. Soil organic matter in temperate agroecosystems: Long-term experiments in North America. CRC Press, Boca Raton, FL.

Paustian, K., W.J. Parton, and J. Persson. 1992. Modeling soil organic matter in organic-amended and nitrogen-fertilized long-term plots. Soil Sci. Soc. Am. J. 56:476–488.

Paustian, K., G.P. Robertson, and E.T. Elliott. 1995. Management impacts on carbon storage and gas fluxes (CO_2,CH_4) in mid-latitude cropland and grassland ecosystems. p. 69–84. *In* R. Lal et al. (ed.) Soil management and greenhouse effect. Advances in soil science. CRC Press, Boca Raton, FL.

Paustian, K., E. Levine, W.M. Post, and I.M. Ryzhova. 1997. The use of models to integrate information and understanding of soil C at the regional scale. Geoderma 79:227–260.

Peterson, G.A., A.D. Halvorson, J.L. Havlin, O.R. Jones, D.J. Lyon, and D.L. Tanaka. 1998. Reduced tillage and increasing cropping intensity in the Great Plains conserves soil C. Soil Till. Res. 47:207–218.

Potvin, M.A., and A.T. Harrison. 1984. Vegetation and litter changes of a Nebraska sandhills prairie protected from grazing. J. Range Manage. 37:55–58.

Rastetter, E.B., A.W. King, B.J. Cosby, G.M. Hornberger, R.V. O'Neill, and J.E. Hobbie. 1992. Aggregating fine-scale ecological knowledge to model coarser-scale attributes of ecosystems. Ecol. Appl. 2:55–70.

Reardon, P.O., and L.B. Merrill. 1976. Vegetative response under various grazing management systems in the Edwards plateau of Texas. J. Range Manage. 29:195–8.

Redmann, R.E. 1975. Production ecology of grassland plant communities in western North Dakota. Ecol. Monogr. 45:83–106.

Reeder, G., E. Schuman, and R.A. Bowman. 1998. Soil C and N changes on Conservation Reserve Program lands in the central Great Plains. Soil Till. Res. 47:339–350.

Rice, E.L., and R.L. Parenti. 1978. Causes of decreases in productivity in undisturbed tall grass prairie. Am. J. Bot. 65:1091–1097.

Risser, P.G., E.C. Birney, H.D. Blocker, S.W. May, W.J. Parton, and J.A. Wiens. 1981. The true prairie ecosystem. US/IBP Synth. Ser. 16. Hutchinson Ross Publ. Co., Stroudburg, PA.

Robles, M.D., and I.C. Burke. 1997. Legume, grass and Conservation Reserve Program effects on soil organic matter recovery. Ecol. Appl. 7:345–357.

Robles, M.D., and I.C. Burke. 1998. Soil organic matter recovery on Conservation Reserve Program fields in southeastern Wyoming. Soil Sci. Soc. Am. J. 62:725–730.

Sims, P.L., J.S. Singh, and W.E. Lauenroth. 1978. The structure and function of ten western North American grasslands. Abiotic and vegetational characteristics. J. Ecol. 66:251–285.

Smeins, F.E., T.W. Taylor, and L.B. Merrill. 1976. Vegetation of a 25-year exclosure on the Edwards Plateau, Texas. J. Range Manage. 29:24–29.

Smeins, F.E., and D.E. Olsen. 1970. Species composition and production of a native northwestern Minnesota tall grass prairie. Am. Mid. Natur. 84:398–410.

Smith, P., D.S. Powlson, M.J. Glendining, and J.U. Smith. 1997. Potential for carbon sequestration in European soils: preliminary estimates for five scenarios using results from long-term experiments. Glob. Change Biol. 3:67–79.

Tomanek, G.W., and F.W. Albertson. 1957. Variations in cover, composition, production, and roots of vegetation on two prairies in Western Kansas. Ecol. Monogr. 27:267–281.

Towne, G., and C. Owensby. 1984. Long-term effects of annual burning at different dates in ungrazed Kansas tallgrass prairie. J. Range Manage. 37:392–397.

Tyson, K.C., D.H. Roberts, C.R. Clement, and E.A. Garwood. 1990. Comparison of crop yields and soil conditions during 30 years under annual tillage or grazed pasture. J. Agric. Sci. 115:29–40.

Soil Conservation Service. 1981. Land resource regions and major land resource areas for the United States. USDA Agric. Handb. 296. U.S. Gov. Print. Office, Washington, DC.

U.S. Department of State. 1994. Climate action report 1994. Rep. U.N. Framework Convent. Climate Change. U.S. Gov. Print. Office, Washington, DC.

Weaver, J.E., and G.W. Tomanek. 1951. Ecological studies in a midwestern range: The vegetation and effects of cattle on its contribution and distribution. Univ. Nebraska Conserv. Surv. Div. Bull. no. 31.

16 The Response of Soil Science to Global Climate Change

Rattan Lal

The Ohio State University
Columbus, Ohio

ABSTRACT

There exists an enormous potential to sequester C in soils of agricultural ecosystems to curtail the rate of enrichment of CO_2 concentration in the atmosphere. About 25 to 75% of the soil organic carbon (SOC) pool in agricultural ecosystems has been lost due to land misuse and soil mismanagement. Conversion to an appropriate land use and adoption of recommended agricultural practices can facilitate regaining the lost organic C in the soil. This strategy requires conducting feasibility studies towards identification of ecoregions and soils with large potential for C sequestration. It also is important to broaden the scope of soil science and reach out to other disciplines for developing interdisciplinary programs. In addition, it is imperative that processes and properties affecting soil C pool and fluxes be understood and the data bank on pedon level improved. With the new paradigm, soil science has a major role to play in addressing two principal challenges of the 21st century: achieving global food security and mitigating the accelerated greenhouse effect.

World soils play an important role in the changing global C cycle. The amount of CO_2 emitted into the atmosphere as CO_2 due to land use change and soil cultivation from 1800 to 1998 is estimated at 136 ± 55 Pg compared with 270 ± 30 Pg emitted by the fossil fuel combustion (IPCC, 2000). Of the 136 ± 55 Pg emitted from land use change, 78 ± 17 Pg came from soil cultivation and depletion of SOC pool (Lal, 1999). So far, world soils have been a source of atmospheric enrichment of CO_2. However, adoption of judicious land use and recommend management practices can make world soils a net sink for atmospheric CO_2.

In most natural ecosystems, the SOC pool is larger than the biotic (vegetation) pool. Whereas the ratio of soil/biotic pool ranges from 1.0 to 1.7 for tropical and temperate forests, respectively, the ratio has a much higher range of 4.0 to 32.78 for boreal forest, tropical savannas and temperate grasslands (Table 16–1). Because of low above-ground biomass and scanty vegetation cover, the ratio is very large for tundra and wetland ecosystems.

The mean SOC pool of 80 Mg C ha^{-1} (Table 16–1) is high for agricultural soils of sub-Saharan Africa, South Asia and other soils of the tropics that have long been cultivated for subsistence farming based on low external input. Even on the

Table 16–1. Biotic and pedologic C pool in principal global biomes (recalculated from IPCC, 2000).

Biome	Area (10^9 ha)	C Pool Vegetation	C Pool Soil†	C Pool Total	Soil/vegetation pool
		Mg ha^{-1}			
Tropical forest	1.76	120.5	122.7	243.2	1.02
Temperate forest	1.04	56.7	96.2	152.9	1.7
Boreal forest	1.37	64.2	343.8	408	5.36
Tropical savannas	2.25	29.3	117.3	146.6	4.0
Temperate grassland	1.25	7.2	236	243.2	32.78
Deserts and semi-desert	4.55	1.8	42	43.8	23.33
Tundra	0.95	6.3	127.4	133.7	20.22
Wetlands	0.35	42.9	642.9	685.8	149.86
Croplands	1.6	1.9	80.0	81.9	42.1
Total	15.16				

†Soil C pool is estimated to 1-m depth.

basis of this high mean, conversion of natural ecosystems to croplands world over have lost SOC pool by an average of 16.2 Mg ha^{-1} for temperate forest, 37.3 Mg ha^{-1} for tropical savannas, 42.7 Mg ha^{-1} for tropical forest, 156 Mg ha^{-1} for temperate grasslands and 263.8 Mg ha^{-1} for boreal forest ecosystems. The relative magnitude of loss is more for cool and moist ecosystems than warm and dry regions.

The depletion of SOC pool from croplands and agricultural soils is exacerbated by soil misuse and attendant degradative processes such as accelerated soil erosion, nutrient depletion, secondary salinization, etc. An important issue is to develop strategies for sequestration of soil C lost due to past land misuse and soil mismanagement. The objective of this chapter is to outline strategies for enhancing SOC pool in agricultural soils and other degraded ecosystems, identify and prioritize researchable issues in soil C sequestration, and outline new paradigms for soil science to meet the challenges of the 21st century.

STRATEGIES OF SOIL CARBON SEQUESTRATION

Soil C sequestration potential is relatively large in degraded and depleted soils of the developing countries of the tropics and subtropics. These soils, managed for subsistence farming with low external input for longtime, have lost a large proportion of their original SOC pool. However, these ecoregions pose a major challenge towards implementation of appropriate strategies for soil C sequestration because of weak institutions and lack of supporting infrastructure. The soil C sequestration strategy involves several steps (Fig. 16–1). The first step is to conduct surveys to identify soils and ecoregions that have a large potential of SOC sequestration, and where institutions and infrastructure also are adequate to facilitate widespread adoption of recommended agricultural practices. The second step is to identify recommended soil and crop management practices, which are specific to each ecoregion, soil type, moisture regime and terrain characteristics. Conducting feasibility studies and establishing demonstration plots in collaboration with agricultural extension and research institutions is the next step in assessing the acceptability of recommended technology by the farming community.

Identification of appropriate policies also is important to encourage farmers to adopt improved technology. Soil scientists and agronomists are to be encouraged to continue inter-disciplinary and on-farm research to improve upon the existing technologies.

Monitoring and verification of SOC pool and fluxes for recommended practices also are an important issue that needs to be addressed. Simple and cost-effective methods are to be developed/adapted to assess temporal changes in SOC pool in relation to management systems.

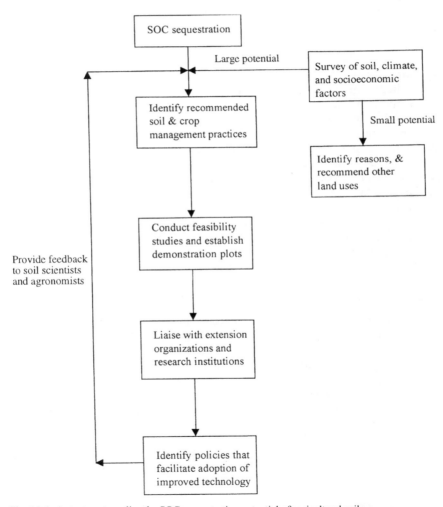

Fig. 16–1. A strategy to realize the SOC sequestration potential of agricultural soils.

A NEW PARADIGM FOR SOIL SCIENCE

Soil science has made important contributions towards achieving global food security, especially during the later half of the 20th century. The mere foundation of the Green Revolution that brought about a quantum jump in food production during 1960s and 1970s, comprised appropriate soil and water management technologies (e.g., irrigation, fertilizer use, residue management and tillage methods, etc.) and improved varieties. Similar to achieving food security, soil science also has an important role to play in improving environment quality especially in enhancing water quality and mitigating the greenhouse effect. Towards this goal, soil scientists have to be proactive in conducting multidisciplinary research in relevant environmental issues.

The era of pedocentrism, studying soil for the sake of soil, if it was ever appropriate, is over. The new paradigm in soil science recognizes the importance of broadening the scope and reaching out to other disciplines for effectively addressing the issues of food security and environment quality (Wild, 1989; Simonson, 1991; Miller, 1993; Bouma, 1994, Bridges & Catizzone, 1996). Some important concerns relevant to environment quality include quality of surface- and groundwater, impact of industrial/urban and mining activities on soil and water quality, soil application of industrial/urban waste, air quality and the greenhouse effect. Addressing these global issues requires a close collaboration with other scientific disciplines. Soils play an important role in moderating the emission of CO_2, CH_4, and N_2O into the atmosphere. Processes and properties affecting the rate and magnitude of these emissions (Fig. 16–2) need to be understood and quantified. There is a strong interaction between soil and the atmospheric processes. Therefore, such studies need to be conducted in collaboration with climatologists, microbiologists, and biochemists.

Importance of soil science in understanding the processes involved and technologies required for mitigating the greenhouse effect cannot be overemphasized. Towards this goal, of expanding the scope and specifically addressing the issue of the accelerated greenhouse effect, soil scientists have to adopt a holistic approach (Fig. 16–3). While continuing to enlarge scientific strength and maintaining professional integrity, soil scientists need to work in close association with:

1. Crop physiologists and plant scientists to understand net primary productivity (NPP) and the fraction returned to the soil for improving SOC pool,
2. Climatologists and hydrologists to understand dynamics of soil temperature and moisture regimes that affect microbial decomposition and oxidation of organic material returned to the soil, and to establish links between soil processes and atmospheric processes, geologists and sedimentologists to understand the impact of erosional and depositional processes on dynamics of soil C redistributed over the landscape, and economists, social scientists and policy makers to assess the value of soil C and to identify policy issues that facilitate adoption of improved technology.

Soil scientists need to work together with researchers from other disciplines to address emerging environmental concerns. There also are numerous new tools

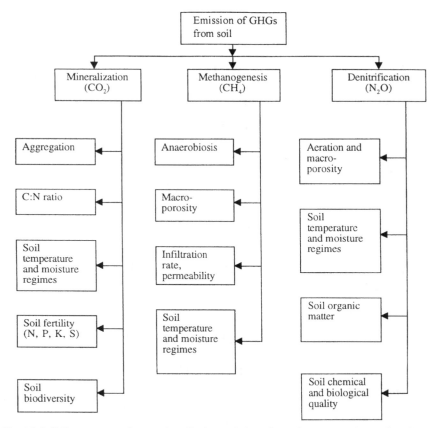

Fig. 16–2. Soil processes and properties affecting emission of greenhouse gases from soil to the atmosphere.

available in other disciplines, which have direct applications in soil science. The study of soil science and its applications to meet the challenge of the greenhouse effect can be revolutionized through the use of geographic information systems (GIS), remote sensing, biotechnology, and modeling and simulation techniques. Use of these techniques allows dynamic and continuous characterization of SOC pool and fluxes in time and space (Bouma & Beek, 1994).

RESEARCHABLE PRIORITIES IN SOIL SCIENCE AND THE GREENHOUSE EFFECT

Two major challenges facing soil science during the 21st century are:

1. Bringing about a quantum jump in food production, especially in developing countries of the tropics and subtropics where 97.5% of the increase in population is occurring, and

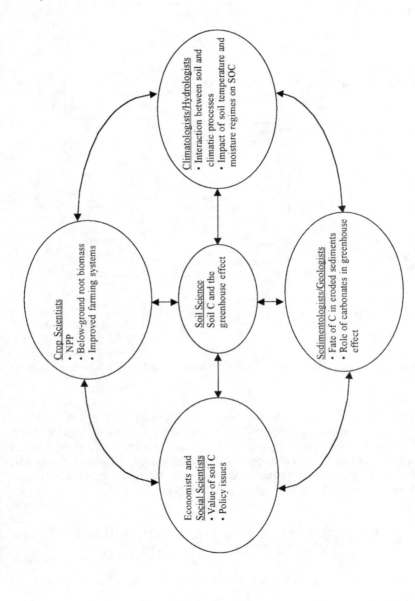

Fig. 16–3. A holistic and multidisciplinary program in soil science in relation to the greenhouse effect.

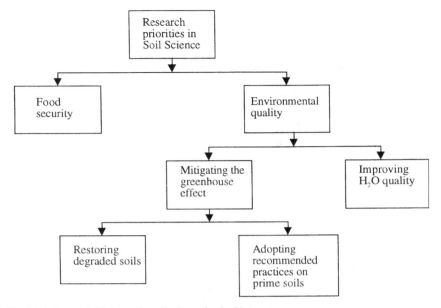

Fig. 16–4. Researchable issues in soil science for the 21st century.

2. Addressing issues of environmental degradation with regard to water quality and the accelerated greenhouse effect (Fig. 16–4). Improving soil quality through enhancement of SOC is a common link in all issues.

Restoring degraded soils, which offers a tremendous potential of soil C sequestration, requires a thorough understanding of energy flow, partitioning of NPP and identification of its pathways, mass balance of water and elements on a watershed scale, and biogeochemical cycles of major elements (e.g., C, N, P, S). Identification and choice of soil restorative measures also requires evaluation of soil resilience and factors affecting it (Lal, 1997). Development of soil quality indicators, based on key soil properties, is important to assessing effectiveness of soil restorative measures.

Adoption of recommended soil and crop management practices has dual purpose of enhancing and sustaining NPP by improving productivity per unit input of energy. Because C sequestration is the goal, sustainability of a land use and soil/crop management system may be assessed on the basis of C balance of the system Eq. [1].

$$I_s = \Sigma \left(\frac{C \text{ output}}{C \text{ input}} \right)_t \qquad\qquad [1]$$

Where I_s is the index of sustainability and t is the time. A non-negative trend in I_s is indicative of a sustainable system.

There are several researchable issues with regards to soil C sequestration that need to be addressed for principal soils and ecoregions. Important among these are the following:

1. Rate of Soil Carbon Sequestration: Quantification of the potential and attainable rate of soil C sequestration is a priority topic. Potential, attainable and actual rates of SOC sequestration need to be quantified in relation to soil and ecological factors, and socioeconomic and political parameters. The SOC sequestration rates need to be determined for a wide range of soil and crop management practices (e.g., conservation tillage, cover crops, soil fertility management, water management, crop rotations, etc.). These rates depend on climate, soil properties, and other site-specific parameters. There are vast areas of degraded soils (Oldeman et al., 1990), especially in arid and semiarid regions. Assessment of the SOC and soil inorganic carbon (SIC) sequestration potential through restoration of these soils is a priority. Nutrient depletion is another issue especially in sub-Saharan Africa (Stoorvogel et al., 1993). Restoration of soil fertility for achieving food security also can lead to SOC sequestration.

2. Fate of Eroded Carbon: Little is known about the fate of C eroded and redistributed over the landscape, deposited in depressional sites and transported into aquatic ecosystems. Multidisciplinary research (involving soil scientists, hydrologists, sedimentologists and biogeochemists) is needed to effectively address this issue.

3. Soil Inorganic Carbon: The role of SIC on the global C cycle is not understood, and little is known about the impact of change in land use and management on SIC. Leaching SIC into groundwater may be an important mechanism of C sequestration, and needs to be studied. The impact of anthropogenic factors on the fate of C in the exposed caliche needs to be studied and quantified.

4. Value of Soil Carbon: The value of soil C (on weight basis as $/ton and area basis as $/ha) needs to be determined by assessing the on-site and off-site benefits of soil C sequestration, and evaluating additional costs incurred to achieve the desired rate and magnitude of C sequestration. In addition to the improvements in soil quality and its positive impact on productivity, ancillary benefits of soil C sequestration include those related to improvement in water quality, reduction in soil erosion, and decrease in risks of damage to waterways and infrastructure caused by siltation.

5. Standardizing Carbon Assessment Methods: It is important to standardize soil sampling, sample preparation, and analyses protocols so that error and costs involved are minimal. Precise measurements of different SOC pools and secondary carbonates (SIC) remain a major challenge.

6. Databank on Soil Carbon: There is an urgent and strong need to strengthen the databank on soil C at the pedon level, especially for principal soils in tropical ecoregions. The available data are severely constrained the lack of information on soil bulk density and C pool for subsoil.

7. Assessment of Carbon Pool at Soilscape and Landscape Levels: Use of remote sensing and models is inevitable. It also is necessary to develop

relationship (pedotransfer functions) between soil C and other properties. Soil C is not randomly distributed over the landscape. It is strongly correlated with clay content and is influenced by soil drainage. It may be useful to determine soil C indices based on management, soil properties and climate.

CONCLUSIONS

The importance of soil and soil management in achieving global food security and improving environment quality cannot be overemphasized. Despite its importance, soil science per se has not made as much impact on the global scene as have its sister sciences of geology, geography, biology and climatology. Yet, soil science has a major role to play in global issues of environment quality, especially the accelerated greenhouse effect, which can be addressed through development of interdisciplinary programs. It is important that soil scientists broaden the scope of their work and reach out to other disciplines for effectively addressing the challenges of 21st century.

Goals of soil science in achieving food security can be strongly augmented by collaboration with biologists (biotechnology), engineers (food processing and waste disposal) and social scientists (policy considerations) and economists (sustainability and profitability). Similarly, achieving the objective of mitigating the greenhouse effect requires a close collaboration between soil scientists and climatologists, geologist, biologist and other relevant disciplines.

The role of soil science in addressing environmental issues related to the accelerated greenhouse effect can be greatly augmented by improving the knowledge base. There is a need to understand soil process and properties that influence soil C pool and fluxes in relation to land use and management options. It is important to standardize soil sampling, sample preparation and analyses protocols for assessment of soil C pool at pedon or soilscape level. Estimating C sequestration rate for innovative practices for major soil of different ecoregions is a high priority.

REFERENCES

Bouma, J. 1994. Sustainable land use and future focus of pedology. Soil Sci. Soc. Am. J. 58:645–646.

Bouma, J., and K.J. Beek. 1994. New techniques and tools. p. 23–30. *In* L.O. Fresco et al. (ed.) The future of the land: Mobilizing and integrating knowledge for land use options. J. Wiley & Sons, Inc., Chichester, UK.

Bridges, E.M., and M. Catizzoned. 1996. Soil science in a holistic framework: Discussion of an improved integrated approach. Geoderma 7:275–287.

Bullock, P. 1994. The need for a paradigm change in soil science in the next century. p. 427–437. *In* Trans. 15th World Congr. Soil Sci. Vol. 9, Acapulco, Mexico. 10–16 July. Int. Soc. Soil Sci., and INEGI, Mexico.

Intergovernmental Panel on Climate Change. 2000. IPCC special report: Land use, land use change and forestry: Summary for policy makers. IPCC, Washington, DC.

Lal, R. 1997. Degradation and resilience of soils. Philos. Trans. R. Soc. Land. (B) 352:997–1010.

Lal, R. 1999. Soil management and restoration for C sequestration to mitigate the accelerated greenhouse effect. Progr. Environ. Sci. 1:307–326.

Miller, F.P. 1993. Soil science: A scope broader than its identity. Soil Sci. Soc. Am J. 57:299, 564.

Oldeman, R.O., R.T.A. Hakkeling, and W.G. Sombroek. 1990. World map of the human induced soil degradation. ISRIC, Wageningen, Holland.

Simonson, R.L. 1991. Soil science: Goals for the next 75 years. Soil Sci. 151:7–18.

Stoorvogel, J. J., E.M.A. Smaling, and B.H. Janssen. 1993. Calculating soil nutrient balances in Africa at different scales. Fert. Res. 35:227–235.

Wild, A. 1989. Soil scientists as a member of the scientific community. J. Soil Sci. 40:209–211.